Mechanical Properties of Self-Compacting Concrete

RILEM STATE-OF-THE-ART REPORTS

Volume 14

RILEM, The International Union of Laboratories and Experts in Construction Materials, Systems and Structures, founded in 1947, is a non-governmental scientific association whose goal is to contribute to progress in the construction sciences, techniques and industries, essentially by means of the communication it fosters between research and practice. RILEM's focus is on construction materials and their use in building and civil engineering structures, covering all phases of the building process from manufacture to use and recycling of materials. More information on RILEM and its previous publications can be found on www.RILEM.net.

The RILEM State-of-the-Art Reports (STAR) are produced by the Technical Committees. They represent one of the most important outputs that RILEM generates—high level scientific and engineering reports that provide cutting edge knowledge in a given field. The work of the TCs is one of RILEM's key functions.

Members of a TC are experts in their field and give their time freely to share their expertise. As a result, the broader scientific community benefits greatly from RILEM's activities.

RILEM's stated objective is to disseminate this information as widely as possible to the scientific community. RILEM therefore considers the STAR reports of its TCs as of highest importance, and encourages their publication whenever possible.

The information in this and similar reports is mostly pre-normative in the sense that it provides the underlying scientific fundamentals on which standards and codes of practice are based. Without such a solid scientific basis, construction practice will be less than efficient or economical.

It is RILEM's hope that this information will be of wide use to the scientific community.

For further volumes:
http://www.springer.com/series/8780

Kamal H. Khayat · Geert De Schutter
Editors

Mechanical Properties of Self-Compacting Concrete

State-of-the-Art Report of the RILEM Technical Committee 228-MPS on Mechanical Properties of Self-Compacting Concrete

Springer

Editors
Kamal H. Khayat
Civil, Architectural and Environmental Engineering
Missouri University of Science and Technology
Rolla, MO
USA

Geert De Schutter
Magnel Laboratory for Concrete Research
Ghent University
Ghent
Belgium

ISSN 2213-204X ISSN 2213-2031 (electronic)
ISBN 978-3-319-34479-9 ISBN 978-3-319-03245-0 (eBook)
DOI 10.1007/978-3-319-03245-0
Springer Cham Heidelberg New York Dordrecht London

Printed on acid-free paper

Springer is part of Springer Science+Business Media (www.springer.com)

State-of-the-Art Report

This report has been prepared by RILEM Technical Committee TC 228-MPS "Mechanical Properties of SCC", consisting of the following members:

- Kamal H. Khayat (Chairman), Missouri University of Science and Technology, USA;
- Geert De Schutter (Secretary), Ghent University, Belgium;
- Sofiane Amziane, Université de Clermont-Ferrand, France;
- Veerle Boel, Ghent University, Belgium;
- Violeta Bokan Bosiljkov, University of Ljubljana, Slovenia;
- Bart Craeye, Artesis University College of Antwerp, Belgium;
- Pieter Desnerck, Center for Infrastructure Engineering Studies, Missouri University of Science and Technology, Rolla, MO 65409-0710, USA; Ghent University, Ghent, Belgium;
- Liberato Ferrara, Politecnico di Milano, Italy;
- Steffen Grünewald, Hurks Beton/TU Delft, the Netherlands;
- Assem Hassan, Memorial University of Newfoundland, Canada;
- Antonios Kanellopoulos, University of Cyprus, Cyprus;
- Michael Khrapko, cbecon, New Zealand;
- Mohamed Lachemi, Ryerson University, Toronto, Canada;
- Andreas Leemann, EMPA, Switzerland;
- Pietro Lura, EMPA, Switzerland;
- Claudio Mazzotti, University of Bologna, Italy;
- Richard Morin, City of Montreal, Canada;
- Bertil Persson, Sweden;
- Wolfram Schmidt, BAM Berlin, Germany;
- Pedro Serna, UPV Valencia, Spain;
- Caijun Shi, Hunan University, China;
- Kosmas Sideris, Democritus University of Thrace, Greece;
- Mohammed Sonebi, Queen's University Belfast, UK;
- Petra Van Itterbeeck, BBRI, Belgium;
- Manuel Vieira, LNEC, Lissabon, Portugal;
- Olafur Wallevik, ICI Rheocenter, Iceland.

All committee members have contributed to this STAR-report, by providing relevant information to lead chapter authors of the chapters, by providing some sections summarizing and contributing to the discussions of the Committee.

Kamal H. Khayat
Geert De Schutter

RILEM Publications

RILEM Publications are presented in six collections, corresponding to the five clusters of active RILEM Technical Committees, sorted by fields of expertise, and a sixth multi-thematic collection dedicated to journals and compendiums:

A. Mechanical Performance and Fracture
B. Test Methods, Materials Characterization and Processing
C. Service Life and Design
D. Durability and Deterioration Mechanisms
E. Bitumen, Masonry and Timber
F. Journals and Compendiums

Each publication is assigned to one of the following series: reports (REP), proceedings (PRO), compendiums (COMP) and journals. The former CD-ROM series is now included in one of these series.
Each publication is available in at least one of the three following editions: print (PR), CD or DVD-ROM (CD), or online (OL).
Online editions are available through our web site, at http://www.rilem.net
The RILEM DVD-ROM, gathering several thousands of online articles, is also published and updated each year (internal publication, circulation restricted to RILEM Benefactor Members).
The following list is presenting our global offer, sorted by series.

RILEM Proceedings

PRO 1: Durability of High Performance Concrete (ISBN: 2-912143-03-9); *Ed. H. Sommer*
PRO 2: Chloride Penetration into Concrete (ISBN: 2-912143-00-04); *Eds. L.-O. Nilsson and J.-P. Ollivier*
PRO 3: Evaluation and Strengthening of Existing Masonry Structures (ISBN: 2-912143-02-0); *Eds. L. Binda and C. Modena*
PRO 4: Concrete: From Material to Structure (ISBN: 2-912143-04-7); *Eds. J.-P. Bournazel and Y. Malier*

PRO 5: The Role of Admixtures in High Performance Concrete (ISBN: 2-912143-05-5); *Eds. J. G. Cabrera and R. Rivera-Villarreal*
PRO 6: High Performance Fiber Reinforced Cement Composites—HPFRCC 3 (ISBN: 2-912143-06-3); *Eds. H. W. Reinhardt and A. E. Naaman*
PRO 7: 1st International RILEM Symposium on Self-Compacting Concrete (ISBN: 2-912143-09-8); *Eds. Å. Skarendahl and Ö. Petersson*
PRO 8: International RILEM Symposium on Timber Engineering (ISBN: 2-912143-10-1); *Ed. L. Boström*
PRO 9: 2nd International RILEM Symposium on Adhesion between Polymers and Concrete ISAP '99 (ISBN: 2-912143-11-X); *Eds. Y. Ohama and M. Puterman*
PRO 10: 3rd International RILEM Symposium on Durability of Building and Construction Sealants (ISBN: 2-912143-13-6); *Eds. A. T. Wolf*
PRO 11: 4th International RILEM Conference on Reflective Cracking in Pavements (ISBN: 2-912143-14-4); *Eds. A. O. Abd El Halim, D. A. Taylor and El H. H. Mohamed*
PRO 12: International RILEM Workshop on Historic Mortars: Characteristics and Tests (ISBN: 2-912143-15-2); *Eds. P. Bartos, C. Groot and J. J. Hughes*
PRO 13: 2nd International RILEM Symposium on Hydration and Setting (ISBN: 2-912143-16-0); *Ed. A. Nonat*
PRO 14: Integrated Life-Cycle Design of Materials and Structures—ILCDES 2000 (ISBN: 951-758-408-3); (ISSN: 0356-9403); *Ed. S. Sarja*
PRO 15: Fifth RILEM Symposium on Fibre-Reinforced Concretes (FRC)—BEFIB'2000 (ISBN: 2-912143-18-7); *Eds. P. Rossi and G. Chanvillard*
PRO 16: Life Prediction and Management of Concrete Structures (ISBN: 2-912143-19-5); *Ed. D. Naus*
PRO 17: Shrinkage of Concrete—Shrinkage 2000 (ISBN: 2-912143-20-9); *Eds. V. Baroghel-Bouny and P.-C. Aïtcin*
PRO 18: Measurement and Interpretation of the On-Site Corrosion Rate (ISBN: 2-912143-21-7); *Eds. C. Andrade, C. Alonso, J. Fullea, J. Polimon and J. Rodriguez*
PRO 19: Testing and Modelling the Chloride Ingress into Concrete (ISBN: 2-912143-22-5); *Eds. C. Andrade and J. Kropp*
PRO 20: 1st International RILEM Workshop on Microbial Impacts on Building Materials (CD 02) (e-ISBN 978-2-35158-013-4); *Ed. M. Ribas Silva*
PRO 21: International RILEM Symposium on Connections between Steel and Concrete (ISBN: 2-912143-25-X); *Ed. R. Eligehausen*
PRO 22: International RILEM Symposium on Joints in Timber Structures (ISBN: 2-912143-28-4); *Eds. S. Aicher and H.-W. Reinhardt*
PRO 23: International RILEM Conference on Early Age Cracking in Cementitious Systems (ISBN: 2-912143-29-2); *Eds. K. Kovler and A. Bentur*
PRO 24: 2nd International RILEM Workshop on Frost Resistance of Concrete (ISBN: 2-912143-30-6); *Eds. M. J. Setzer, R. Auberg and H.-J. Keck*
PRO 25: International RILEM Workshop on Frost Damage in Concrete (ISBN: 2-912143-31-4); *Eds. D. J. Janssen, M. J. Setzer and M. B. Snyder*
PRO 26: International RILEM Workshop on On-Site Control and Evaluation of Masonry Structures (ISBN: 2-912143-34-9); *Eds. L. Binda and R. C. de Vekey*

PRO 27: International RILEM Symposium on Building Joint Sealants (CD03); *Ed. A. T. Wolf*
PRO 28: 6th International RILEM Symposium on Performance Testing and Evaluation of Bituminous Materials—PTEBM'03 (ISBN: 2-912143-35-7; e-ISBN: 978-2-912143-77-8); *Ed. M. N. Partl*
PRO 29: 2nd International RILEM Workshop on Life Prediction and Ageing Management of Concrete Structures (ISBN: 2-912143-36-5); *Ed. D. J. Naus*
PRO 30: 4th International RILEM Workshop on High Performance Fiber Reinforced Cement Composites—HPFRCC 4 (ISBN: 2-912143-37-3); *Eds. A. E. Naaman and H. W. Reinhardt*
PRO 31: International RILEM Workshop on Test and Design Methods for Steel Fibre Reinforced Concrete: Background and Experiences (ISBN: 2-912143-38-1); *Eds. B. Schnütgen and L. Vandewalle*
PRO 32: International Conference on Advances in Concrete and Structures 2 vol. (ISBN (set): 2-912143-41-1); *Eds. Ying-shu Yuan, Surendra P. Shah and Heng-lin Lü*
PRO 33: 3rd International Symposium on Self-Compacting Concrete (ISBN: 2-912143-42-X); *Eds. Ó. Wallevik and I. Níelsson*
PRO 34: International RILEM Conference on Microbial Impact on Building Materials (ISBN: 2-912143-43-8); *Ed. M. Ribas Silva*
PRO 35: International RILEM TC 186-ISA on Internal Sulfate Attack and Delayed Ettringite Formation (ISBN: 2-912143-44-6); *Eds. K. Scrivener and J. Skalny*
PRO 36: International RILEM Symposium on Concrete Science and Engineering—A Tribute to Arnon Bentur (ISBN: 2-912143-46-2); *Eds. K. Kovler, J. Marchand, S. Mindess and J. Weiss*
PRO 37: 5th International RILEM Conference on Cracking in Pavements—Mitigation, Risk Assessment and Prevention (ISBN: 2-912143-47-0); *Eds. C. Petit, I. Al-Qadi and A. Millien*
PRO 38: 3rd International RILEM Workshop on Testing and Modelling the Chloride Ingress into Concrete (ISBN: 2-912143-48-9); *Eds. C. Andrade and J. Kropp*
PRO 39: 6th International RILEM Symposium on Fibre-Reinforced Concretes—BEFIB 2004 (ISBN: 2-912143-51-9); *Eds. M. Di Prisco, R. Felicetti and G. A. Plizzari*
PRO 40: International RILEM Conference on the Use of Recycled Materials in Buildings and Structures (ISBN: 2-912143-52-7); *Eds. E. Vázquez, Ch. F. Hendriks and G. M. T. Janssen*
PRO 41: RILEM International Symposium on Environment-Conscious Materials and Systems for Sustainable Development (ISBN: 2-912143-55-1); *Eds. N. Kashino and Y. Ohama*
PRO 42: SCC'2005—China: 1st International Symposium on Design, Performance and Use of Self-Consolidating Concrete (ISBN: 2-912143-61-6); *Eds. Zhiwu Yu, Caijun Shi, Kamal Henri Khayat and Youjun Xie*

PRO 43: International RILEM Workshop on Bonded Concrete Overlays (e-ISBN: 2-912143-83-7); *Eds. J. L. Granju and J. Silfwerbrand*
PRO 44: 2nd International RILEM Workshop on Microbial Impacts on Building Materials (CD11) (e-ISBN: 2-912143-84-5); *Ed. M. Ribas Silva*
PRO 45: 2nd International Symposium on Nanotechnology in Construction, Bilbao (ISBN: 2-912143-87-X); *Eds. Peter J. M. Bartos, Yolanda de Miguel and Antonio Porro*
PRO 46: ConcreteLife'06—International RILEM-JCI Seminar on Concrete Durability and Service Life Planning: Curing, Crack Control, Performance in Harsh Environments (ISBN: 2-912143-89-6); *Ed. K. Kovler*
PRO 47: International RILEM Workshop on Performance Based Evaluation and Indicators for Concrete Durability (ISBN: 978-2-912143-95-2); *Eds. V. Baroghel-Bouny, C. Andrade, R. Torrent and K. Scrivener*
PRO 48: 1st International RILEM Symposium on Advances in Concrete through Science and Engineering (e-ISBN: 2-912143-92-6); *Eds. J. Weiss, K. Kovler, J. Marchand, and S. Mindess*
PRO 49: International RILEM Workshop on High Performance Fiber Reinforced Cementitious Composites in Structural Applications (ISBN: 2-912143-93-4); *Eds. G. Fischer and V.C. Li*
PRO 50: 1st International RILEM Symposium on Textile Reinforced Concrete (ISBN: 2-912143-97-7); *Eds. Josef Hegger, Wolfgang Brameshuber and Norbert Will*
PRO 51: 2nd International Symposium on Advances in Concrete through Science and Engineering (ISBN: 2-35158-003-6; e-ISBN: 2-35158-002-8); *Eds. J. Marchand, B. Bissonnette, R. Gagné, M. Jolin and F. Paradis*
PRO 52: Volume Changes of Hardening Concrete: Testing and Mitigation (ISBN: 2-35158-004-4; e-ISBN: 2-35158-005-2); *Eds. O. M. Jensen, P. Lura and K. Kovler*
PRO 53: High Performance Fiber Reinforced Cement Composites—HPFRCC5 (ISBN: 978-2-35158-046-2); *Eds. H. W. Reinhardt and A. E. Naaman*
PRO 54: 5th International RILEM Symposium on Self-Compacting Concrete (ISBN: 978-2-35158-047-9); *Eds. G. De Schutter and V. Boel*
PRO 55: International RILEM Symposium Photocatalysis, Environment and Construction Materials (ISBN: 978-2-35158-056-1); *Eds. P. Baglioni and L. Cassar*
PRO56: International RILEM Workshop on Integral Service Life Modelling of Concrete Structures (ISBN 978-2-35158-058-5); *Eds. R. M. Ferreira, J. Gulikers and C. Andrade*
PRO57: RILEM Workshop on Performance of cement-based materials in aggressive aqueous environments (e-ISBN: 978-2-35158-059-2); *Ed. N. De Belie*
PRO58: International RILEM Symposium on Concrete Modelling—CONMOD'08 (ISBN: 978-2-35158-060-8); *Eds. E. Schlangen and G. De Schutter*
PRO 59: International RILEM Conference on On Site Assessment of Concrete, Masonry and Timber Structures—SACoMaTiS 2008 (ISBN set: 978-2-35158-061-5); *Eds. L. Binda, M. di Prisco and R. Felicetti*

PRO 60: Seventh RILEM International Symposium on Fibre Reinforced Concrete: Design and Applications—BEFIB 2008 (ISBN: 978-2-35158-064-6); *Ed. R. Gettu*

PRO 61: 1st International Conference on Microstructure Related Durability of Cementitious Composites 2 vol., (ISBN: 978-2-35158-065-3); *Eds. W. Sun, K. van Breugel, C. Miao, G. Ye and H. Chen*

PRO 62: NSF/ RILEM Workshop: In-situ Evaluation of Historic Wood and Masonry Structures (e-ISBN: 978-2-35158-068-4); *Eds. B. Kasal, R. Anthony and M. Drdácký*

PRO 63: Concrete in Aggressive Aqueous Environments: Performance, Testing and Modelling, 2 vol., (ISBN: 978-2-35158-071-4); *Eds. M. G. Alexander and A. Bertron*

PRO 64: Long Term Performance of Cementitious Barriers and Reinforced Concrete in Nuclear Power Plants and Waste Management—NUCPERF 2009 (ISBN: 978-2-35158-072-1); *Eds. V. L'Hostis, R. Gens, C. Gallé*

PRO 65: Design Performance and Use of Self-consolidating Concrete—SCC'2009 (ISBN: 978-2-35158-073-8); *Eds. C. Shi, Z. Yu, K. H. Khayat and P. Yan*

PRO 66: 2nd International RILEM Workshop on Concrete Durability and Service Life Planning—ConcreteLife'09 (ISBN: 978-2-35158-074-5); *Ed. K. Kovler*

PRO 67: Repairs Mortars for Historic Masonry (e-ISBN: 978-2-35158-083-7); *Ed. C. Groot*

PRO 68: Proceedings of the 3rd International RILEM Symposium on 'Rheology of Cement Suspensions such as Fresh Concrete (ISBN 978-2-35158-091-2); *Eds. O. H. Wallevik, S. Kubens and S. Oesterheld*

PRO 69: 3rd International PhD Student Workshop on 'Modelling the Durability of Reinforced Concrete (ISBN: 978-2-35158-095-0); *Eds. R. M. Ferreira, J. Gulikers and C. Andrade*

PRO 70: 2nd International Conference on 'Service Life Design for Infrastructure' (ISBN set: 978-2-35158-096-7, e-ISBN: 978-2-35158-097-4); *Ed. K. van Breugel, G. Ye and Y. Yuan*

PRO 71: Advances in Civil Engineering Materials—The 50-year Teaching Anniversary of Prof. Sun Wei' (ISBN: 978-2-35158-098-1; e-ISBN: 978-2-35158-099-8); *Eds. C. Miao, G. Ye, and H. Chen*

PRO 72: First International Conference on 'Advances in Chemically-Activated Materials—CAM'2010' (2010), 264 pp, ISBN: 978-2-35158-101-8; e-ISBN: 978-2-35158-115-5; *Eds. Caijun Shi and Xiaodong Shen*

PRO 73: 2nd International Conference on 'Waste Engineering and Management—ICWEM 2010' (2010), 894 pp, ISBN: 978-2-35158-102-5; e-ISBN: 978-2-35158-103-2; *Eds. J. Zh. Xiao, Y. Zhang, M. S. Cheung and R. Chu*

PRO 74: International RILEM Conference on 'Use of Superabsorsorbent Polymers and Other New Addditives in Concrete' (2010) 374 pp., ISBN: 978-2-35158-104-9; e-ISBN: 978-2-35158-105-6; *Eds. O. M. Jensen, M. T. Hasholt, and S. Laustsen*

PRO 75: International Conference on 'Material Science—2nd ICTRC—Textile Reinforced Concrete—Theme 1' (2010) 436 pp., ISBN: 978-2-35158-106-3; e-ISBN: 978-2-35158-107-0; Ed. W. Brameshuber

PRO 76: International Conference on 'Material Science—HetMat—Modelling of Heterogeneous Materials—Theme 2' (2010) 255 pp., ISBN: 978-2-35158-108-7; e-ISBN: 978-2-35158-109-4; Ed. W. Brameshuber

PRO 77: International Conference on 'Material Science—AdIPoC—Additions Improving Properties of Concrete—Theme 3' (2010) 459 pp., ISBN: 978-2-35158-110-0; e-ISBN: 978-2-35158-111-7; *Ed. W. Brameshuber*

PRO 78: 2nd Historic Mortars Conference and RILEM TC 203-RHM Final Workshop—HMC2010 (2010) 1416 pp., e-ISBN: 978-2-35158-112-4; *Eds. J. Válek, C. Groot, and J. J. Hughes*

PRO 79: International RILEM Conference on Advances in Construction Materials Through Science and Engineering (2011) 213 pp., e-ISBN: 978-2-35158-117-9; *Eds. Christopher Leung and K. T. Wan*

PRO 80: 2nd International RILEM Conference on Concrete Spalling due to Fire Exposure (2011) 453 pp., ISBN: 978-2-35158-118-6, e-ISBN: 978-2-35158-119-3; *Eds. E. A. B. Koenders and F. Dehn*

PRO 81: 2nd International RILEM Conference on Strain Hardening Cementitious Composites (SHCC2-Rio) (2011) 451 pp., ISBN: 978-2-35158-120-9, e-ISBN: 978-2-35158-121-6; *Eds. R. D. Toledo Filho, F. A. Silva, E. A. B. Koenders and E. M. R. Fairbairn*

PRO 82: 2nd International RILEM Conference on Progress of Recycling in the Built Environment (2011) 507 pp., e-ISBN: 978-2-35158-122-3; *Eds. V. M. John, E. Vazquez, S. C. Angulo and C. Ulsen*

PRO 83: 2nd International Conference on Microstructural-related Durability of Cementitious Composites (2012) 250 pp., ISBN: 978-2-35158-129-2; e-ISBN: 978-2-35158-123-0; *Eds. G. Ye, K. van Breugel, W. Sun and C. Miao*

PRO 85: RILEM-JCI International Workshop on Crack Control of Mass Concrete and Related issues concerning Early-Age of Concrete Structures—ConCrack 3—Control of Cracking in Concrete Structures 3 (2012) 237 pp., ISBN: 978-2-35158-125-4; e-ISBN: 978-2-35158-126-1; *Eds. F. Toutlemonde and J.-M. Torrenti*

PRO 86: International Symposium on Life Cycle Assessment and Construction (2012) 414 pp., ISBN: 978-2-35158-127-8, e-ISBN: 978-2-35158-128-5; *Eds. A. Ventura and C. de la Roche*

PRO 87 draft: UHPFRC 2013—RILEM-fib-AFGC International Symposium on Ultra-High Performance Fibre-Reinforced Concrete (2013), ISBN: 978-2-35158-130-8, e-ISBN: 978-2-35158-131-5; *Eds. F. Toutlemonde*

PRO 88: 8th RILEM International Symposium on Fibre Reinforced Concrete (2012) 344 pp., ISBN: 978-2-35158-132-2, e-ISBN: 978-2-35158-133-9; *Eds. Joaquim A. O. Barros*

PRO 90: 7th RILEM International Conference on Self-Compacting Concrete and of the 1st RILEM International Conference on Rheology and Processing of Construction Materials (2013) 396 pp, ISBN: 978-2-35158-137-7, e-ISBN: 978-2-35158-138-4; *Eds. Nicolas Roussel and Hela Bessaies-Bey*

RILEM Reports

Report 19: Considerations for Use in Managing the Aging of Nuclear Power Plant Concrete Structures (ISBN: 2-912143-07-1); *Ed. D. J. Naus*
Report 20: Engineering and Transport Properties of the Interfacial Transition Zone in Cementitious Composites (ISBN: 2-912143-08-X); *Eds. M. G. Alexander, G. Arliguie, G. Ballivy, A. Bentur and J. Marchand*
Report 21: Durability of Building Sealants (ISBN: 2-912143-12-8); *Ed. A. T. Wolf*
Report 22: Sustainable Raw Materials—Construction and Demolition Waste (ISBN: 2-912143-17-9); *Eds. C. F. Hendriks and H. S. Pietersen*
Report 23: Self-Compacting Concrete state-of-the-art report (ISBN: 2-912143-23-3); *Eds. Å. Skarendahl and Ö. Petersson*
Report 24: Workability and Rheology of Fresh Concrete: Compendium of Tests (ISBN: 2-912143-32-2); *Eds. P. J. M. Bartos, M. Sonebi and A. K. Tamimi*
Report 25: Early Age Cracking in Cementitious Systems (ISBN: 2-912143-33-0); *Ed. A. Bentur*
Report 26: Towards Sustainable Roofing (Joint Committee CIB/RILEM) (CD 07) (e-ISBN 978-2-912143-65-5); *Eds. Thomas W. Hutchinson and Keith Roberts*
Report 27: Condition Assessment of Roofs (Joint Committee CIB/RILEM) (CD 08) (e-ISBN 978-2-912143-66-2); *Ed. CIB W 83/RILEM TC166-RMS*
Report 28: Final report of RILEM TC 167-COM 'Characterisation of Old Mortars with Respect to Their Repair (ISBN: 978-2-912143-56-3); *Eds. C. Groot, G. Ashall and J. Hughes*
Report 29: Pavement Performance Prediction and Evaluation (PPPE): Interlaboratory Tests (e-ISBN: 2-912143-68-3); *Eds. M. Partl and H. Piber*
Report 30: Final Report of RILEM TC 198-URM 'Use of Recycled Materials' (ISBN: 2-912143-82-9; e-ISBN: 2-912143-69-1); *Eds. Ch. F. Hendriks, G. M. T. Janssen and E. Vázquez*
Report 31: Final Report of RILEM TC 185-ATC 'Advanced testing of cement-based materials during setting and hardening' (ISBN: 2-912143-81-0; e-ISBN: 2-912143-70-5); *Eds. H. W. Reinhardt and C. U. Grosse*
Report 32: Probabilistic Assessment of Existing Structures. A JCSS publication (ISBN 2-912143-24-1); *Ed. D. Diamantidis*
Report 33: State-of-the-Art Report of RILEM Technical Committee TC 184-IFE 'Industrial Floors' (ISBN 2-35158-006-0); *Ed. P. Seidler*
Report 34: Report of RILEM Technical Committee TC 147-FMB 'Fracture mechanics applications to anchorage and bond' Tension of Reinforced Concrete Prisms—Round Robin Analysis and Tests on Bond (e-ISBN 2-912143-91-8); *Eds. L. Elfgren and K. Noghabai*
Report 35: Final Report of RILEM Technical Committee TC 188-CSC 'Casting of Self Compacting Concrete' (ISBN 2-35158-001-X; e-ISBN: 2-912143-98-5); *Eds. Å. Skarendahl and P. Billberg*
Report 36: State-of-the-Art Report of RILEM Technical Committee TC 201-TRC 'Textile Reinforced Concrete' (ISBN 2-912143-99-3); *Ed. W. Brameshuber*

Report 37: State-of-the-Art Report of RILEM Technical Committee TC 192-ECM 'Environment-conscious construction materials and systems' (ISBN: 978-2-35158-053-0); *Eds. N. Kashino, D. Van Gemert and K. Imamoto*
Report 38: State-of-the-Art Report of RILEM Technical Committee TC 205-DSC 'Durability of Self-Compacting Concrete' (ISBN: 978-2-35158-048-6); *Eds. G. De Schutter and K. Audenaert*
Report 39: Final Report of RILEM Technical Committee TC 187-SOC 'Experimental determination of the stress-crack opening curve for concrete in tension' (ISBN 978-2-35158-049-3); *Ed. J. Planas*
Report 40: State-of-the-Art Report of RILEM Technical Committee TC 189-NEC 'Non-Destructive Evaluation of the Penetrability and Thickness of the Concrete Cover' (ISBN 978-2-35158-054-7); *Eds. R. Torrent and L. Fernández Luco*
Report 41: State-of-the-Art Report of RILEM Technical Committee TC 196-ICC 'Internal Curing of Concrete' (ISBN 978-2-35158-009-7); *Eds. K. Kovler and O. M. Jensen*
Report 42: 'Acoustic Emission and Related Non-destructive Evaluation Techniques for Crack Detection and Damage Evaluation in Concrete'—Final Report of RILEM Technical Committee 212-ACD (e-ISBN: 978-2-35158-100-1); *Ed. M. Ohtsu*

Preface

Dear Colleagues,

We are pleased to provide you with this state-of-the-art report on the mechanical properties of Self-Compacting Concrete (SCC). Given to the fast growth of research output in the field of SCC and the growing acceptance of this high-performance material around the world in both precast and cast-in-place applications, the members of the RILEM Technical Committee TC 228-MPS "Mechanical Properties of SCC" have worked diligently to provide a concise summary of key topics pertaining to the mechanical properties of SCC. This report comprises of eight chapters covering various topics pertaining to mechanical properties of SCC. This includes compressive, tensile, flexural, and shear strengths and elastic modulus, stress–strain behaviour, and mechanical properties at elevated temperatures. The report also discusses creep, drying and autogenous shrinkage behaviours of SCC and the applicability of various models to SCC. Bond strength to reinforcement, existing concrete, and freshly cast SCC are discussed. A review of the main structural behaviour characteristics of SCC used in various structural elements is provided. A comprehensive review of the behaviour of fibre-reinforced SCC (FR-SCC) is provided. Finally, the report discusses the properties of specialty SCC including lightweight and heavyweight SCC, preplaced aggregate SCC, underwater concrete, self-levelling concrete, and self-compacting repair mortar and concrete.

I would like to acknowledge the contributions of all 26 members of this TC 228-MPS who have worked diligently since 2008 to gather a large body of technical knowledge literature that is synthesized in report. Particular thanks go to the lead authors of the various chapters listed below for championing the efforts in analysing large body of technical literature and synthesizing the findings and the Committee's option in their respective chapters. Special thanks are extended to the Secretary of the Committee, Prof. Geert De Schutter, for his help in coordinating the work of the Committee and the preparation of this state-of-the-art report on the mechanical properties of SCC.

Kamal H. Khayat

Contents

Authors

Veerle Boel Ghent University, Ghent, Belgium

Bart Craeye Artesis University College of Antwerp, Antwerp, Belgium

Geert De Schutter Ghent University, Ghent, Belgium

Pieter Desnerck Center for Infrastructure Engineering Studies, Missouri University of Science and Technology, Rolla, MO 65409-0710, USA; Ghent University, Ghent, Belgium

Liberato Ferrara Politecnico di Milano, Milan, Italy

Assem Hassan Memorial University of Newfoundland, Newfoundland, Canada

Kamal H. Khayat Civil, Architectural and Environmental Engineering, Missouri University of Science and Technology, Rolla, MO 65409-0710, USA

Mohamed Lachemi Ryerson University, Toronto, Canada

Andreas Leemann EMPA, Dübendorf, Switzerland

Pietro Lura EMPA, Dübendorf, Switzerland

Claudio Mazzotti University of Bologna, Bologna, Italy

Caijun Shi Hunan University, Changsha, China

Mohamed Sonebi Queen's University Belfast, Belfast, UK

Petra Van Itterbeeck BBRI, Brussels, Belgium

Manuel Vieira LNEC, Lissabon, Portugal

Chapter 1
Introduction and Glossary

Geert De Schutter and Kamal H. Khayat

1.1 Introduction

Self-compacting concrete (SCC) [1], also known as self-consolidating concrete [2], was developed in the late 1980's, although earlier 'look-alikes' surely exist, though not defined as such. In comparison with conventional concrete, referred to in this report as vibrated concrete (VC), SCC can be considered on the one hand as a new type of high-performance material of a different approach to mix design and rheological characteristics. On the other hand, SCC can be seen as a new approach to casting concrete enabled by adjusted fresh concrete properties. In reality, SCC is a combination of both approaches that has enabled to push the boundaries of concrete technology to a new area.

Several methods exist for the mix design of SCC, as explained in [3]. In various parts of the world, different concepts might be followed for the proportioning of SCC and are referred to as 'powder-type SCC', 'VMA-type SCC', or 'mixed-type SCC'. This makes it complicated to present general conclusions and recommendations concerning the mix design and specific characteristics of SCC. Since the introduction of 'modern' SCC, RILEM has been very active in producing state-of-the-art reports related to this innovative class of high-performance cementitious material. At first, focus was aimed at describing the mix design and workability of SCC [3] then on mixing and casting of SCC [4]. Later on, other aspects of SCC were targeted, including durability [5]. Also, the American Concrete Institute (ACI) Committee 237 "SCC" is working on producing a comprehensive state-of-the-art report dealing with all aspects of SCC [6].

G. De Schutter (✉)
Ghent University, Ghent, Belgium
e-mail: geert.deschutter@ugent.be

K. H. Khayat
Missouri University of Science and Technology, Rolla, MO, USA
e-mail: khayatk@mst.edu

K. H. Khayat and G. De Schutter (eds.), *Mechanical Properties of Self-Compacting Concrete*, RILEM State-of-the-Art Reports 14,
DOI: 10.1007/978-3-319-03245-0_1, © RILEM 2014

Although SCC has been used in actual structures for the last 20 years around the world, until recently it was not fully clear whether existing design codes for structural design can be fully applied when using SCC. Some initial problems have been reported in the past, e.g. bond and shear behaviour of SCC. Due to the lower aggregate content, aggregate interlock was expected to play a limited role in comparison with VC, which might result in a reduced bond and shear strengths. Another point of discussion was the strain-softening behaviour of SCC under compression, and linked to that, it was not clear to which extent the traditional parabola-rectangular constitutive law used for VC would be applicable to SCC. Creep and shrinkage of SCC have also been extensively discussed in the literature, leading to contradictory conclusions (see Chap. 3).

In recent years, many research groups have been investigating the mechanical properties and structural behaviour of SCC. Recent publications have answered a lot of questions, showing that the mechanical behaviour of SCC is similar to that of VC. Although the contribution of aggregate interlock to the shear resistance might be lower in case of SCC, the greater intrinsic quality of the matrix and of the interfacial transition zone (ITZ) with the aggregate seem to compensate this negative effect.

The compressive stress-strain relation for SCC seems to be fundamentally of the same nature as for VC; however, with slightly lower Young's elastic modulus, and with slightly higher peak strain. This latter issue seems to depend on the powder type, and still needs further fundamental evaluation. The toughness of SCC seems to be slightly higher than VC of similar compressive strength, which is mainly due to the effect of the higher peak strain value, as the strain softening curve is quite comparable to the case of VC.

In spite of some early problems related to relatively low bond strength to reinforcement when using SCC, which are probably related to segregating of the SCC mixtures, bond between SCC and reinforcing steel and prestressing strands was found to be as good as in the case of VC. For small bar diameters, bond of SCC can be significantly higher than that of VC of similar compressive strength. When the SCC mixture is designed to exhibit proper static stability, the top-bar effect of SCC can be similar or even lower than VC.

One aspect which will certainly be further developed in the future is the incorporation of fibers in SCC. It has already been shown that excellent properties can be obtained when adding steel fibers to SCC, while maintaining appropriate rheological properties needed to remain self-compaction properties. The alignment of the rigid steel fibers, following the flow patterns of the SCC, when cast into relatively thin formwork section, could be better exploited by applying advanced Computational Fluid Dynamics (CFD) techniques. Casting operations could be optimized to obtain the desired fiber orientation, depending on the mechanical actions in the finalized structure.

The previous paragraphs have pointed to some existing discussions related to the mechanical properties of SCC, and to some challenges and remaining research areas. In order to give a detailed answer to these and other questions and discussions, based on thorough understanding of various mechanical properties of this new class of high-performance cementitious material, this report presents the state-of-the art review of the mechanical properties of SCC.

In the sequel of Chap. 1, a glossary is presented. In Chap. 2, an overview of mechanical properties is given on the material level. Chap. 3 discusses creep and shrinkage behaviour, also evaluates the application of existing creep and shrinkage code models to SCC. A detailed study of bond properties between reinforcement and SCC as well as between multi layers of SCC and bond to existing concrete is presented in Chap. 4. The link between materials and structures is given in Chap. 5, discussing the structural behaviour of various types of SCC elements. FRSCC is reported in Chap. 6, including the important issue of fiber orientation and structural performance of FRSCC. The mechanical properties of some special SCC are summarized in Chap. 7, including lightweight and heavyweight SCC, preplaced aggregate SCC, underwater concrete, self-levelling concrete, and self-compacting repair mortar and concrete. Finally, a summary and general conclusions are presented in Chap. 8.

1.2 Glossary

Addition

A fine-grained inorganic material. Includes two types [EN206]: inert or nearly inert addition (Type I) and pozzolanic or latent hydraulic addition (Type II). (See also supplementary cementitious materials, fines, and fine fillers).

Admixture

A material, other than water, aggregate, hydraulic cement, and fiber reinforcement, used as an ingredient of a cementitious mixture to modify its freshly mixed, setting, or hardened properties and that is added to the batch before or during its mixing.

Air content

Volume of air-voids in fresh or hardened concrete usually expressed as a percentage of the total volume of the mix. Vibrated concrete is usually considered to be fully compacted when its air content is below 1.5 %, excluding any entrained air.

Air-entrainment

Intentional introduction of air uniformly distributed in small air bubbles into concrete, primarily to improve frost resistance. Volume of entrained air in SCC usually varies between 5–8 % of the total volume of concrete.

Anti-washout agents (AWA)

Stabilizer/thickener applied in the case of underwater concreting.

Autogenous shrinkage

The external-macroscopical (bulk) dimensional reduction (volumetric or linear) of the cementitious system, which occurs under sealed, isothermal, and unrestrained conditions.

Binder

A cementitious material, either a hydrated cement or reaction product of cement or lime and reactive siliceous material; the kind of cement and curing conditions governing the characteristics of the product formed. Also materials such as asphalt, resins, and other materials forming the matrix of concrete, mortar, and sanded grout.

Bingham fluid

A material which when subjected to shear stress behaves as an elastic solid until the yield stress is reached, after which there is a linear variation between shear stress and rate of strain (velocity gradient). The slope of the linear relationship refers to the plastic viscosity. Fresh cement-based materials at relatively low shear rate values and early age typically behave as Bingham fluids.

Bleed water

The water that rises to the surface subsequent to the placing of the concrete. The rise of mixing water within, or its emergence from, newly placed concrete, can be caused by consolidation and settlement of the plastic concrete.

Bleeding test

The standard test for determining the relative quantity of mixing water that will bleed from a sample of freshly mixed concrete.

Blocking

The condition in which coarse aggregate particles combine to form elements large enough to obstruct the flow of the fresh concrete among closely spaced reinforcement and/or other obstructions in the formwork.

Bond

The attachment of a material at the interface of two surfaces.

Cohesiveness

The tendency of the concrete constituent materials to adhere together, resulting in resistance to segregation, settlement, and bleeding.

Compaction

The process in which the volume of entrapped air in fresh concrete is reduced below 1–2 % of total volume of concrete, usually by mechanical means, such as vibration. Interchangeable with "consolidation". In SCC, compaction (consolidation) is achieved by gravity flow of the material without the need of mechanical vibration, rodding, or tamping.

Compressive strength

The capacity to withstand axially directed pushing stresses. When the limit of compressive strength is reached, the concrete is crushed.

Consolidation

See compaction.

Conventional vibrated concrete
See vibrated concrete.

Consistency
The relative mobility or ability of freshly mixed concrete or mortar to flow.

Creep
Time-dependent deformation due to sustained load.

Drying shrinkage
Shrinkage occurring in a specimen that is exposed to the environment and allowed to dry.

Ductility
The extent to which materials can be deformed without failure.

Fatigue
The progressive structural damage that occurs when a material is subjected to cyclic loading.

Fiber
A slender or threadlike component sometimes used to strengthen cementitious materials, e.g. metal, synthetic, or natural.

Fillers
Finely divided inert materials, such as limestone powder, silica, or colloidal substances, sometimes added to Portland cement, paint or other materials to reduce shrinkage, improve workability, or act as an extender or material used to fill an opening in a form.

Filling ability
The ability of SCC to flow into the formwork and fill completely all spaces within the formwork under its own weight, also referred to as deformability or non-restricted deformability. It is sometimes referred to as 'flowability'.

Fines and fine fillers
Includes powder (see Powder) and materials finer than 0.125 mm (No. 100 sieve).

Flexural strength
Highest tensile stress experienced within the material at its moment of rupture in flexure.

Flowability
See Filling ability.

Flow-rate
The speed at which a sample of a fresh concrete mixture spreads horizontally outwards from the base of a slump cone until it reaches a selected radial distance, usually a concentric ring of 500 mm diameter. It is indicated by a 'flow-time', such as t_{50}, interchangeable with t_{500}, measured in second.

Flow-time

Usually the time taken for the concrete flow in an Orimet, V-funnel, or O-funnel test; an indication of the level of filling ability and rate of flow. Flow-time can be used to assess plastic viscosity and risk of blocking of SCC.

Fluidity

The ease by which fresh concrete flows under gravity. In rheology, fluidity is the reciprocal value of viscosity.

Fly ash

The finely divided residue that results from the combustion of ground or powdered coal and that is transported by flue gasses from the combustion zone to the particle removal system. Because of its spherical shape, fineness, and pozzolanic activity, fly ash can modify the fresh and hardened properties of concrete.

Formwork pressure

Lateral pressure acting on vertical or inclined formed surfaces, resulting from the fluid-like behaviour of the unhardened concrete confined by the forms.

Fracture energy

The energy needed to create a fracture surface in a material.

Ground granulated blast-furnace slag (GGBFS, also known as GGBS)

A fine granular, mostly latent hydraulic binding material that can be added to concrete to modify its fresh and hardened characteristics. A waste product in the manufacture of pig iron and chemically a mixture of lime, silica, and alumina. GGBFS is also referred to in some cases as slag cement.

High-range water-reducing admixture (HRWRA)

A water-reducing admixture capable of producing large water reduction or greater flowability, also known as superplasticiser.

J-ring test

Test used to determine the passing ability of SCC, or the degree to which the passage of concrete through the bars of the J-Ring apparatus is restricted. Results are expressed as blocking 'step' BJ (mm), spread SFj (mm) or flow-time t_{J50} (second) (= t_{J500}).

J-Ring flow

The distance of lateral flow of concrete using the J-Ring in combination with a slump cone.

L-Box test

Test used to assess the confined flow of SCC and the extent to which it is subject to blocking by reinforcement.

Metakaolin

Mineral admixture used as binding material (supplementary cementitious material) in concrete. A highly processed reactive aluminosilicate pozzolan, a finely-divided material that reacts with slaked lime at ordinary temperature and in the presence of moisture to form strong slow-hardening cement. It is formed by calcining purified kaolinite, generally between 650–700 °C in an externally fired rotary kiln.

Mixture robustness
The characteristic of a mixture that encompasses its tolerance to variations in constituent characteristics and quantities, variations during concrete mixing, transport, and placement, as well as environmental conditions.

Modulus of elasticity
The slope of the stress-strain curve in the elastic deformation region.

Mortar
The fraction of concrete with maximum aggregate size $\leq$4 mm. The maximum size can be different (e.g. 4.75 mm as per ASTM, ACI, and CSA specifications).

Newtonian fluid
A fluid, which shows a linear relationship between shear rate and shear stress, passing through the origin in a simple shear flow. Fresh SCC mixtures of high filling ability can display this type of behaviour. The slope of the linear relationship is the Newtonian viscosity (or viscosity).

Passing ability
The ability of SCC to flow under its own weight (without vibration) and completely fill all spaces within intricate formwork, containing obstacles, such as reinforcement.

Paste
Mixture of cementitious materials and fillers with water, with or without chemical admixtures.

Paste volume
Volume of paste in concrete or in mortar; it can be expressed as volume percent of the entire mixture.

Plastic settlement
Settlement of a concrete after casting while the material is still in fresh state.

Plastic viscosity
The resistance of the plastic material to undergo a given flow. It is computed as the slope of the shear stress versus shear rate curve measurements. Concrete with higher plastic viscosity takes longer to flow. It is closely related to t50 and V-funnel time (higher plastic viscosity: higher t50 and V-funnel time). (See also Bingham fluid).

Powder
Includes cement, fillers, and additions.

Powder-type SCC
SCC mixtures that rely extensively on the amount and character of the fines and powder included in the mixture for meeting workability performance requirements (passing ability and stability).

Pump-ability
The ability of a SCC mixture to be pumped without significant degradation of workability or blockage.

Rate of flow
See flow-rate.

Rheology
The science of flow and deformation of matter. It includes studies of the handling and placing of freshly mixed concrete, and the behaviour of slurries, pastes, and the like. In the context of SCC, rheology refers to the evaluation of yield stress, plastic viscosity, and thixotropy to achieve the desired level of filling ability, passing ability, and segregation resistance. Rheology is also used to evaluate deformation of hardened concrete.

Rheological properties
Properties dealing with the deformation and flow of matter.

Robustness
See mixture robustness.

Segregation
The differential concentration of the components of mixed concrete, aggregate, or the like, thus resulting in non-uniform proportions in the mass. In the case of SCC, segregation may occur during transport, flow into the form, or after placement when the concrete is in a plastic state. This results in non-uniform distribution of in situ properties of the concrete.

Segregation resistance
The ability of concrete to remain uniform in terms of composition during placement and until setting. Segregation resistance encompasses both dynamic and static stability.

Self-compacting concrete (SCC)
Fresh concrete which has an ability to flow under its own weight, fill the required space or formwork completely and produce a dense and adequately homogeneous material without a need for mechanical compaction.

Self-consolidating concrete (SCC)
See self-compacting concrete.

Self-leveling concrete
SCC with the ability to attain a fully levelled surface after casting.

Service life

The time during which the structure performs its design functions with and without unforeseen maintenance or repair.

Settlement

The condition in which the aggregates in SCC tend to sink to the bottom of the form resulting in non-homogeneous concrete. Surface settlement can also be caused by bleeding of free water and loss of air as well as movement of aggregate particles within fresh concrete (consolidation).

Shear strain

A deformation of a material in which parallel internal surfaces slide past one another.

Shear thinning (also known as pseudo-plasticity)

A material is shear thinning when the viscosity is decreasing at increasing shear rate.

Shear thickening

A material is shear thickening when the viscosity is increasing at increasing shear rate.

Shear strength

The strength of a material when loaded in shear.

Shear stress

The stress component acting tangentially to a plane.

Shrinkage

Contraction of concrete due to loss of water, e.g. due to the hydration process (autogenous shrinkage) or due to water loss to the environment (drying shrinkage).

Shrinkage-reducing agent

Admixture added to cementitious materials to reduce the shrinkage.

Sieve segregation test

Test for assessment of resistance to (static) segregation. Result is a Segregation index SI indicating the proportion of mortar, which separates through a sieve within a given period of time.

Silica fume

Very fine amorphous silica produced in electric arc furnaces as a by-product of the production of elemental silicon or alloys containing silicon. Silica fume can be added to SCC to improve the rheological and mechanical properties.

Slump flow retention

The ability of concrete to maintain its slump flow over a given period of time.

Slump flow spread

The distance of lateral flow of concrete during the slump-flow test. Slump flow spread is the numerical value of flow determined as the average diameter of the circular deposit of SCC at the conclusion of the slump flow test.

Slump-flow test

A test for filling ability of SCC using a 300 mm tall conical mould (as for the slump test). The primary result is the average horizontal spread diameter (SF).

Slump test

A test for consistency of fresh traditional vibrated concrete. Result is expressed as the depth of the vertical settlement of concrete after removal of a standard (Abrams) mould 300 mm high.

Speed of flow

See flow-rate.

Spread

The result of the 'Slump-flow' (SF) or J-ring (SFJ) tests.

Stability

The ability of a concrete mixture to resist segregation of the paste from the aggregate.

Stability, Dynamic – The resistance to segregation when external energy is applied to concrete - namely during placement.

Stability, Static – The resistance to segregation when no external energy is applied to concrete - namely from immediately after placement and until the onset of hardening.

Stabilizer

Admixture applied to improve the stability of the mixture (See al-so viscosity-modifying admixture).

Strain

Geometrical measure of deformation representing the relative displacement between particles in a material.

Stress

Amount of force exerted per unit area.

Superplasticiser

An admixture producing much-increased consistency of a fresh mixture without significant retardation or air-entrainment. Also known as 'high-range water reducing admixture—HRWRA'.

Supplementary cementitious materials

See Additions; [EN206]: Type II pozzolanic or latent hydraulic addition.

t_{50} measurement
(also referred to as the T-20 in.)—The time for the concrete to reach the 500 mm (20 in.) diameter spread drawn on the slump plate, after starting to raise the slump cone.

Tensile strength
Stress at which a material breaks when loaded in tension.

Texture
The pattern or configuration apparent in an exposed surface, as in concrete and mortar, including roughness, streaking, striation, or departure from flatness.

Thixotropy
The property of a material that enables it to stiffen in a short period while at rest (structural build-up), but to acquire a lower viscosity when mechanically agitated (structural break-down); the process being time-dependent and reversible. The material having this property is termed thixotropic.

Top-bar effect
The effect that the bond between a reinforcing bar and concrete can be of a lower quality when the bar is located in the upper zone of the concrete element.

Traditional vibrated concrete
See vibrated concrete.

Transfer length
The length needed before a reinforcing bar or fiber in concrete is exposed to full stress.

Transportability
The ability of concrete to be transported from the mixer to the placement site while remaining in a homogeneous condition.

V-funnel test
Device used to determine the time for a given volume of concrete to flow out through a funnel opening.

Vibrated concrete
Concrete, which requires vibration to achieve adequate compaction or consolidation.

Viscosity (plastic)
See plastic viscosity.

Viscosity-modifying admixture (VMA)
An admixture used for enhancing the rheological properties of cement-based materials in the plastic state to reduce the risk of segregation and washout. Also known as thickening agent, stabilizer, or viscosity-enhancing agent (VEA).

Visual Stability Index (VSI)

A test that involves the visual examination of the SCC slump flow spread resulting from performing the slump flow test. The test can also be conducted on exposed concrete surfaces, such as in wheel barrels.

Water-cement ratio (W/C)

The ratio of the mass of water, exclusive only of that absorbed by the aggregate, to the mass of cement in concrete, mortar, or grout.

Water-cement ratio, corrected ***(W/C_{corr})***—The ratio of the mass of water, exclusive only of that absorbed by the aggregate and including the water content in the admixtures, to the mass of cement in concrete, mortar, or grout.

Water-cementitious materials ratio (W/CM)

The ratio of the mass of water, exclusive only of that absorbed by the aggregate, to the mass of cementitious materials (cement and addition Type II) in concrete, mortar, or grout. Also referred to as water-binder ratio (W/B).

Water-powder ratio (W/P)

The ratio of the mass of water, exclusive only of that absorbed by the aggregate, to the mass of cement and other powder additions (Types I and II) in concrete, mortar, or grout.

Workability

That property of freshly mixed concrete or mortar that determines the ease with which it can be mixed, placed, consolidated, and finished into a homogenous condition. For SCC, workability encompasses filling ability, passing ability, and segregation resistance, and it is affected by rheology. (See also consistency).

Yield stress

The minimum shear stress required to initiate (static yield stress) or maintain (dynamic yield stress) flow. The dynamic yield stress is closely related to slump flow of SCC (lower yield stress results in higher slump flow). Dynamic yield stress is calculated as the intercept of the shear stress versus shear rate plot from rheometer flow curve measurements. Also known as yield value, the stress on the stress-strain or shear stress – rate of shear strain diagram which corresponds to the change from elastic to plastic behaviour of a solid material or static to dynamic behaviour of a Bingham fluid.

References

1. De Schutter, G., Bartos, P.J.M., Domone, P., Gibbs, J.: Self-Compacting Concrete. Whittles Publishing, CRC Press, Taylor and Francis Group, Caithness (2008)
2. Daczko, J.A.: Self-Consolidating Concrete: Applying What we Know. Spon Press, New York (2012)
3. Skarendahl, A., Petersson, O. (eds.): 'Self-compacting concrete', State-of-the-art report of RILEM Technical Committee 174-SCC. RILEM Publications, Bagneux (2000)

4. Skarendahl, A., Billberg, P. (eds.): 'Casting of self-compacting concrete', State-of-the-art report of RILEM Technical Committee 188-CSC. RILEM Publications, Bagneux (2006)
5. De Schutter, G., Audenaert, K. (eds.): 'Durability of self-compacting concrete', State-of-the-art report of RILEM Technical Committee 205-DSC. RILEM Publications, Bagneux (2007)
6. ACI Committee 237: 'Self-Consolidating Concrete', 237R-07 Self-Consolidating Concrete. RILEM Publications, Bagneux (2007)

Chapter 2
Mechanical Properties

Pieter Desnerck, Veerle Boel, Bart Craeye and Petra Van Itterbeeck

2.1 Introduction

For more than 20 years, self-compacting concrete (SCC) has been applied in the construction industry. In this period, a lot of research has been performed with regards to the applicability, mix design, pump-ability, durability, rheology, etc. of SCC. In the nineties, little attention was devoted to the mechanical properties of the material and to its structural performance. Recently, an increasing amount of research has been dedicated to these mechanical properties of SCC [1].

Research projects often include data on compressive strength, tensile strength, and Young's modulus, although the main focus remained on other aspects of the concrete as mentioned above. In the past, some authors, e.g. Domone [2] and Holschmacher [3], have presented surveys on the mechanical properties of SCC based on available literature. Due to a scarce amount of available test results, these studies were based on a limited set of data. In this way they did not include, neglect or generalise the influence of some major parameters for the mechanical behaviour, such as type of aggregate. Therefore, it was decided, in the scope of this chapter to develop an extensive database with results on fresh and hardened properties of SCC reported between 1990 and 2011, originating from numerous journal and conference papers.

P. Desnerck (✉)
Center for Infrastructure Engineering Studies, Missouri University of Science and Technology, Rolla, MO 65409-0710, USA

P. Desnerck · V. Boel
Ghent University, Ghent, Belgium

B. Craeye
Artesis University College of Antwerp, Antwerp, Belgium

P. Van Itterbeeck
BBRI, Brussels, Belgium

K. H. Khayat and G. De Schutter (eds.), *Mechanical Properties of Self-Compacting Concrete*, RILEM State-of-the-Art Reports 14,
DOI: 10.1007/978-3-319-03245-0_2,

Table 2.1 Ranges of reported fresh properties for SCC

Property		# measurements	Min.	10 % centile	Mean	90 % centile	Max.
Slump-flow	[mm]	1,545	320	600	690	780	933
V-funnel	[s]	741	1.0	3.5	9.3	16.0	57.2
L-box	[-]	563	0.20	0.73	0.86	0.97	1.12
Sieve stability	[%]	141	0.20	2.24	8.35	15.51	29.30
Air content	[%]	488	0.4	1.3	3.2	6.1	12.8

To allow for a full assessment of the mechanical performance of SCC this database also includes information on the mixture proportion and design, the fresh concrete properties and testing conditions (e.g. type and geometry of specimens). Mixture composition information included in the database are water content, cement content and type (Portland, binary types, etc.), addition content and type (limestone, fly ash, ground granulated blast-furnace slag, silica fume), viscosity modifying agent (VMA) content, superplasticiser content and type (polycarboxylic, etc.), air entrainment content, aggregate type, aggregate content and maximum aggregate size, etc.

To be able to quantify the self-compacting properties of the reviewed mixtures that are reported in the literature, results of fresh concrete property tests such as slump-flow (SF), V-funnel, L-Box, U-Box, sieve segregation, etc. were collected as well. With respect to the mechanical properties, information on the test age, test standards, specimens sizes and types (cylinders, cubes, prisms), compressive strength, direct tensile strength, splitting tensile strength, flexural tensile strength, and Young's modulus were gathered.

With more than 1,500 mixtures reviewed in the database that cover a wide range of SCC types (e.g. powder-type SCC, VMA-type SCC, Combination-type SCC) [4]—taken from more than 200 papers [5–238] (35 different countries)—an extensive dataset is available for thorough investigation of the influences of different parameters on the mechanical properties of SCC's. Based on the data entered in the database, the influence of some parameters, which could be of major interest to users, designers, etc. are derived (see following sections), including W/C, W/B, W/P, S/A, addition content, and ratio of cylinder to cube compressive strength.

In most studies, the fresh properties of the mixtures are characterized by means of the V-funnel, L-box, etc. The most commonly reported fresh property is the slump-flow. The ranges of fresh properties of the so claimed SCCs are listed in Table 2.1. For the V-funnel measurements, different dimensions of the outlet can be used. However, in papers the authors do not often include the applied type. Therefore, all V-funnel measurements are combined regardless of the dimensions.

As can be seen from Table 2.1, the slump-flow values range from 320 to 933 mm. The majority of the test results (95.1 %) are within the range of 550–850 mm defined by EN 206-9 [239] for SCC (Fig. 2.1).

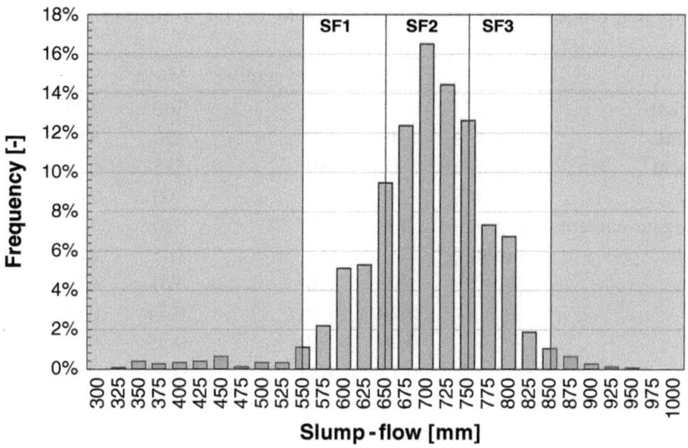

Fig. 2.1 Histogram of reported SF-values for SCC mixtures (based on 1,551 mixtures)

The committee believes that concrete mixtures with slump-flow values of 320 mm cannot be classified as self-compacting. Therefore, for the analysis of the hardened properties, only concrete mixtures with slump-flow values within the limits defined by EN206 (550–850 mm) were considered in this chapter.

As the database contains results from over 250 papers, it can provide insights on the ranges of SCC properties. These total ranges are of importance to determine classes of 'low', 'mean' and 'high' values of the different parameters which on their turn can be used in the analysis, as it will become clear in the following sections. SCC can be developed according to the powder-type SCC, VMA-type SCC or combination type SCC approach [4]. This is noted in the database as well. Based on the amount of mixtures in which a VMA is used, the powder-type approach was chosen in 83 % of the cases, whereas the VMA- or combination-type SCC mixtures were utilised in 17 % of the reported research projects.

Table 2.2 presents an overview of some of the major mixture properties of the different SCC mixtures that are considered in the database. Besides the maximum and minimum values, the mean values and standard deviations (SD) are provided as well as 10 % and 90 % centile values. Values of the water-to-cement ratio (W/C) are mentioned as well as the corrected water-to-cement ratio $(W/C)_{corr}$. The $(W/C)_{corr}$ considers the water content that is included in the superplasticiser as in some mixtures it can be as high as 14 l/m^3, which could lead to a significant contribution to the overall water content.

The cement, sand, and coarse aggregate contents generally vary in over a wide range, covering almost the entire range found for VC. With respect to the W/C, values between 0.19 and 2.73 are found with a mean value of 0.54. The highest reported W/C is relatively high, but it has to be emphasized that, in general, also other types of binders are used. All types of cement are applied in SCC mixtures: CEM I, CEM II, CEM III (according to European standards [240]), ASTM Type I,

Table 2.2 Applied ranges of mixture properties for SCC (1,474 mixtures out of 214 reported studies)

Property		Min.	10 % centile	Mean	90 % centile	Max.
Cement content	[kg/m^3]	83	250	360	490	700
Binder content	[kg/m^3]	150	300	420	550	750
Powder content[a]	[kg/m^3]	275	410	535	625	1272
Sand content	[kg/m^3]	135	600	800	940	1190
Coarse aggregate content	[kg/m^3]	283	650	820	1035	1740
Water content	[kg/m^3]	77	160	185	220	434
Paste volume	[l/m^3]	151	320	370	425	709
W/C	–	0.19	0.38	0.54	0.80	2.73
$(W/C)_{corr}$	–	0.20	0.39	0.55	0.81	2.81
W/B	–	0.15	0.33	0.46	0.63	1.33
W/P	–	0.14	0.27	0.36	0.45	0.65
C/P[b]	–	0.15	0.50	0.69	0.95	1.00
S/A[c]	–	0.10	0.37	0.50	0.60	0.81
Max. grain size	mm	5	10	16	20	25

[a] Powder = cement + SCMs + fillers
[b] C/P = cement to powder ratio by mass
[c] S/A = sand to aggregate ratio by mass

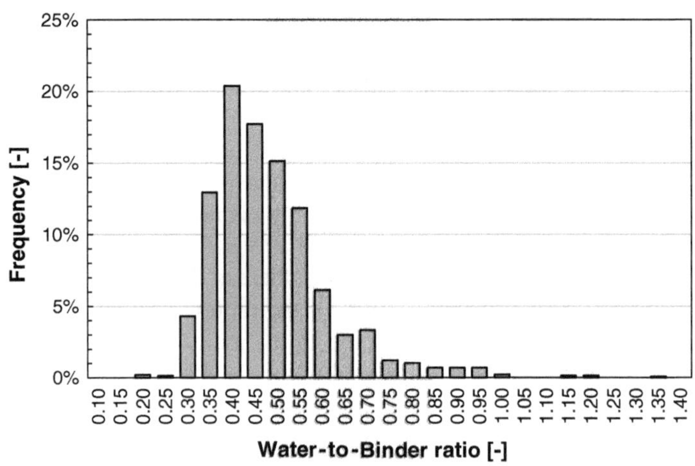

Fig. 2.2 Histogram of the applied W/B in SCC mixtures (based on 1,474 mixtures)

Type II, CEM Type GU (according to ASTM Standards [241]), etc. The W/B for the investigated SCC ranges from 0.15 to 1.33 with a mean value of 0.46. The majority of the mixtures were composed with a W/B between 0.33 and 0.63 (Fig. 2.2).

The higher powder content needed for SCC is obtained by increasing the cement content or more often by increasing the amount of additions (powder-type SCC). The addition can be composed of one or more types. As addition material,

Table 2.3 Origin of main addition materials (based on 1,352 mixtures)

Addition class	Amount (%)	Addition class	Amount (%)
Basalt	0.2	Natural pozzolan	0.2
Fly ash	35.1	Quartz	1.8
Granite	0.3	Rubble powder	0.2
Kiln dust	0.1	Silica	9.2
Limestone	40.7	Slag	9.0
Metakaolin	0.7	Vegetal ash	2.4

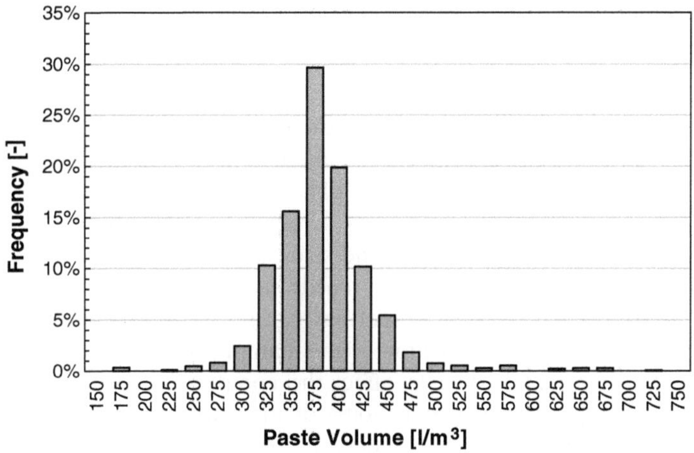

Fig. 2.3 Histogram of applied paste volumes in SCC mixtures (based on 1,474 mixtures)

inert as well as pozzolanic materials are widely used. The most used materials are limestone filler, fly ash, ground granulated blast-furnace slag, and silica fume. Less common fillers but nonetheless used in some cases, we note marble powder, glass powder, rice husk ash, metakaolin, volcanic ash, and granite powder.

In Table 2.3 the main addition type is mentioned together with the frequency of occurrence, as listed in the database. As SCC often is composed of larger amounts of fines compared to vibrated concrete, the powder content is higher as well. According to the database common powder contents for SCC mixtures are situated in the range of 410–625 kg/m^3.

Another important property influencing the mechanical properties of SCC is the paste volume. The total paste volume of a SCC mixture is, in most cases, higher than that of a VC mixture, which could have its repercussions on mechanical properties. The applied paste volumes (excluding air) in SCC mixtures can be seen in Fig. 2.3. Values of 150 l/m^3 up to 710 l/m^3 were found in the database with a mean value around 370 l/m^3, whereas for VC the mean value is situated around 290 l/m^3 [4].

The following paragraphs summarize the findings concerning SCC on the compressive strength, stress-strain relationship, modulus of elasticity, mechanical behaviour at elevated temperatures, and in situ properties. When applicable, the codes Eurocode 2 [257], ACI 318-08 [253], and CEB/FIP Model Code 2010 (MC2010) [268] are taken into consideration. In addition, more information regarding the application of other provisions can be found in NCHRP Report 628 [243].

2.2 Compressive Strength

2.2.1 Introduction

Design rules for concrete structures are often based on the compressive strength of the concrete. In Eurocode 2 concrete is classified solely on the basis of its compressive strength, in accordance with EN 206-1 [242] where cylinders 150/300 mm and cubes 150 mm are used as a reference. In other standards such as AASHTO and ASTM different types and sizes of specimens are also used. Many of the other mechanical properties (e.g. tensile strength, modulus of elasticity, and compressive strain) and physical properties (e.g. related to durability) of concrete are moreover expressed as a function of this parameter.

Many parameters are known to affect the compressive strength of concrete, most of them being interdependent. Some of the important parameters that may affect the compressive strength of concrete are the W/C, cement compressive strength, properties of the aggregates (shape, grading, surface texture mineralogy, strength, stiffness, and maximum grain size), air-entrainment, curing conditions, testing parameters, specimen parameters, loading conditions, and test age [244, 245]. Based on the discussion of the Sect. 2.1, the influence of the paste volume (air content excluded) will also be investigated. In the following section, curing and testing of the specimens are in accordance to international standards and are as such not within the scope of this chapter.

2.2.2 Compressive Strength at 28 days

As mentioned previously, only the SCC-mixtures are taken into account for which the slump-flow values are between 550 and 850 mm, thus meeting the requirements of the slump-flow classes defined in [246]. This results in 1,476 mixtures with 28-day compressive strength results out of 3,739 SCC mixtures. The limits of some parameters in the following analysis are based on the mean values and 10/90 percentiles listed in Table 2.2.

Table 2.4 Strength ratios $f_{ccub,x}/f_{ccub,150}$ [249–253]

Side x [mm]	$f_{ccub,x}/f_{ccub,150}$ [-]				
	Neville	NBN	UNESCO	AASHTO	ACI
100	1.02	1.06	1.00	–	–
120	1.01	1.04	1.00	–	–

Table 2.5 Strength ratios $f_{ccyl,d}/f_{ccyl,150}$ (h/d = 2) [249–253]

Diameter d [mm]	Height h [mm]	$f_{ccyl,d}/f_{ccyl,150}$ [-]				
		Neville	NBN	UNESCO	AASHTO	ASTM
100	200	1.03	1.03	1.03	1.00	1.00
110	220	1.02	1.03	1.02	1.00	–

Table 2.6 Strength ratios $f_{ccyl,150}/f_{ccub,150}$ [249–253]

Neville	NBN	UNESCO	AASHTO	ACI
0.81	0.79	0.80	–	–

2.2.2.1 Strength Ratio

In practice different standards can be followed for the compressive testing of concrete, e.g. EN 12390-3 [247], ASTM C39 [248], and AASHTO T 22 [249]. These standards allow the testing of a wide variety of specimens with different shapes and sizes.

In the case of VC the influence of specimen size and shape on the measured compressive strength is well documented in various papers and books [244, 245], this is however not yet the case for SCC. Tables 2.4, 2.5 and 2.6 list some strength conversion factors utilised in practice originating from different sources [249–253]. The values $f_{ccub,x}$ and $f_{ccyl,d}$, respectively represent the compressive strength determined on cubes side x and cylinders with diameter d. In MC 2010, it is mentioned that in case test specimens other than cylinders 150/300 mm are used to specify the concrete compressive strength, the conversion factors should either be determined experimentally or, when given in national codes used accordingly for a given category of testing equipment [MC 2010] [268].

Because of the differences in mixture composition, changes in the conversion formulas might be expected when comparing Z and VC.

According to Domone [2], the ratio f_{ccyl}/f_{ccub} increases from 0.8 to near 1.0 with increasing strength, which is a wider range than the 0.9–1.0 range reported by Klug and Holschemacher in [254].

In [255] results originating from own research (based on characteristic strength values) and from literature (based on mean strength values) are compared with the EC2 conversion factors. A tendency for the development of higher conversion factors in the case of SCC was noticed. The first CEB International

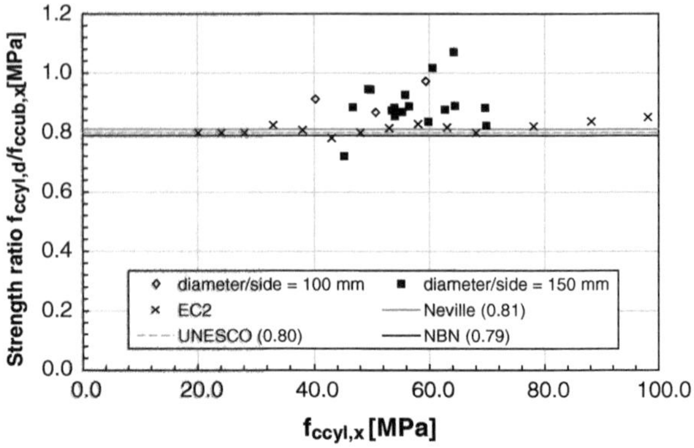

Fig. 2.4 Strength ratio $f_{ccyl,100}/f_{ccub,100}$ and $f_{ccyl,150}/f_{ccub,150}$ versus cylinder compressive strength $f_{ccyl,x}$ (20 results from 9 papers)

Recommendations from 1964 state that experimental conversion factors of f_{ccyl}/f_{ccub} for VC are generally situated within the region of 0.70–0.90. Even taking into account the band of variation found for VC, still ±50 % of the experimental results for SCC was found beyond the upper limit of 0.90.

In Fig. 2.4 experimental strength conversion factors $f_{ccyl,d}/f_{ccub,x}$ for SCC from the database are plotted in function of the cube compressive strength and compared to conversion factors of Tables 2.4, 2.5 and 2.6 and Eurocode 2 [257]. The analysis is based on cubes of 100 and 150 mm and the corresponding cylinders 100/200 and 150/300 mm. Figure 2.4 gathers results from in total nine papers [26, 60, 61, 122–124, 154, 206, 224]. It needs to be mentioned that these conversion factors were calculated through the mean cylinder and cube compressive strength values since in the original publications no information could be found with regard to the scatter present on the SCC experimental results. The EC2 (and EN 206), however, refers to the characteristic compressive strength values. The mean value of the strength ratio $f_{ccyl,d}/f_{ccub,x}$ found with these results is 0.90 with a standard deviation of 0.07. Even though the information is based on only a limited amount of results and further research on this matter is still needed to further confirm this conclusion, it can nonetheless be stated that a first observation seems to indicate that the conversion factor may be higher for SCC than reported for VC [249–253]. This is in accordance with the results of [2, 255]. Taking into account the limited amount of data, no dependency on the compressive strength was noticed, however, which was the case in [2].

Recently, strength results of 25 powder-type SCC mixtures and 2 VC mixtures have been investigated in three different laboratories [256]. The study was performed in order to analyse the effect on the strength ratio of several parameters, including the type of cement (CEMI 52.5 N, CEM III/A 42.5 N LA, CEM I 52.5 R HES), filler (limestone fillers with different fineness, quartz filler, and fly ash), aggregate (gravel, calcareous rubble, and porphyry rubble), and superplasticiser type (two types of polycarboxylate ether with different solid concentrations (PCE1 35 % vs. PCE2 30 %), NFS: naphthalene formaldehyde sulphonate, and MFS: melamine formaldehyde sulphonate).

The influence of the W/C, the C/P, and the powder content (P) is investigated as well. The strength results from literature and those found experimentally by the authors tend towards the same conclusion regarding the ratio $f_{ccyl,150}/f_{ccub,150}$, i.e. a value of 0.90 (±0.06) can be retained to make the conversion between cubes side 150 mm and cylinders diameter 150 mm (h/d = 2). Regarding the ratio $f_{ccub,100}/f_{ccub,150}$ a value, 1.04 (±0.05), is found in agreement with those proposed for VC (Fig. 2.4). Besides, the influence of the W/C, the C/P, the powder content, the type of cement, filler, superplasticiser, and aggregate on the compressive strength seemed to be insignificant or too scattered to be considered.

Among the influencing factors of the strength ratio mentioned in [244], three factors might be responsible for the difference between VC and SCC. The denser microstructure of SCC and enhanced bonding to the aggregates [31, 258] may lead to a more uniform stress distribution during compression. Lower stress concentrations could also decrease the chance of premature failure. This might influence the effect of the specimen size on the strength ratio. In the case of VC lateral stresses are known to affect the stress state over a depth of 0.866 d of a cone- or pyramid shaped region from each end of the specimen. Due to the considerable difference in mixture composition between SCC and VC there might be a different effect of the multi-axial stresses in the cylinders and cubes. SCC contains less coarse aggregates compared to VC. Thus, the wall effect might become less important in the case of SCC. For VC it is known that changing the aggregate grading affects cube strength more than cylinder strength. This effect might be enhanced in the case of SCC.

In [2] the results of the studies of Schiessl and Zilch [259] are mentioned explaining the difference between SCC and VC based on the contribution of aggregate interlock to the shear strength of cracked sections. They found that SCC exhibited 10 % lower shear strength for any given normal stress compared to VC. This is believed to be caused by the smoother crack surfaces in the SCC due to a lower coarse aggregate content. During the compression test in cubes shear movement has a more significant effect on the ultimate average stress. This can explain lower cube strengths and higher f_{ccyl}/f_{ccub} ratios.

As a result of the different standards mentioned earlier the compressive strength results contained in the database used for the analysis of the mechanical properties of SCC consists of a wide scope of compressive strength specimens. In order to be able to incorporate as many collected results as possible in the parametric studies in this chapter, it is important that as many results as possible can be converted to a reference compressive strength (either cylinder or cube compressive strength). In the database, 109 strength results are available from cylinders 150/300 mm, whereas 659 results are available from cubes 150 mm. As the compressive strength $f_{ccyl,150}$ is more common used the conversion is performed to cylinders 150/300 mm, i.e. the equivalent cylinder compressive strength $f_{ccyl,150,eq}$.

Based on the database and Tables 2.4, 2.5 and 2.6 the following conversion factors were derived for SCC. In this way, considering the equivalent cylinder compressive strength, 1,374 results can be analysed.

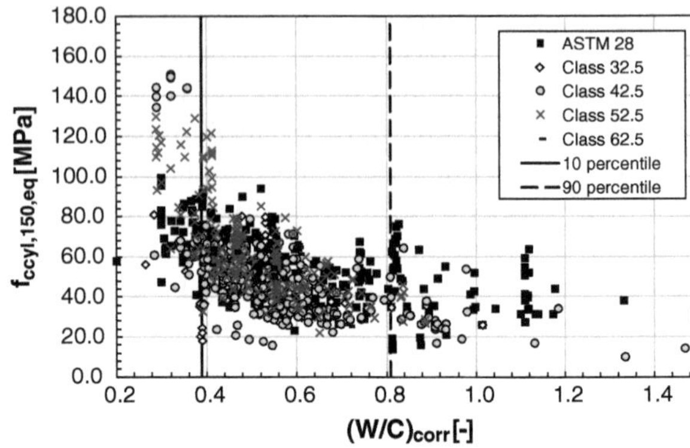

Fig. 2.5 Equivalent cylinder compressive strength $f_{ccyl,150,eq}$ versus W/C_{corr} (1,163 results from 165 papers)

$$f_{ccub,100} \approx f_{ccub,120} \approx f_{ccub,150} \tag{2.1}$$

$$f_{ccyl,100} \approx f_{ccyl,110} \approx f_{ccyl,150}(h/d = 2) \tag{2.2}$$

$$f_{ccyl,150,eq} \approx 0.9 \text{ x } f_{ccub,150} \tag{2.3}$$

The equivalent cylinder compressive strengths $f_{ccyl,150,eq}$ from the database range from 10 to 150 MPa.

2.2.2.2 Cement Strength Class

According to [260] and Feret's law [261] the 28-day cement compressive strength is a decisive key parameter with regards to the concrete strength. All results were divided into five groups of strength classes based on the actual measured 28-day cement compressive strength f_{cem28}, according to [240, 262] or on the cement class if experimental data was not available.

ASTM 28: $f_{cem28} \geq 28$ MPa
Class 32.5: 32.5 MPa $\leq f_{cem28} <$ 42.5 MPa (f_{cem28} known) or CEM 32.5
Class 42.5: 42.5 MPa $\leq f_{cem28} <$ 52.5 MPa (f_{cem28} known) or CEM 42.5
Class 52.5: 52.5 MPa $\leq f_{cem28} <$ 62.5 MPa (f_{cem28} known) or CEM 52.5
Class 62.5: 62.5 MPa $\leq f_{cem28} <$ 72.5 MPa (f_{cem28} known)

In Fig. 2.5 the equivalent cylinder compressive strength $f_{ccyl,150,eq}$ is plotted against the W/C_{corr}, in which the contribution of water content in the admixtures is taken into consideration. Regarding the results for which the actual measured cement compressive strength is unknown, there is a large uncertainty with respect to the group in which they are classified. Therefore, in Fig. 2.6 only the results for which f_{cem28} are known are presented. The 10/90 percentiles from Table 2.2 are indicated in both figures.

Cylinder compressive strengths higher than 90 MPa are reported in six publications (43 results) from which in only one paper (26 results) the f_{cem28} is known,

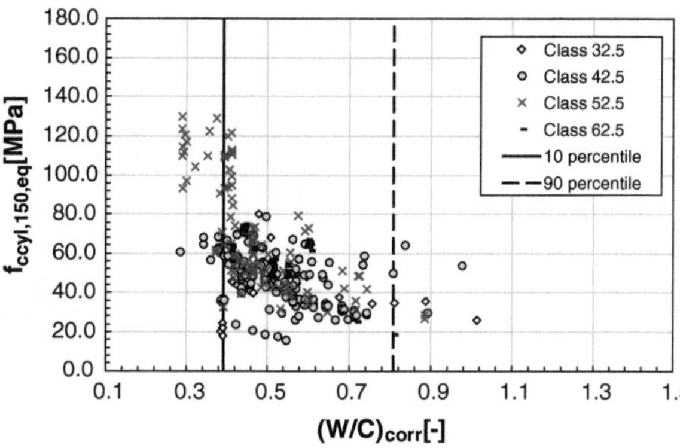

Fig. 2.6 Equivalent cylinder compressive strength $f_{ccyl,150,eq}$ versus corrected W/C, known f_{cem28} (320 results from 41 papers)

i.e. 53.9 MPa [8, 77, 140, 141, 182, 186]. The W/C_{corr} is lower than 0.52 and the W/P lower than 0.46. Silica, vegetal ash, fly ash, and quartz were used as additions.

Beneath the value of 90 MPa, there is a large scatter among the different cement groups, but still the dependency on the W/C can be observed. Also some upward shift can be observed for cement groups with f_{cem28} higher than 42.5 MPa. Within the 10/90 percentiles the results obtained from ASTM cements lie within the same strength region of the European cements, which is not the case above the 90 percentile. No more explicit notable shift of the cement groups was observed when considering the W/P instead of W/C_{corr}. The large scatter is due to the many different parameters varying in the mixtures (e.g. aggregate type, D_{max}, filler type, etc.).

2.2.2.3 Addition Type

As mentioned in the introduction, the most common used addition types are limestone, fly ash, ground granulated blast-furnace slag, and silica. The following analysis is based on the main addition types, secondary addition types are not taken into account. In Fig. 2.7 the equivalent cylinder compressive strength $f_{ccyl,150,eq}$ is plotted against the corrected W/C. The strengths obtained with limestone and fly ash lie within the same region. When slag is used an upward shift can be mentioned compared to the use of limestone and fly ash. Strengths of 90 MPa and higher were only achieved when silica (39 of 41 results) or fly ash (2 of 41 results) was used, and were reported in six publications [8, 77, 140, 141, 182, 186].

In [193] nine mixtures were developed with equal constituents, except from the addition type. With each addition type three concrete strengths were obtained ($f_{ccyl,150,eq}$ of 28, 45, and 64 MPa). The cement content was kept constant for each strength. In the case of fly ash and natural pozzolana a higher water content was needed to obtain the same strength as in the case of limestone was used. As such one could say larger strengths (at 28 days) will be obtained with fly ash and natural pozzolana compared to limestone with equal cement content and a constant C/P. Larger differences in strength will be noticed at higher $(W/C)_{corr}$.

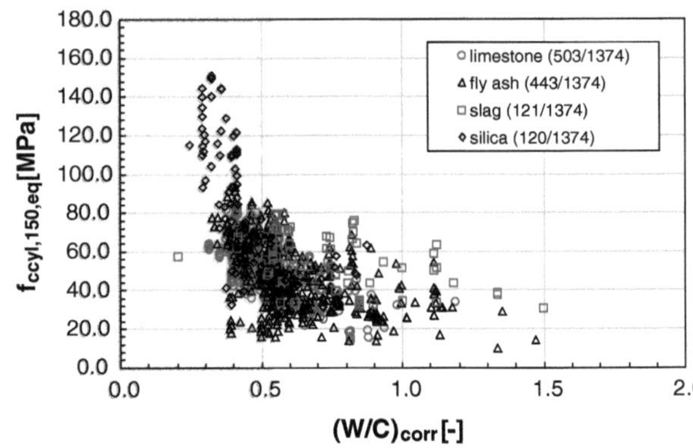

Fig. 2.7 Influence of addition type (1,187 results from 178 papers)

In [220] two identical mixtures were made with the only difference being the addition type, i.e. limestone and marble. The marble powder particles were finer than those of the limestone filler. However, the strength was almost the same.

In the study of [218], the potential of limestone, basalt and marble powder as partial replacement of Portland cement is investigated. The specific surface area of the fines is 4,000 cm^2/g for cement, 8,900 cm^2/g for marble powder, 2,500 cm^2/g for limestone powder, and 6,300 cm^2/g for basalt powder. The powder content was kept constant at 550 kg/m^3, the W/P at 0.33. The C/P was varied from 1.0 to 0.7. When compared to the control mixture where only cement was used increasing amounts of mineral admixtures generally decrease the strength, except when the cement was partially replaced by marble powder. No important difference was found between limestone powder and basalt powder. The strength increase in case of replacement of marble powder was only noticed with a C/P of 0.8 or higher.

Uysal and Yilmaz [218] reported that only limestone and basalt powders act like inert mineral admixtures reducing the compressive strength in case of partial replacement of cement. The good performance with the marble powder is believed to be due to the physical nature of better packing since it is the finest powder used. A denser matrix is achieved. Furthermore, the surface of the marble powder is believed to act as nucleation site for the early reaction products of CH and CSH. The effect of nucleation on the strength is dependent on the mineral admixture's affinity to cement hydrates, and it increases with fineness and specific surface area of the mineral admixture [218].

According to [2], limestone powder which is a common addition to SCC mixtures makes a contribution to strength gain, mainly in the beginning of the hardening phase. In [31], by means of mercury porosimetry on hardened cement pastes, it was illustrated that limestone filler, apart from any chemical influence, also helps realizing a more dense structure by the filling effect, with positive influence on the compressive strength at 28 days. Furthermore this effect is found to influence the development of the pore structure of the hardened cement paste with limestone following Power's model [237].

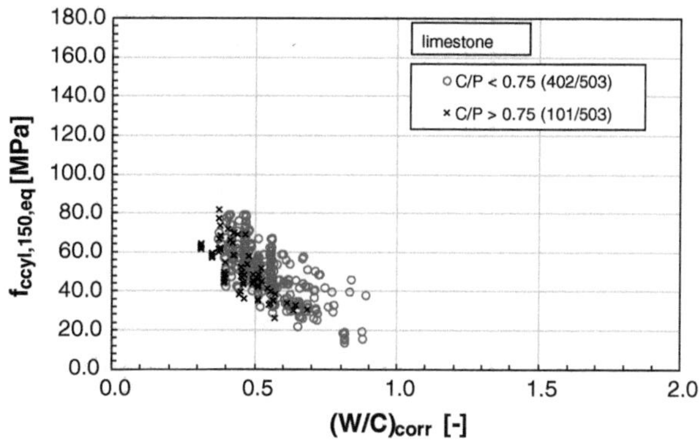

Fig. 2.8 Influence of C/P in case of limestone as addition type (503 results from 100 papers)

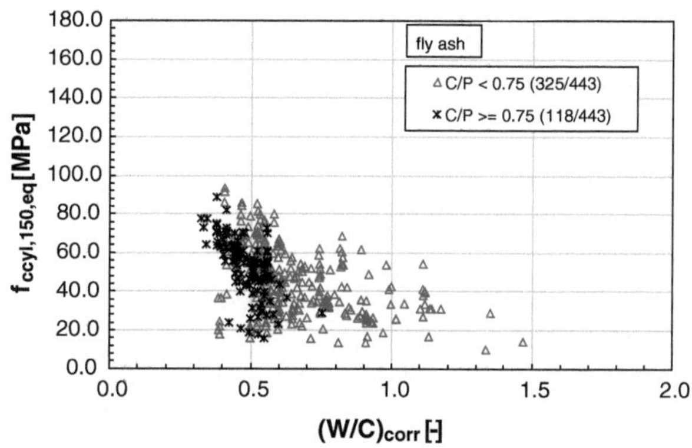

Fig. 2.9 Influence of C/P in case of fly ash as addition type (443 results from 81 papers)

It has to be emphasized that the Blaine finesses of the filler can have a significant effect on the compressive strength. In Some cases, high Blaine finesses of the limestone filler plays a major role in the obtained mechanical properties [273].

2.2.2.4 C/P and Cement Content

The influence of the C/P is investigated for various addition types: limestone, fly ash, ground granulated blast-furnace slag, and silica. In Figs. 2.8 and 2.9 the strength results obtained with limestone and fly ash are presented.

Values of C/P higher than or equal to 0.75 are positioned in the lower left part of the point cloud. At a constant W/C a higher C/P leads to lower strengths. This was also found varying the C/P in [238]. This is probably due to the fact that increasing the cement content also requires the increase in water content to maintain the W/C, thus leading to a higher W/P. More water in the mixture leads to a higher capillary porosity and lower compressive strength. Also less addition will be present to contribute to the strength (e.g. fly ash). No further conclusions could be made by considering the different groups of cement strength. In the case of

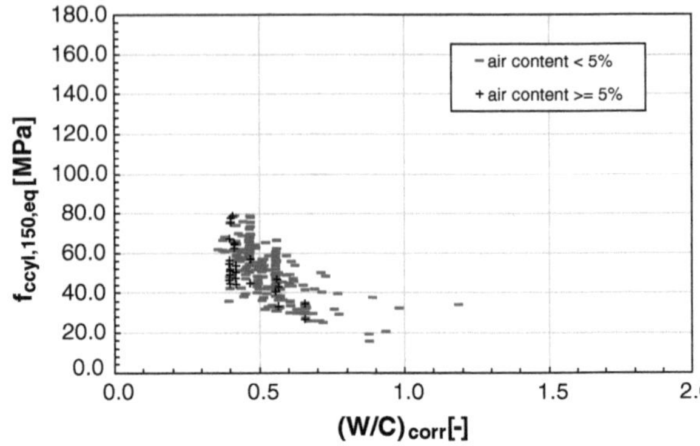

Fig. 2.10 Influence of air content in case of limestone as addition type (261 results from 51 papers)

ground granulated blast-furnace slag and silica no differences could be observed due to the variation of the C/P, probably due to the limited amount of results for this purpose.

2.2.2.5 Air Content

In [186] it was observed that the increased air content decreased the compressive strength of SCC. The reduction in compressive strength was about 4 MPa per 1 % increase in air content.

Considering the database, the influence of the air content was investigated, but no clear trends could be observed since many parameters intervene at the same time. When variation of some of those parameters (C/P, addition type, and aggregate surface/origin) is excluded, there is a small trend of a decrease in compressive strength with the increase in air content. In Fig. 2.10 the influence of the air content is illustrated for mixtures with limestone as addition type.

2.2.2.6 Coarse Aggregate Type/Origin and Maximum Aggregate Size (D_{max})

Coarse aggregates can have an influence on the compressive strength due to their shape, nominal maximum size, surface texture, and origin. According to [263], crushed rock was used over three times greater than gravel (uncrushed) aggregates, primarily reflecting local availability. In [2], crushed aggregate mixtures have higher cube strengths. This is similar to the behaviour of VC, but the average difference between the two best-fit curves was found to be small (4 MPa). The opposite behaviour was mentioned regarding the cylinder strengths, but fewer data were available, and there was only a partial overlap of the W/C's of SCC and VC.

The effect of the surface texture and the origin of coarse aggregates on the compressive strength are presented in Figs. 2.11 and 2.12. In Fig. 2.11, the results related to crushed (63 %) and uncrushed (37 %) coarse aggregates are shown. In

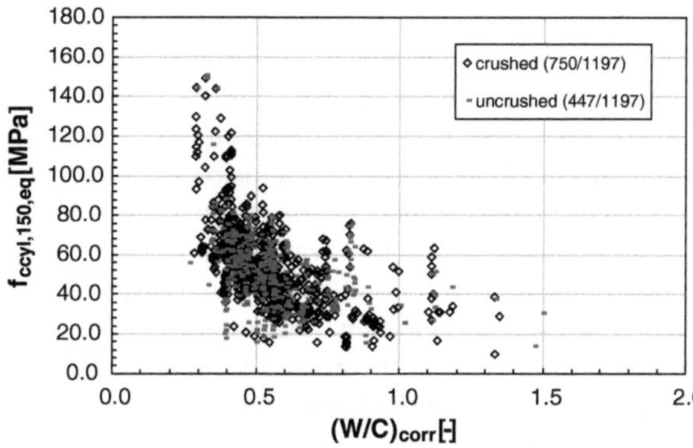

Fig. 2.11 Influence of coarse aggregate—crushed versus uncrushed (1,197 results from 171 papers)

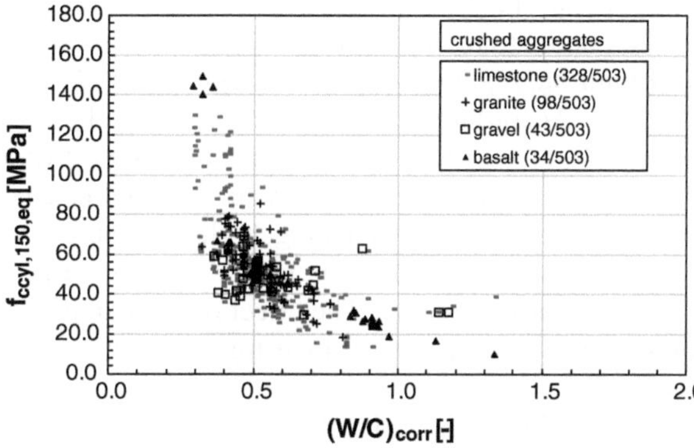

Fig. 2.12 Influence of coarse aggregate—crushed—type of origin (503 results from 71 papers)

the range of 30–90 MPa no obvious difference can be noticed between SCC made with crushed or uncrushed aggregates. The crushed aggregates, however, cover a larger strength range compared to the uncrushed aggregates. Results above 90 MPa are found in six papers. Silica or fly ash was used. A rougher surface could contribute to a compressive strength higher than 90 MPa, but the appropriate addition types have to be applied. In general, rougher surfaces will enhance the bond between the aggregate and the cement matrix which benefits the compressive strength. In [29] two mixtures have been compared with the only difference being the type of coarse aggregate. The mixture with gravel (2/8 and 8/16) had an equivalent cube compressive strength of 50.8 MPa whereas the mixture with crushed limestone (4/14) reached a strength of 67.2 MPa.

In Fig. 2.12 the actual origin of the crushed aggregates is depicted. Besides the fact that quite a lot aggregate origins are not specified in the papers it can be observed that crushed aggregates often consist firstly of limestone and secondly of granite and, uncrushed aggregates of gravel (371 of 447). The origin of the coarse aggregates depends on the local availability. No clear trend could be observed regarding the origin of the aggregates. The higher the compressive strength the more important becomes the aggregate compressive strength and as such the origin. At normal strength the compressive strength is mainly governed by the strength of the cement matrix. Consequently, as most strength results are lower than 90 MPa no effect of the aggregate origin is noticed on the compressive strength. In [215, 217] a comparison was made between two types of crushed aggregate. In [215] no clear conclusion could be drawn when the strength results of mixtures with crushed limestone or crushed basalt were studied. A 5.4 MPa higher compressive strength was found for the mixture with limestone aggregates in cases limestone filler and fly ash were used as additions. When only limestone filler was used as addition, the mixture with limestone aggregates showed a 3.6 MPa lower strength compared to the mixture with basalt aggregates. In [217] the application of crushed limestone and pumice was investigated on eight different mixtures. For all four comparisons the mixtures with pumice showed much lower strengths (only 40 % of the strength of the mixtures with limestone).

De Larrard [264] emphasizes of the need to reduce the maximum aggregate size to attain high strength. In [263] 70 % of the investigated SCC mixtures used had a maximum aggregate size in the range of 16–20 mm (value depending on local practice). According to the database, only 47.5 % of the aggregate sizes lie between 16 and 20 mm. A mean value for the maximum aggregate size was found to be 16 mm. There is a tendency to use smaller coarse aggregate sizes, probably linked to the desire of reaching higher compressive strength. However, it was observed that small coarse aggregate sizes were also found in combination with low compressive strength. The effect of the maximum aggregate size, D_{max}, on the compressive strength was investigated, but as such no clear influence was observed.

A lower D_{max} also contributes to a better segregation resistance and passing ability.

2.2.2.7 Strength Model

Over the years, several strength models have been proposed for VC. In a first step, some of those models are applied on the available database. Subsequently, an adapted model for SCC, as proposed by Poppe [265], is applied.

Lambotte combined the research results of Waltz with the strength model proposed by Bolomey into a new formula (Eq. 2.4) [266]. In this internationally known model the cement compressive strength f_{cem28} and the W/C are the only parameters.

$$f_{ccub,150} = 0.483\ f_{cem28}/(W/C)\ [MPa] \tag{2.4}$$

In Fig. 2.13 all the results with known f_{cem28} are used to calculate the cube compressive strength as proposed by Lambotte ($f_{ccub,150,cal}$). The corrected W/C was used for the calculation. The calculated values are plotted in function of the equivalent cube compressive strength ($f_{ccub,150,eq}/0.9$). A 95 % confidence interval has been plotted. Several calculated results are close to $f_{ccub,150,eq}$. Although some

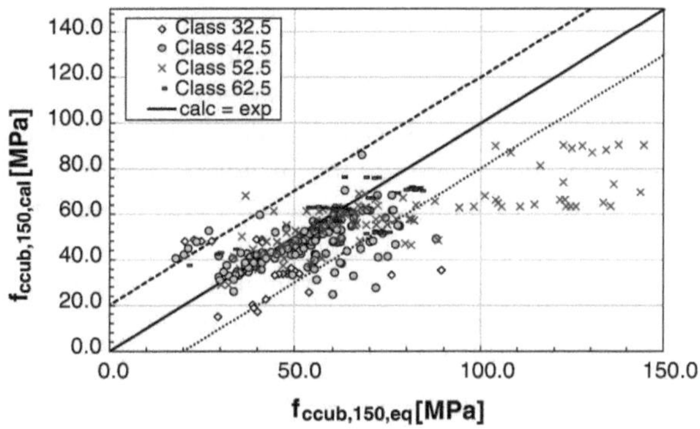

Fig. 2.13 Calculated cube compressive strength according to Lambotte versus equivalent cube compressive strength (320 results from 41 papers)

calculated values overestimate the actual strength, most of the calculated values underestimate the strength. This is probably due to the use of higher powder contents by incorporating additions and the use of superplasticisers. More dispersed cement leads to higher hydration, and the additions realize a more dense structure by the filling effect or, even in some cases influencing the formation of hydration products [31, 265]. Also, this model does not incorporate the effect of the W/P or C/P. The effect becomes even more important for higher compressive strengths. The model of Lambotte is valid for fully VC, i.e. with an air content of 1 %. The air content mentioned in this analysis varies from 0.6 % till 9.0 %. Although there should be (based on the limited amount of results for which the air content of the fresh concrete is known), no clear effect of the air content on the Lambotte model could be found. This is probably due to the contribution of other parameters.

After evaluating the different influencing parameters on compressive strength, it is believed that an improvement of Lambotte's model can be obtained by adding parameters related to the C/P, addition type, and the use of superplasticiser, as proposed by Poppe [265]. In this latter reference, it is suggested to use Eq. 2.5 to calculate the cube compressive strength on cubes 150 mm.

$$f_{ccub,150} = \frac{0.483.\mathrm{f}_{\mathrm{cem28}}(\mathrm{a}_1.\frac{\mathrm{sp}}{\mathrm{C}} + 1)}{(\frac{\mathrm{W}}{\mathrm{C}} + \mathrm{a}_2.\left(\frac{\mathrm{C}}{\mathrm{P}} - 1\right))} \tag{2.5}$$

with sp/C the dosage of superplasticiser, by mass of cement, a_1 a coefficient depending on the superplasticiser type and a_2 a coefficient depending on the addition type. Also in this case, the corrected W/C_{corr} will be used for the analysis. As a-values are depending on the addition and the superplasticiser type, a further analysis is performed on the data results with limestone, silica fume, or fly ash as first or main addition type, and polycarboxylate ether as superplasticiser. However, when investigating the effect of the superplasticiser no clearly defined regions could be observed of the mixtures with the same superplasticiser type. The results were plotted according to Poppe, with different a-values for different addition types. A better fitting was the result, but due to the fact that besides the main addition type other addition types are often used, there is still an important distribution on the results. Also the derived a-values have no scientific meaning. For this reason, the graphs are not included. However, when mixtures from individual experimental programs were used, it was found that it was possible to find valid models.

In [198], based on experimental data, genetic programming (GP) was used to model and formulate the fresh and hardened properties of SCC containing pulverised fuel ash (PFA). Twenty-six mixtures were made with W/P of 0.38–0.72, 183–317 kg/m^3 of cement, 29–261 kg/m^3 of fly ash, and 0–1 % total liquid mass of polycarboxylate ether superplasticiser by mass of powder. Graded crushed basalt aggregate with a D_{max} of 20 mm was used, together with well-graded quartzite sand with a fineness modulus of 2.74. The mass of coarse aggregate was fixed for all mixtures at 837 kg/m^3 to enhance the filling ability, passing ability, and the resistance to segregation. The formula for the compressive strength at 28 days is given by Eq. 2.6.

$$
\begin{aligned}
\text{Compressive strength} = & \left[\left(\frac{\mathrm{C}+\mathrm{CA/sand}-70}{\mathrm{W/P}}-\mathrm{C}-\mathrm{SP}\right)\right. \\
& \left.\times\left(\frac{3\times 28-\mathrm{PFA}+\mathrm{W/P}+\mathrm{C}\times\mathrm{W/P}+\mathrm{C}+26.1}{4}\right)\right]^{1/5} \\
& \times\left(\frac{\sqrt{[5]28}}{(\mathrm{CA/sand})^4}+\ln(28)\right)
\end{aligned}
\tag{2.6}
$$

With cement (C) content (kg/m^3), pulverised fuel ash (PFA) content (kg/m^3), water-to-powder ratio (W/P), superplasticiser (SP) content, and ratio of coarse aggregate to sand (CA/sand), respectively.

Results showed that the derived GP model is capable of accurately predicting the compressive strength at 28 days used in the training process. The model can also effectively predict compressive strength at 28 days for new mixtures designed within the practical range with variation of the mix ingredients. The performance of the GP-model is illustrated in Fig. 2.14.

Also this model cannot be generalized for the database, since many parameters are involved (filler type, superplasticiser type, aggregate type, etc.). However, it is believed that, when mixtures from individual experimental programs will be used, valid models will be found based on Eq. 2.5. Probably the model could also be optimised incorporating the actual measured 28-day cement compressive strength f_{cem28}, in case different cement types are used.

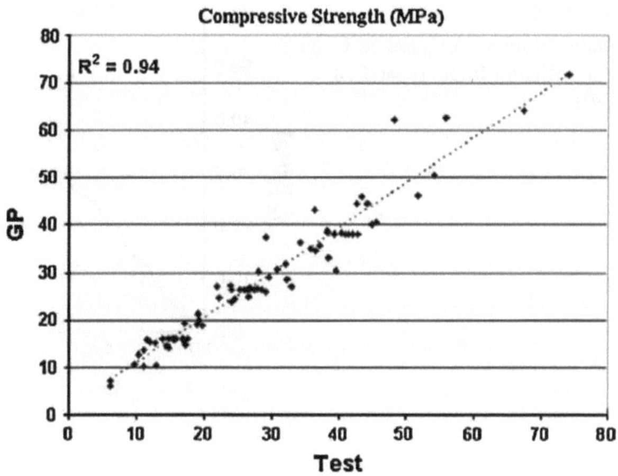

Fig. 2.14 Performance of GP model versus test results for compressive strength [198]

2.3 Stress-Strain Relationship

Until now, very little is known about the differences between the stress-strain behaviour of SCC and that of VC. Most research programs include tests to determine the compressive strength and some include Young's modulus as well, but the stress-strain relationship is almost never recorded.

Above that, accurate measurements of the stress-strain relationship require a high level of care and an appropriate steering signal when executing the tests [267]. Therefore, different conclusions can be found in the literature just because research programs were performed with different test set-ups and more or less attention to side-effects.

2.3.1 Ascending Part of the Stress-Strain Relationship

Georgiadis [79] performed tests on cubes (sides 100 mm) made of SCC of two strength classes and with several filler types. The modulus of elasticity for all SCC's was smaller compared to VC but the compressive strengths were slightly improved which they attribute to the improved ITZ quality.

When SCC was produced with limestone filler, cement kiln dust or ground granulated blast-furnace slag, the toughness was higher than the one recorded for VC. However, when glass filler was used, the peak strain increased (Fig. 2.15). No additional details are given about the properties of the filler materials (grading curves, chemical composition, etc.) or test set-up and the recorded strains seem to be on the high side as one should expect values of the peak strain ε_{c1} in the range of 1.8 till 2.7 ‰ [254].

Anagnostopoulos et al. [14] looked into the influence of fire on the stress-strain relationship of SCC. They ran tests after heating to 300 and 600 °C, but also on

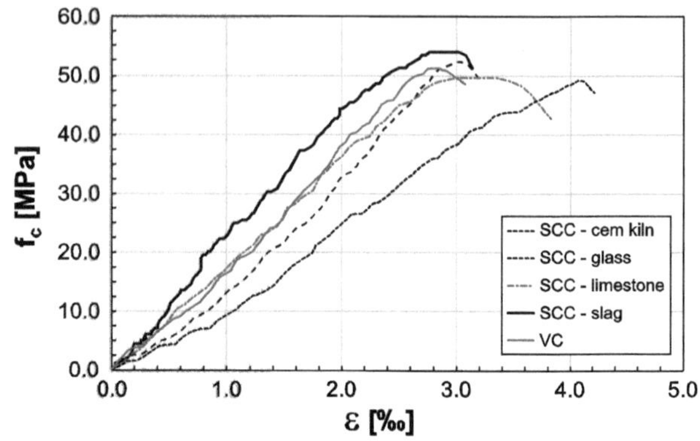

Fig. 2.15 Stress-strain relationships of VC and SCC with different fillers (based on [79])

reference specimens stored at 20 °C. All SCCs contained limestone filler and had a total powder content of 475 kg/m^3. To obtain different compressive strengths the ratio between cement and filler in the total powder content was varied. No noticeable difference between SCC and VC was recorded with respect to the shape of the stress-strain curves. Comparing the curves between the two applied strength classes, it can be mentioned that the higher the strength class, the steeper the slope of the ascending branch of the stress-strain curve and the more linear the descending one. The peak strain decreased at increasing compressive strength.

The stress-strain behaviour of SCC was measured on cylindrical specimens with slenderness ratios (ratio of height to diameter) of 3.0 or 3.6 by Desnerck [60]. During the experiments the specimens were subjected to a continuously increasing deformation rate of 0.003 mm/s. The axial deformations in the central part of the specimen, the circumferential deformation and the stress applied to the specimen were recorded. In total, four SCC mixtures containing limestone filler and three VC mixtures were tested. The 28-day cylinder compressive strength of the SCC mixtures varied between 45 and 60 MPa.

From the obtained results it can be seen that the peak strain at different ages (3 days up to 3 months) and concrete strengths for SCC (range of 20–70 MPa) is higher than for VC (Fig. 2.16).

Besides the influence of limestone filler, the use of blast furnace slag, fly ash, and silica fume has been investigated. For almost all tested compositions, the peak strain ε_1 of SCC was higher than those of VC. The largest normalised peak strains were measured for SCC containing limestone filler. The values for the concrete compositions with fly ash, silica fume, or blends of these SCMs were slightly lower. The lowest values were found when blast furnace slag was used as addition (Fig. 2.17).

In their study, the Belgian Building Research Institute (BBRI) [26] used cylindrical specimens (diam. 100 mm and height 200 mm). The tests were displacement-controlled at a speed of 60 μm/min. A friction reducing sheet was applied.

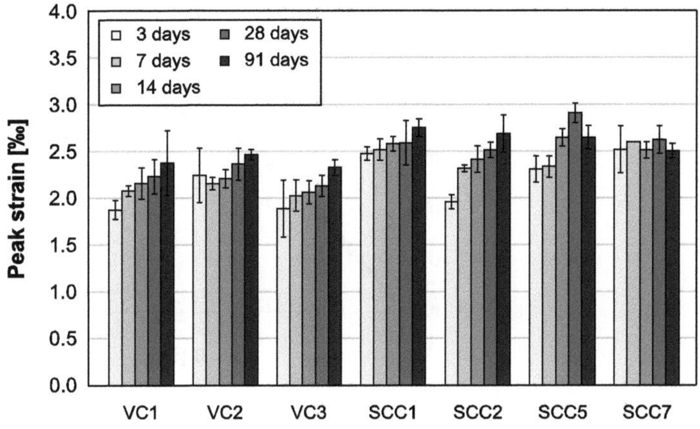

Fig. 2.16 Measured peak strain values for limestone SCCs and VCs at different concrete ages [60]

Fig. 2.17 Normalised peak strain values for SCCs with different additions and different concrete ages [60] (*LS* Limestone, *BFS* Blast furnace slag, *FA* Fly ash, *SF* silica fume)

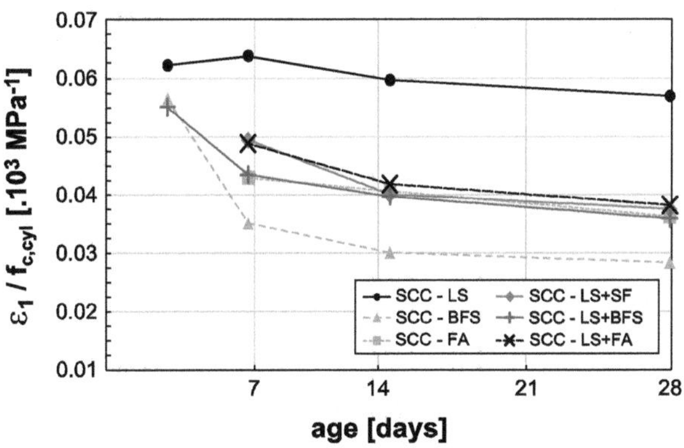

In total eight SCCs and one VC were tested. It was found that the scatter on the obtained stress-strain relationships was rather small (see Fig. 2.18). Concretes with lower compressive strengths (40–50 MPa) showed a more ductile behaviour than higher strengths SCCs (50–60 MPa). When high amounts of fly ash were combined with blast furnace slag, the behaviour was most brittle. As reported by Desnerck [60] as well, the peak strain is increasing at increasing compressive strength.

2.3.2 Strain-Softening Behaviour

Desnerck [60] investigated the stress-strain behavior of SCC paying special attention (by applying the proper steering signal) to the descending branch of the stress-strain diagram to evaluate the so called 'strain-softening' behaviour of the

Fig. 2.18 Comparison of measured stress-strain curves for SCC and VC mixture [26]

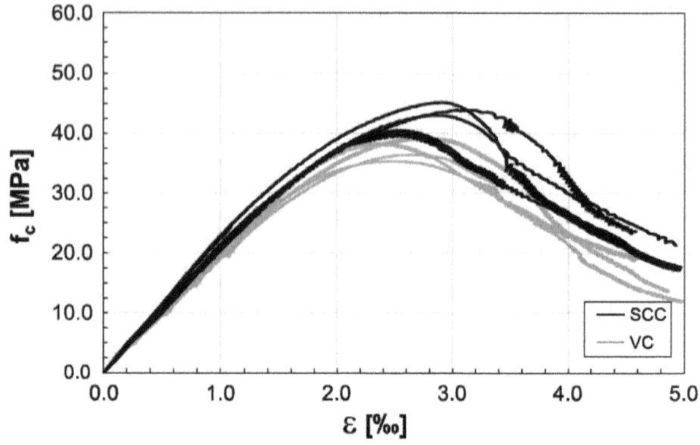

Fig. 2.19 Comparison of normalised strain-softening behaviour of limestone SCC mixtures at 28 days [60]

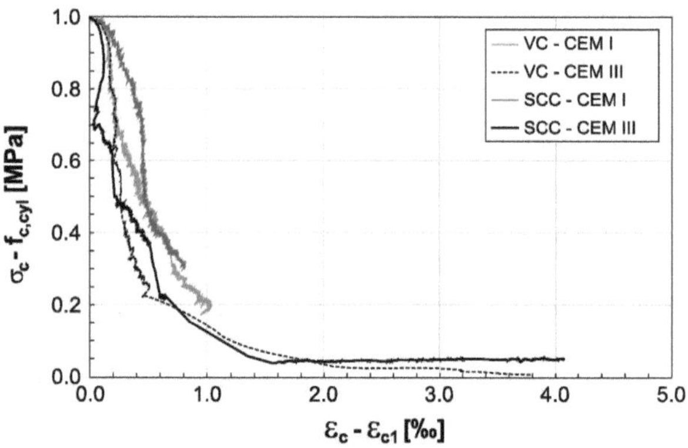

material. He concluded that the strain-softening behaviour of SCC and VC are comparable for the same compressive strength level (Fig. 2.19). Based on the total stress-strain relationship measured during the experiments, the toughness was determined (area underneath the entire stress-strain diagram). The toughness of the tested limestone-containing SCC was slightly higher than that for VC. The largest difference was found for the ascending branch, whereas the area underneath the descending branch was comparable. However, BBRI [26] found that the normalised toughness's did not differ significantly when comparing SCC specimens to VC specimens or the Eurocode2 [254] provisions.

2.4 Tensile Strength

For the evaluation of the tensile strength of SCC, and concrete in general, three well-known methods are being used: (i) the direct tensile test, (ii) the splitting tensile test and (iii) the bending tensile test (3-point or 4-point). Due to the difficult

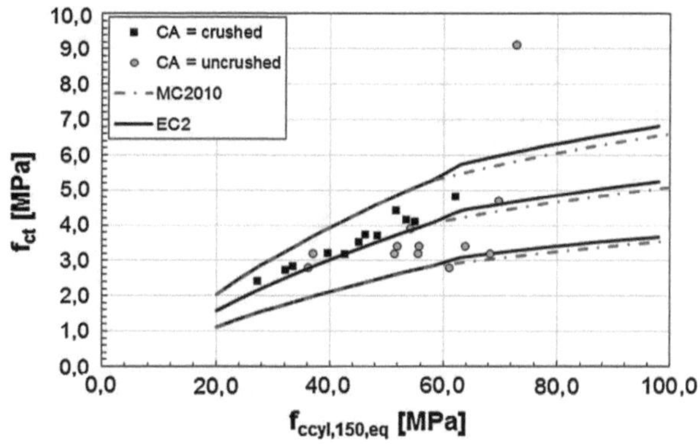

Fig. 2.20 Direct tensile strength versus equivalent cylinder compressive strength of SCC mixtures and the effect of the coarse aggregates (CA) shape (25 results from 6 studies)

test setup direct tensile strength tests are rather scarce. Therefore, the tensile strength is quite often derived from test results originating from splitting tensile or bending (e.g. flexural) tensile strength tests. By means of models offered by e.g. Eurocode 2 (EC2) [257] or Model Code 2010 (MC2010) [268], it is possible to convert these test results. However, it is not quite clear whether these conversion factors can still be used for SCC. Note that the upper and lower limit curves provided in the figures are the 5 % and 95 percentile provided by EC2 and MC2010.

2.4.1 Direct Tensile Strength

For the evaluation of the direct tensile strength of SCC, a limited amount of test results can be found in the literature (25 data results originating from 6 studies). This is mainly due to the complexity of the test equipment and execution of the experiments. The test results are obtained at different ages of the samples (14–90 days). The most common shape of the test specimens was cylindrical. In one study [177], the dog-bone shape was used for preferential rupture.

In Fig. 2.20, the correlation between the direct tensile strength and the equivalent cylinder compressive strength is expressed. All compressive test results were converted into $f_{ccyl,150,eq}$ by using Eq. 2.3 obtained by means of the database.

An increasing trend of direct tensile strength is noticed for increasing cylinder compressive strength (Fig. 2.20). For this limited amount of data, the results tend to follow the mean relationship proposed by the EC2 and MC2010, especially if crushed coarse aggregates (limestone or calcareous) are being used. On the other hand, the test results originating from mixtures based on uncrushed coarse aggregates (gravel) lie between the 5 percentile range and the mean value proposed by EC2 and MC2010.

Based on limited data, it can be stated that no significant effect of paste volume or filler type on the correlation between direct tensile strength and compressive

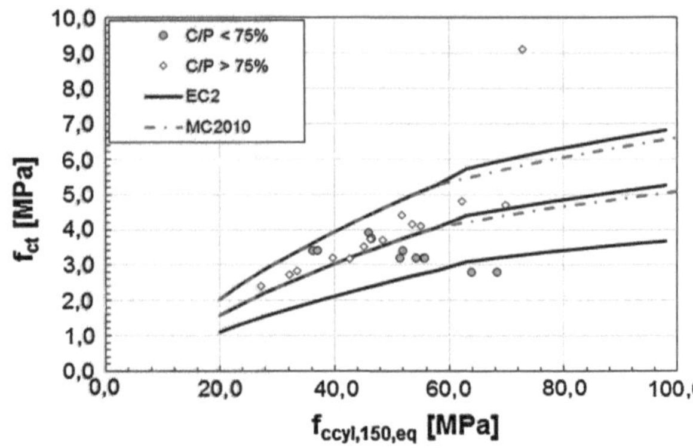

Fig. 2.21 Direct tensile strength versus equivalent cylinder compressive strength of SCC mixtures and the effect of C/P (25 results from 6 studies)

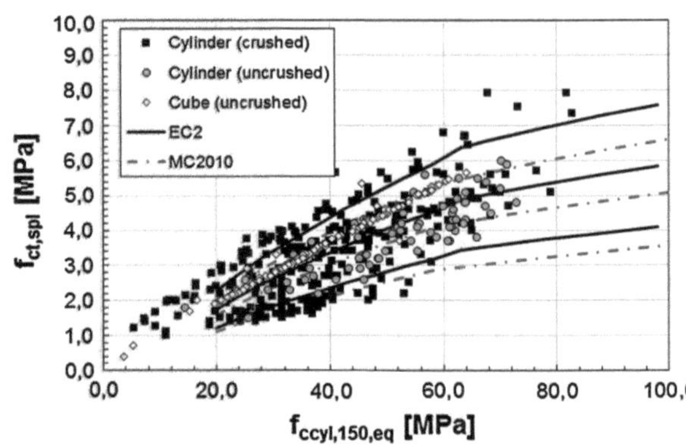

Fig. 2.22 Splitting tensile strength (sample shape: cylinder or cube) versus equivalent cylinder compressive strength of SCC mixtures and the effect of coarse aggregates shape (crushed vs. uncrushed) (536 results from 49 studies)

strength can be found from the database. However, the effect of the C/P is noticeable, as depicted in Fig. 2.21. SCC mixtures with a C/P less than 0.75 tend to lie beneath the mean values proposed by the EC2 and MC2010.

Overall, more experimental results are necessary to obtain reliable conclusions regarding the direct tensile strength of SCC.

2.4.2 Splitting Tensile Strength

For the evaluation of the splitting tensile strength of SCC, a considerable number of test results can be found in literature (608 data results originating from 60 studies) with test ages varying from 1 to 365 days. Mostly, the shape of the test specimens is cylindrical (399 from 43 studies) or cubic (137 results from 6 studies). The correlation between splitting tensile strength (of cubes and cylinders) and cylinder compressive strength is given in Fig. 2.22.

Note that all compressive test results have been converted into $f_{ccyl,150,eq}$ by using the conversion factors proposed in Sect. 2.2 (Eq. 2.3). In EC2 and MC2010

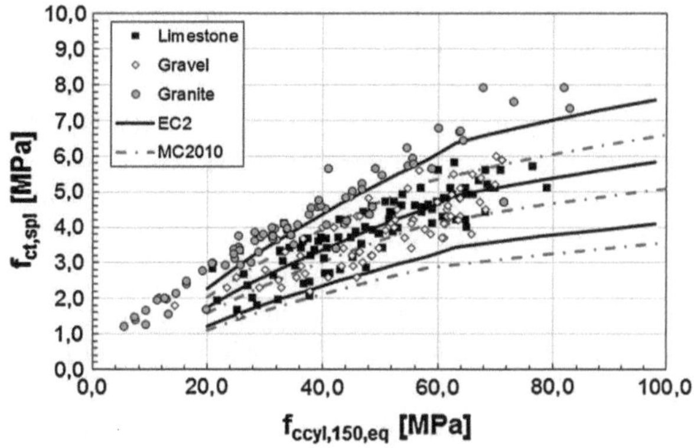

Fig. 2.23 Cylinder splitting tensile strength versus equivalent cylinder compressive strength of SCC mixtures and the effect of coarse aggregates type (granite vs. limestone vs. gravel) (256 results from 27 studies)

it is mentioned that the relationship between splitting tensile strength and compressive strength is based on the cylinder compressive strength.

According to [2] cylinder splitting results fall in the 5–95 percentile ranges given in EC2, with the majority being the upper half of this, and a few above the upper limit. An explanation can be found in a possible underestimation of the used conversion factor between cylinder and cube compressive strength, as mentioned in [26]. No significant difference between SCC and VC is found.

An increasing trend of splitting tensile strength is noticed for increasing cylinder compressive strength (Fig. 2.22). The behaviour of cubes under splitting tensile stress tends to follow the mean relationship proposed by EC2 (for f_{ccub} lower than 40 MPa), and the maximum range proposed by MC2010. A linear trend is identified between $f_{ct,spl}$ of cubes and $f_{ccyl,150,eq}$ (originating from one study) which is likely to be confirmed by Topçu and Uygunoglu [210].

Although higher splitting tensile strength results can be expected when cubic samples are used (10 % higher values compared to cylinders, according to EN 12390-6), this cannot be confirmed by the data of Fig. 2.22. A rather high percentage of the widely scattered splitting tensile strength results of cylinders lie between the 5 and 95 percentile or maximum ranges (or even above) of EC2, and mostly in the upper half of the MC2010 range. An explanation was found in the improved bond between the paste and the aggregates in the SCC mixtures.

Splitting tensile strength results of SCC using granite coarse aggregates lie in the upper half or above the ranges proposed by EC2 and MC2010 (Fig. 2.23). When limestone or gravel aggregates are used, this is not the case: the data tend to follow the mean relation proposed by EC2. The cylinder splitting tensile strength as a function of the cube compressive strength was also investigated on SCC and VC mixtures produced with limestone and siliceous aggregates by [14]. The splitting tensile strength of SCC mixtures was slightly lower (decreased by 6–8 %) than that of the VC mixtures produced with the same W/C and cement content when the mixtures were produced with quartz aggregate. When crushed limestone aggregate was used, SCC mixtures had similar splitting tensile strength as the reference VC mixtures. The ratio between the splitting tensile strength measured for SCC and the splitting tensile strength measured for VC of equal W/C is influenced by the aggregate type [14].

Further analysis of the database shows that there is not a significant effect of coarse aggregate size or paste volume on the correlation between splitting tensile strength and compressive strength. However, according to [11] larger maximum aggregate size and higher coarse aggregates volume (thus, smaller paste volume) leads to higher splitting tensile strength. The influence of C/P or filler type is not significant according to the database.

It can be concluded that using models proposed by the EC2 and MC2010 for estimating the tensile strength of SCC can result in safe values for the prediction of splitting tensile strength.

A limited amount of test results can be found in literature dealing with the evolution of splitting tensile strength in time. These results can be valuable for the evaluation of the early-age (thermal) cracking behaviour of massive concrete structures, as internal stress build up due to the thermal gradient and can exceed the tensile strength of the concrete at young age. According to a study performed by [255], a rather good prediction (small underestimation) of the splitting tensile strength evolution can be obtained for the SCC mixtures made with Ordinary Portland Cement (OPC) using the model proposed by EC2 (using β_{cc} factor). In case fly ash is added to the OPC, a slower strength evolution is noticed; however, the EC2 does not account for the use of fly ash.

Data originating from one study [255] (6 SCC mixtures with limestone filler or fly ash, W/C = 0.45–0.61 and C/P = 0.50–0.67) indicates that the conversion factor A_{sp} to correlate the splitting tensile strength and the direct tensile strength, is slightly lower (0.84 ± 0.04) than the values proposed by EC2 and MC90 (A_{sp} = 0.9) or by MC2010 (A_{sp} = 1.0). Thus, when determining f_{ct} by means of spitting tensile strength results and based on the previously mentioned codes, an overestimation of the tensile strength is obtained. The conversion factor for bending A_{sfl} (0.59 ± 0.10) was significantly lower than the 0.69 prescribed in EC2 and MC2010.

2.4.3 Flexural Tensile Strength

For the evaluation of the flexural tensile strength of SCC, a limited amount of test results can be found in literature (4-point loading: 27 data results originating from 4 studies, 3-point loading: 78 data results from 7 studies, all data have a ratio specimen depth/span length between 3 and 4). The correlation between bending tensile strength and cube compressive strength is given in Fig. 2.24. Note that all compressive test results have been converted into $f_{ccyl,150,eq}$ by using the conversion factors proposed in Sect. 2.2 (Eq. 2.3).

An increase in flexural tensile strength is noticed when cylinder compressive strength is increased (Fig. 2.24). As could be expected, higher values of $f_{ct,fl}$ are

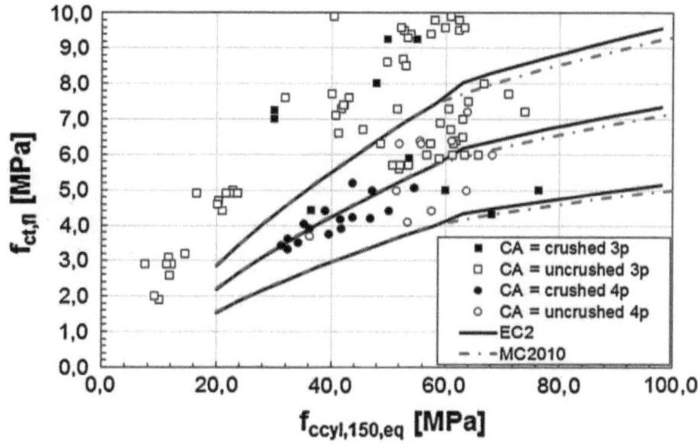

Fig. 2.24 Flexural tensile strength (3-point or 4-point bending tests) versus equivalent cube compressive strength of SCC and the effect of coarse aggregate shape (crushed vs. uncrushed) (105 results from 11 studies)

found when the 3-point bending test is used: most of the results lie above the mean value and even above the upper range proposed by the EC2 and MC2010.

In case of the 4-point bending test, the data tend to follow the mean value proposed by the EC2 and MC2010. All of the data are located between the upper and lower ranges proposed by the codes.

The determination of $f_{ct,fl}$ by means of 3-point testing method results in an overestimation of the flexural tensile strength. Therefore, it is advised to use the 4-point bending test to obtain realistic values for the flexural tensile strength.

The influence of the shape of the coarse aggregate is not noticeable. Also no significant tendencies were found with regards to aggregate type, paste volume, C/P, or filler type, mainly due to a limited amount of available flexural data.

2.5 Modulus of Elasticity

Due to the considerable contribution of aggregates to the overall stiffness of concrete, it is often assumed that SCC—with its higher paste content—is characterised by a lower modulus of elasticity (E_c). Some studies [1–3] have reported that the modulus of elasticity of tested SCC mixtures was lower than that of companion VC mixtures, with a similar compressive strength. Pineaud et al. [167] studied the effect of the paste volume and W/C. By varying the paste volume between 359 and 452 l/m^3 a decrease in E_c was found for an increasing paste volume. The influence of the W/CM on the modulus of elasticity was found to be comparable to that observed on the compressive strength. A survey by Domone [2] seems to indicate that the difference between SCC and VC in the modulus of elasticity is greater for lower compressive strengths. According to Klug and Holschemacher [254] the scatter present on the SCC results is smaller, causing all results to remain within an acceptable band for design using CEB-fib Model Code 1990 [269], validating as such still the use of the common relationship between the modulus of elasticity and the characteristic compressive strength. The survey by

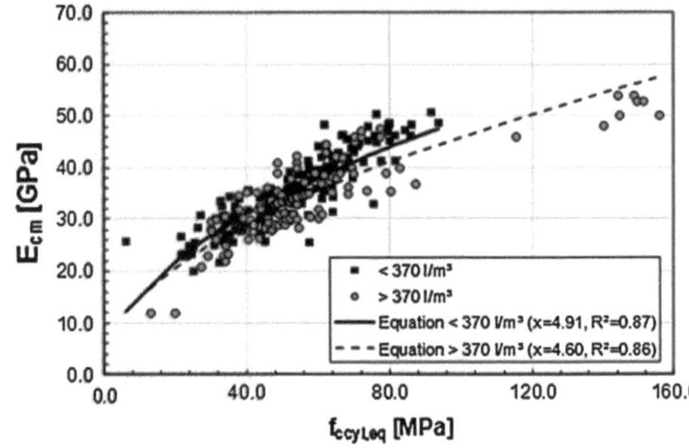

Fig. 2.25 Influence of paste volume on E-modulus (314 results)

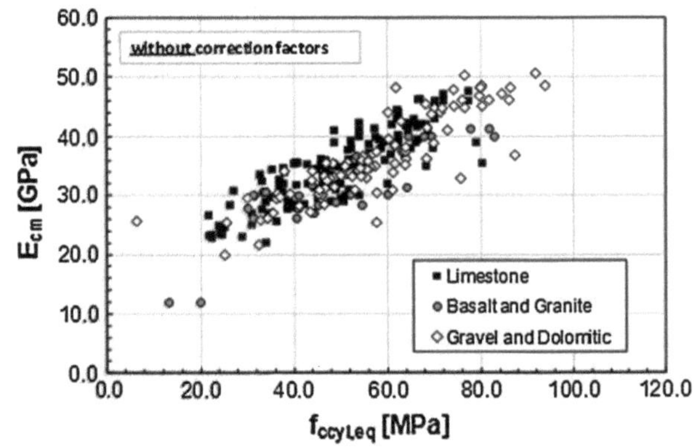

Fig. 2.26 Influence of aggregate type on E-modulus without taking into account the correction factor according to EC2 (276 results)

Van Itterbeeck et al. [255] also found that the E-modulus results for SCC conformed well to the Eurocode 2 predictions and remained well within the band of the CEB-fib Model Code [270]. A similar overall scatter was also observed within this study for SCC and VC.

Since the RILEM database includes a large variety of SCC mixtures with a wide range of paste volumes (results in Fig. 2.25, paste volume range from 280 to 560 l/m^3) the results were first analysed as a function of the paste volume. All results were expressed in function of an equivalent cylinder (diameter 150 mm length 300 mm) compressive strength. Only the secant E-modulus results were included in this analysis. Figure 2.25 could perhaps indicate a 'slight' influence of paste volume. However, the scatter on the results is more important and thus cannot be solely related to paste volume.

When analysing the results in function of the coarse aggregate origin (see Fig. 2.26) a clear influence on the E-modulus cannot be observed. Data scatter appears to increase when the EC2 correction factors are applied, taking into account the effect of aggregate type (Fig. 2.27). Therefore, no conversion factors will be used in this paragraph.

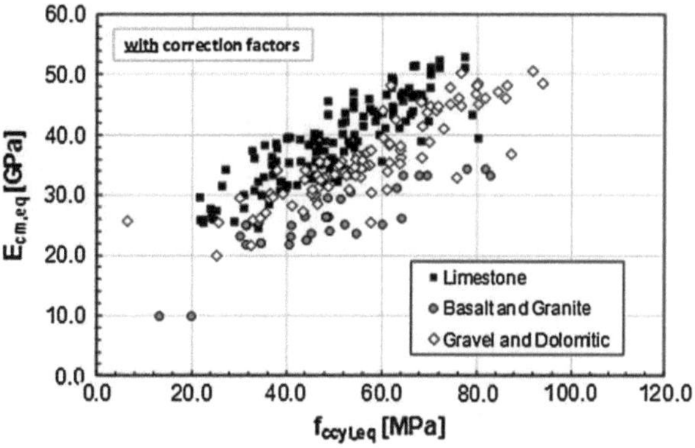

Fig. 2.27 Influence of aggregate type on E-modulus with taking into account the correction factor according to EC2 (276 results)

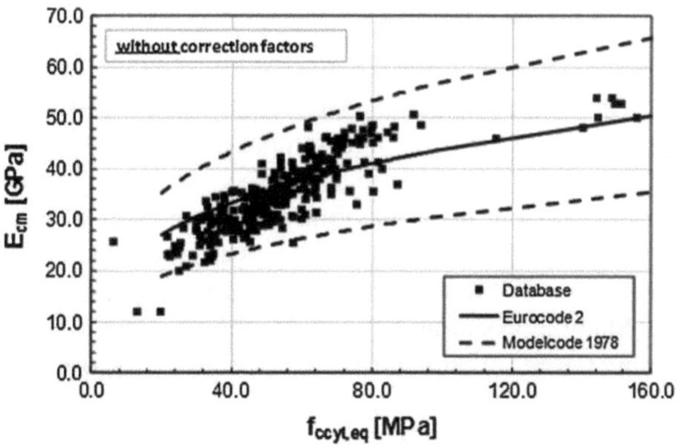

Fig. 2.28 Comparison of E-modulus results from RILEM database with Eurocode 2 provisions (314 results)

In Figs. 2.28 and 2.29 the E-modulus results from the database are compared with the Eurocode 2 [254] and ACI318-11 [253] predictions. The results were not compared with the ModelCode 2010 since, in contrast with the Eurocode 2 and ACI 318-08, it provides information with regard to the tangent E-modulus instead of the secant E-modulus. All results seem to conform to the Eurocode 2 [254] and fit well into the band acceptable for design using the CEB-fib Model Code [270]. A general observation which could thus be made with regard to the modulus of elasticity of SCC after viewing the graph presented in Fig. 2.28 is that it seems to be similar to that for VC, with an important but similar scatter present on the results for both types of concrete. Figure 2.29 indicates that the ACI318-08 captures the relationship between compressive strength and E-modulus better, but seems to provide an underestimation of the modulus of elasticity.

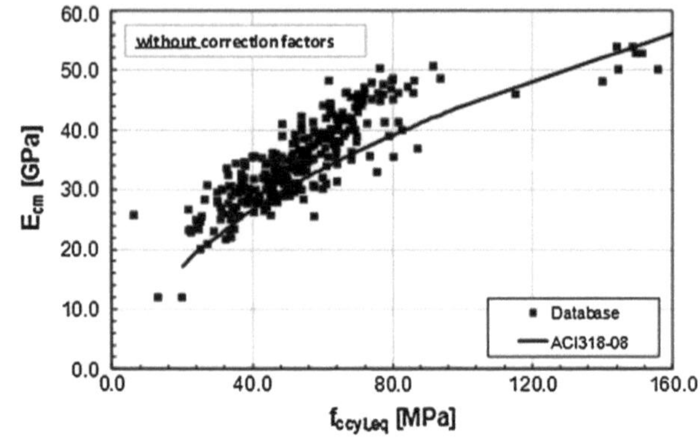

Fig. 2.29 Comparison of E-modulus results from RILEM database with ACI318-08 provisions (314 results)

2.6 Mechanical Behaviour of SCC at Elevated Temperatures

For the evaluation of the fire resistance of a structural system, knowledge on the properties of SCC at high temperatures is essential: thermal, mechanical, deformation properties, and material specific phenomena, such as fire induced spalling. In this section, the behaviour of the mechanical properties of plain SCC (without fibers) is discussed and (where possible) compared to the behaviour of VC. Similar properties of FRSCC are discussed in Chap. 6.

Many physical and chemical changes occur in concrete when subjected to high temperature. When the cement paste is exposed to increasing temperature, the following temperature regions can be identified by means of thermo-gravimetric investigations [271–273]:

- At a temperature of 100 °C: the cement paste loses weight due to the evaporation of free water present in the capillary pores. This process is accompanied by a slight expansion of the cement paste.
- At temperatures between 100 and 200 °C: a continued evaporation of the free water is identified. At 180 °C the dehydration of the calcium silicate hydrates (C–S–H) begins and leads to the breakdown of the chemically bound water which results in a non-linear weight reduction. This process is accompanied by a significant shrinkage of the cement paste. At a temperature of 200 °C, the weight loss of SCC is about 12 % (comparable to HPC).
- At temperatures between 200 and 400 °C: the weight loss continues at a decreased rate mainly due to the loss of water from the gel pores and the further dehydration of the C–S–H. At 400 °C, the weight loss of SCC is about 15 % (17 % for HPC) [271].
- At temperatures between 400 and 600 °C: the weight loss increases non-linearly and is caused by the decomposition of portlandite (especially at 450–500 °C). The total weight loss at 500 °C is 18 % for SCC (20 % for HPC), whereas at 600 °C this is 19 % for SCC (compared to 21 % for HPC). At 500 °C, approximately 70 % of the CSH gel is decomposed.

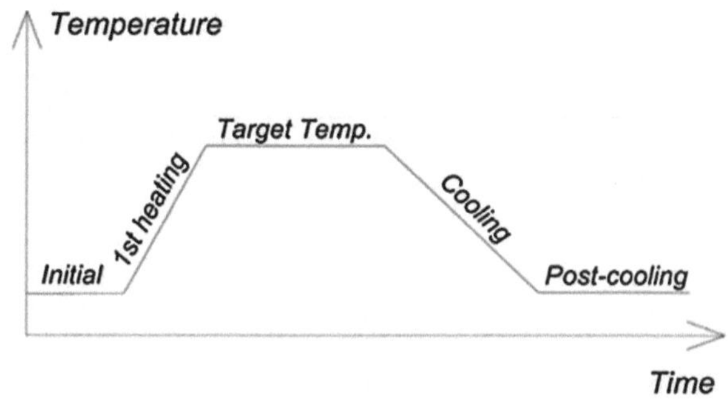

Fig. 2.30 Different temperature stages of fire influencing the remaining strength [271]

- At temperatures between 600 and 800 °C: the decarbonation of the calcite starts: $CaCO_3 \rightarrow CaO + CO_2$. In SCC the calcite is part of the unhydrated cement grains or limestone filler of the cement paste. According to [272], the weight loss between 730–770 °C is defined as the loss due to decarbonation. At 800 °C, the total weight loss reaches 30 % for SCC, compared to 23 % for HPC. According to [31] this peak in mass loss is not identified for VC were for SCC a steep mass reduction was determined at 750 °C. The lack of limestone filler (calcite) in the VC explains the absence of the decarbonation process and the mass loss peak.
- At temperatures exceeding 800 °C: the chemical conversion of the CSH is completed at 850 °C, leading towards a marginal reduction of the weight. However, at about 800 °C ceramic binding may form in the cement paste, which increases only the residual strength, not the hot strength. OPC paste starts to melt around 1,100 and 1,350 °C.

A number of studies can be found in literature dealing with the behaviour of SCC exposed to elevated (up to 100 °C) or high temperatures (up to 800 °C), or on the mechanical behaviour of SCC exposed to fire. In order to evaluate the residual strength of SCC members exposed to fire, it is necessary to determine the residual or hot strength properties. Four strength stages can be identified according to [274]: initial strength (before fire), hot strength (during fire), residual strength (<1 day after fire) and post-cooling strength (>1 day after fire) (Fig. 2.30).

To obtain those strength characteristics, a different experimental set-up is needed. For the evaluation of the hot strength and the residual (or post-cooling) strength two types of tests can be used: residual and hot tests. The difference lies in the different scenarios to which concrete can be subjected mechanically. While data of hot methods are obtained by testing the samples at maintained elevated temperatures (which simulates the behaviour of concrete in a structure under fire), residual and post-cooling test results are obtained by testing the samples after cooling (which simulates the scenario of a concrete element cooled after fire).

Table 2.7 Overview of research on mechanical properties of heated SCC with temperatures up to 700°

Study	[14]	[271]	[276]	[70]	[195]
T-range	20–600 °C	20–600 °C	20–105 °C	20–600 °C	20–700 °C
Mechanical property	$f_{ccub150}$ $f_{ct,spl}$	$f_{ccub,150}$	$f_{ccub150}$ $f_{ct,spl}$	f_{ccyl} $f_{ct,fl}$ E_c	$f_{ccub100}$ $f_{ct,spl}$
Residual or hot	Residual	Residual	Residual	Residual	Residual
Cement type	CEM II/A	CEM I	CEM I	CEM I CEM II/B	CEM II/A
Filler type	Limestone BFS Glass filler	Limestone	Limestone	Limestone	Silica fume
Aggregate type	Granite Limestone	Siliceous	Limestone	Siliceous	Granite
W/B	0.40–0.55	0.48	0.50	0.57–0.61	0.45–0.57
$f_{ccub150,eq\ init}$ [MPa]	37.1–54.0	65.0	57.5	41.1–60.0	33.7–73.2

2.6.1 Mechanical Properties of SCC at Elevated Temperatures

2.6.1.1 Introduction

Due to its higher paste volume and due to the presence of a higher content of additions compared to VC, the mechanical behaviour of SCC under or after thermal load is expected to be different. The dense microstructure of SCC seems to be a disadvantage when exposed to fire. The few studies on SCC subjected to high temperatures show that differences exist between the properties of SCC and VC after being subjected to high temperature (Tables 2.7 and 2.8). The obtained test results indicate a decrease in mechanical properties (compressive strength, tensile strength and modulus of elasticity), comparable or higher than VC. According to [275] plain SCC and VC show similar behaviour with regard to either the relative decay of mechanical properties with respect to room temperature conditions and the higher sensitivity to temperature exposure with increasing strength.

2.6.1.2 Residual or Hot Compressive Strength of Heated SCC

Indications of comparable residual compressive strength losses of SCC and VC (up to 16 %) due to elevated temperatures up to 105 °C were found by [276]. SCC and VC had comparable W/C (0.5) and the same raw materials were used (Table 2.7). The samples were tested immediately after the removal from the heated environment. Explanations are found in an increase in capillary porosity, measured by means of fluorescence microscopy, in a microscopical crack width

Table 2.8 Overview of research on mechanical properties of heated SCC with temperatures above 700°

Study	[275]	[162]	[207]	[273, 278]
T-range	20–800 °C	20–800 °C	20–800 °C	20–800 °C
Mechanical property	f_{ccyl} $f_{ct,spl}$ E_c	$f_{ccyl100/200}$ E_c	f_{ccyl}	$f_{ccub100}$
Residual or hot	Hot	Hot residual	Hot	Residual
Cement type	CEM I ASTM C-150	CEM I CEM II	CEM I	CEM I
Filler type	FA BFS	Limestone	Limestone	FA BFS Limestone Marble Basalt
Aggregate type	Limestone	Gneiss Granite	Limestone	Limestone
W/B	0.28	0.40–0.70	0.64	0.33–0.47
$f_{ccub150,eq\ init}$ [MPa]	67.8	44.4–97.8	24.6	60.0–75.0

enlargement through the aggregates and in higher weight losses compared to the reference samples (20 °C).

A study performed by [70] showed that for SCC heated from 20 to 150 °C, a small loss of strength can be observed, associated to an evaporation of free water as well as to an increase in porosity of the tested concrete, which is confirmed by [279]. This porosity increase is an expansion of the pores diameter and therefore leads to an increase in permeability. Between 150 and 300 °C, an increase in compressive strength is observed for SCC (Fig. 2.31). This increase could be attributed to a modification of the bonding properties of the cement paste hydrates (rehydration of the paste due to the migration of water in the pores). Beyond 300 °C, the residual compressive strength of the tested concretes decreased quickly, which is linked to the appearance of micro-cracking and consequently an increase in porosity and permeability of the tested concretes.

Heated SCC of different strength classes was studied in [14]: at 300 °C there is a residual compressive strength reduction for the SCC C25/30 between 12 and 15 %, while for the SCC C30/37 the strength loss percentage reaches 18 %. The corresponding strength loss percentages for the equivalent VC are 18 % for both strength classes. At 600 °C the reduction in compressive strength ranges from 52 to 57 % for all mixtures (SCC and VC).

Residual mechanical properties were determined in [195]: the residual compressive strength of SCC mixtures linearly decreased at increasing temperature, and the strength reduction is parallel to that observed on VC mixtures of the same strength category. Only the VC C25/30 mixture showed a steeper strength reduction than the SCC C25/30 mixture when heated at temperatures above 300 °C. At all tested temperatures, the residual compressive strength of the SCC

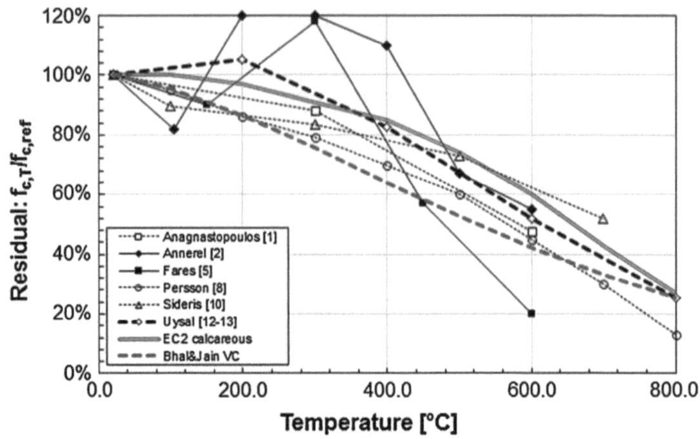

Fig. 2.31 Variation in residual compressive strength of heated SCC as function of temperature (strength determined at room temperature)

mixtures was higher than the one measured on the VC of the same class. This is in accordance with the findings of [279], who reported that concrete specimens produced with SCC had a greater stability below 700 °C compared to VC specimens prepared with the same W/C.

Strength grade has an effect on the hot strength loss of concrete, especially in the temperature range below 400 °C [207].

In Figs. 2.31 and 2.32 an overview of available experimental data on the ratio of (residual and hot) compressive strength at a certain temperature to that at reference temperature (20 °C) as a function of temperature for heated SCC is given.

Strength decreases of SCC can be expected due to thermal load. Comparable degradation behaviour of the residual compressive strength of SCC can be found, which is lower compared to the curves for residual VC given by [280]. In comparison with the curves given by NBN EN 1992-1-2 (Eurocode 2, for hot calcareous VC), the degradation of residual compressive strength of SCC is higher, except for the results by [271, 273, 275, 278] for temperatures between 100 and 400 °C.

The reduction in residual and hot compressive strengths with temperature observed by Persson [162] is gradual without any peaks or outliers. The strength of SCC at fire temperature (hot strength) and the residual or post-cooling strength more or less follow the strength decrease by Eurocode 2 and by [280]. The decrease of the strength of SCC is less than those of HPC with temperature increase.

The effects of elevated temperature on the properties of SCC produced with blast furnace slag (BFS), fly ash (FA) and other mineral additions (limestone powder LP, marble powder MP and basalt powder BP) is examined by [273, 278] (Fig. 2.33). By gradually replacing the cement (CEM I) by BFS or FA, the residual

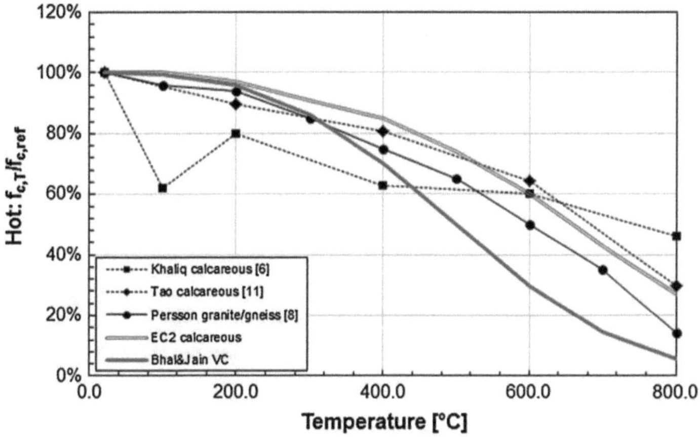

Fig. 2.32 Variation in hot compressive strength of heated SCC as function of temperature (strength determined at room temperature)

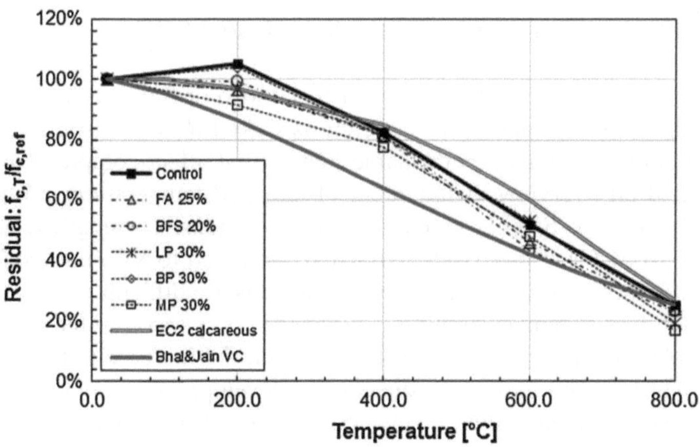

Fig. 2.33 Influence of mineral additions on the residual strength of heated SCC as function of temperature (strength determined at room temperature)

or hot compressive strength loss increases especially when temperatures exceed 400 °C. The increases in the BFS or FA replacement ratios do not significantly affect registered weight losses, however, higher weight losses were observed in the FA series compared to the BFS series. On the other hand, the BFS series showed a higher reduction in the pulse velocity (procedure described in EN 12504-2) compared with the FA series. BP series showed a better performance than other series for all heating cycles. At higher replacement levels of LP, MP and BP further weight losses are observed. Higher weight losses were observed in LP series compared to BP and MP series. Compared to Eurocode 2 and [280] the residual compressive strengths tend to follow the proposed curves of VC.

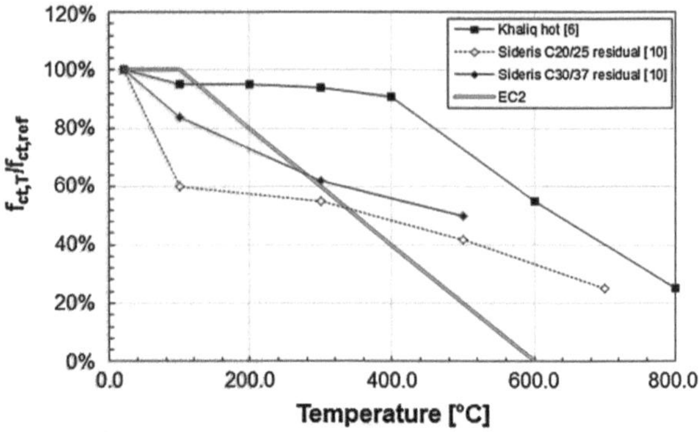

Fig. 2.34 Variation in hot and residual tensile strength of heated SCC as function of temperature

According to [14], SCC produced with limestone filler generally performs better compared to mixtures prepared with ladle furnace slag and glass filler [14] and maintains a greater residual strength after exposure to high temperature.

2.6.1.3 Residual or Hot Tensile Strength of Heated SCC

Also the tensile behaviour of SCC under thermal load or fire exposure is discussed in the literature, as the spalling of concrete is linked to the tensile strength. The tensile strength is dependent on the same factors that affect compressive strength.

In Fig. 2.34 an overview of the effect of elevated temperatures on the tensile strength behaviour of SCC is given. Variations in tensile strength results under thermal load of SCC are observed, also when compared to VC (Eurocode 2). Due to the limited amount of data available, conclusions are difficult to draw. It can be stated that the loss in tensile strength is higher in comparison to the more gradual loss that is obtained for compressive strength with the increase in temperature, as reported by [195]. This is mainly due to the development of micro-cracks as a result of thermal incompatibility between the hydration products and the inert aggregate skeleton of concrete, which is also confirmed by [281, 282].

A similar behaviour was observed for both SCC and VC mixtures of the lower strength categories C20/25 and C25/30: the relative splitting tensile strength is lower after reaching 300 °C. With regard to higher strength categories (C30/37 and C50/60), SCC can have a smaller degree of reduction of relative tensile strength than VC of the same strength class. However, using the Eurocode 2 to predict the residual tensile strength of SCC, heated up to temperatures inferior to 300 °C, can lead to an overestimation.

The effect of temperature on the tensile strength of SCC becomes more pronounced at higher temperatures since the increase in pore pressure due to further temperature increase causes rapid loss of tensile strength.

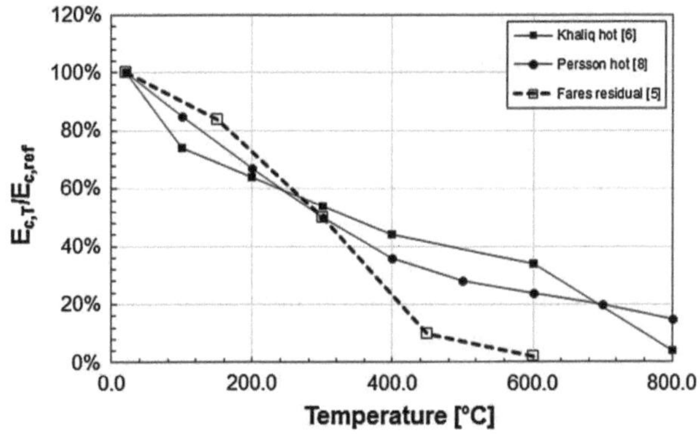

Fig. 2.35 Variation in hot and residual modulus of elasticity of heated SCC as function of temperature

Even at temperatures up to 105 °C, higher residual splitting tensile strength losses were found for SCC (up to 10 %), whereas for VC no significant strength loss was observed [276].

According to [277], the loss in splitting tensile strength of SCC measured at elevated temperatures is gradual and varies almost in a linear fashion up to 400 °C, followed by a sharp reduction in tensile strength between 400 and 800 °C. This sharp reduction in splitting tensile strength beyond 400 °C can be attributed to the development of excessive micro- and macro-cracks resulting from thermal stresses and thermal incompatibility within the concrete. These results were not compared to VC samples, but they were compared to the Eurocode 2. The test curve is significantly higher than the curve of predicted from the code provision (Fig. 2.34).

Up to 800°, a continuously decrease of the flexural strength of SCC is found by Fares et al. [70]. This is associated with the evaporation of bound water and corresponding large mass loss.

2.6.1.4 Residual or Hot Modulus of Elasticity of Heated SCC

The fire response of concrete structures influences the modulus of elasticity of concrete. As shown in Fig. 2.35, a sharp decrease of the modulus of elasticity of SCC of temperature can be obtained with temperature increase. This is mainly due to the disintegration of the hydration products. This behaviour depends on the thermal transient creep, the aggregate type, and the weight loss [277]. At temperatures higher than 300 °C, the hot residual modulus of elasticity seems to be higher than the residual one.

Lower elastic moduli at temperatures encountered during fire were observed in SCC than in VC [162]. The residual modulus of elasticity decreases continuously up to 800 °C due to the evaporation of bound water, corresponding to a large mass loss [70].

2.6.2 Material Specific Phenomena of Heated SCC: Fire-Induced Spalling

Concrete elements exposed to a rapid temperature increase (e.g. fire) may suffer severely from spalling of the outer concrete layers of the elements. According to [271], spalling is the violent or non-violent breaking off of layers or pieces of concrete from the surface of a structural element when it is heated rapidly. Four types of spalling can be identified: aggregate spalling, explosive spalling, surface spalling, and corner/sloughing-off spalling. Many factors have been identified from experiments as influencing spalling of concrete in fire: the heating rate (especially above 3 °C/min), the permeability of the material (a denser concrete is more sensitive for pore pressure build up, for example HPC and SCC compared to VC), pore saturation level (especially above 2–3 % moisture content by weight of concrete), aggregate size/type, the presence of reinforcement (fibers), the shape/size of the samples and the level of externally applied load [271, 272].

The origin of spalling can be explained as a combination of the stresses formed by the thermal expansion and the vapour pressure within the concrete and stresses due to phase changes of some aggregates [283, 284]. The main reason for concrete spalling at elevated temperatures is considered to be the internal pore pressure that builds up due to the vaporization of the free and chemically bound water [195].

In concrete mixtures with finer pore structure, such as HPC or SCC, this internal pressure is not released sufficiently, thus it can lead to severe spalling of the surface layers.

The probability of spalling of SCC is much higher compared to VC, which can be explained by the slower drying rate of SCC due to its lower permeability [285]. However, according to [195], SCC has an explosive spalling tendency similar to the one of VC of the same strength class. Spalling is only dependent on the cement and water content of the mixture and not on the way of its compaction: the self-compacting samples and the samples with VC of the lower strength classes tested (C20/25 and C25/30) did not spall at any temperature up to 700 °C. SCC and VC of the strength class C30/37 suffered from explosive spalling at temperatures between 500 and 580 °C, whereas for the highest strength category mixtures (C50/60) spalling already occurred in the temperature range of 380–458 °C.

A study performed by [14] found that explosive spalling occurs in SCC and VC at 600 °C. The use of ladle furnace slag as filler material instead of limestone filler leads toward higher initial compressive strength but also towards a higher explosive spalling risk. Overall, SCC is expected to spall more than VC due to the lower permeability and higher moisture content.

Explosive spalling was also observed in columns made with SCC but not in those made with VC, even though the VC columns were cured in exactly the same way as the SCC columns [162]. Spalling occurred at a heating rate of 480 °C/h, at a rate of 240 °C/h no spalling occurred. The explosive fire spalling is mainly dependent on the stress in the concrete, the C/P, and water-cementitious materials ratio. An option to limit the amount of explosive fire spalling in columns with SCC to the same level as in columns with VC is to introduce polypropylene fibers into the concrete mix proportions. Another way is to keep the combination of C/P and W/C sufficiently high.

Steel fibers, polypropylene fibers (PP), and others can be used to improve the residual or hot properties of concrete being exposed to elevated temperatures. The incorporation of PP fibers can reduce the damage as the melted fiber form a connected channel system through which heat and water vapour could escape without causing extensive damage to the concrete [279]. Further discussion of the behaviour of steel fibre reinforced concrete (fibers added to overcome the adverse effects of fire induced spalling) is presented in Chap. 6.

Due to the complex physics behind the spalling behaviour of concrete, no valuable prediction models are available to date.

2.7 In-situ Properties

In literature, several papers dealing with in situ properties of SCC can be found. Domone [286] summarized 68 examples of applications of SCC in practice, in the period from 1993 to 2003, including details and ranges of properties, used materials and mix proportions, details of fresh properties, and mechanical properties. This enables concrete practitioners and researches to draw their own conclusions on the suitability and the wide range of concrete applications. This study is a good starting point in the process of identifying the differences and similarities of (i) in situ properties of SCC and properties of SCC tested in laboratory conditions, and (ii) in situ properties of SCC and VC. Ninety percent of the considered cases used SCC with slump flows of 600 and 750 mm, and compressive strength higher than 40 MPa. The maximum aggregate size lies between 16 and 20 mm. The local availability of the aggregates quite often determines the use of crushed rock versus natural gravel. Limestone filler as additional powder is used in 41 % of the cases. In half of the cases VMA was being used to make the mixtures more robust. The median values of the mixture proportions are determined:

- 31.2 vol % coarse aggregates
- 34.8 vol % paste content
- Powder content 500 kg/m^3
- W/P: 0.34.

As expected, there is no unique composition for SCC for a given application. Different types of applications worldwide are mentioned in the literature: columns in high rise buildings, tunnel linings, wall elements for tunnelling, panels, bridge piers, beams for bridges, slabs, architectural purposes, etc.

Some interesting case studies dealing with in situ mechanical properties of SCC are highlighted in the following section.

2.7.1 Case Studies

A demonstration conducted by Sonebi et al. [287] deals with in situ compressive strength of walls (height: 2 m, width 1.95 m, thickness: 0.25 m) using free falling SCC (2 m high). The composition consists of Portland cement Class 42.5, limestone filler, crushed granite aggregates (maximum aggregate size of 20 mm) and polycarboxylate ether as superplasticiser: precise compositions are mentioned in [287]. Compared to the 28 day cube strength of 72.7 MPa, the in situ strength determined by means of 100 mm drilled cores is between 54.5 and 58.6 MPa, thus on average 22 % lower. Although the shape of the specimens is different, the main reason for the difference in results was associated with the core samples being perpendicular to the casting direction, which is not the case for the control cubes (tested in the direction of casting). Nevertheless, a high degree of strength uniformity of the cores was found along the different heights of the wall.

Khayat et al. [116] also reported on the in situ strength of walls (1.5 m in height) cast with SCC (compressive strength up to 40–50 MPa). Compared to laboratory produced cylinders of the same diameter, cylindrical cores (h/d = 2) had slightly lower compressive strength. Core samples extracted from SCC and VC wall elements exhibited an approximate reduction of compressive strength of 10(maximum aggregate size of 10 mm). Again, this was mainly associated with the fact that the core samples were perpendicular to the casting direction. A strength-reduction of 8 % was obtained between cores drilled from the top and bottom sections.

Khayat et al. [288] carried out investigations on the structural behaviour of congested columns cast with SCC of 40-50-60 MPa compressive strength as well as control columns cast with VC of similar strength. The results indicate that the columns with SCC could develop similar in situ compressive strength and greater ductility than columns cast with VC. The estimated compressive strength of SCC elements tested in compression was reported to be 10 % lower than that determined on control 100 × 200 mm cylinders.

Nunes et al. [154] present the results of a study dealing with the use of SCC and VC in a precast factory. The main goal was to evaluate the feasibility of replacing VC (strength class C45/55) by SCC of the same strength class while maintaining the constituent materials. A large number of specimens (cylinder, cube and prism) were cast along with full-size precast elements. The composition of the SCC consisted of Portland cement class 52.5R, limestone filler, siliceous natural fine and coarse sand, crushed calcareous aggregates (maximum aggregate size of 12.5 mm), and polycarboxylate ether superplasticiser: precise compositions mentioned in [154]. In-situ properties of SCC and VC are obtained by means of compressive strength of drilled cores out of a U-shaped element (two vertical walls and one horizontal slab) at the age of 60 days. The results were compared to cube specimens (150 mm side) and cylinders (300/150 mm) tested at 28 days. Overall, the SCC exhibited an improved mechanical behaviour, mainly due to the lower W/C. More important is the finding that a more uniform strength along the full-size elements (slab vs. wall elements) was obtained for SCC, due to a combination of proper mix-design in combination with controlled mixing and placing. It is concluded that in general SCC exhibits a more compact and homogenous matrix and higher resistance to fluid and chloride ingress [154].

Two non-reinforced concrete columns (6.0 m in height), cast with SCC or a VC, considered for high-level radioactive waste disposal purposes (using Portland cement class 42.5, limestone filler, sand and aggregates and a polycarboxylate ether superplasticiser (precise compositions mentioned in [289]) were tested to determine in situ properties of SCC and VC [290]. Compressive strength of drilled cores and precast cubes at 28 days, direct tensile strength of drilled cores at 56 days, and secant modulus of elasticity of drilled cores at 28 days are reported. Cores were drilled out of the columns at different heights (high, mid, and low levels at 5, 3, and 1 m, respectively, above the bottom surface). Three cubes were stored at 20 °C (>90 % RH). It is shown that the height of the extruded cores is of a great importance for SCC columns, which is less prominent for columns out of VC. Due to the self-compacting capacity, a greater degree of compaction was obtained for the SCC cores at the lower levels, hence resulting in increased compressive strength (3.6 % higher for cores at the lower level) and modulus of elasticity (10.8 % higher for cores at lower level). Cores from the lower parts also registered 3.8 % higher ultrasonic pulse velocity (NBN EN 12504-4) and significantly lower water permeability (NBN B 15-222) compared to cores from the upper parts of the wall elements. For the tensile behaviour, the position of the cores had no significant effect.

Full-scale reinforced columns (3.0 m in height) applying VC and SCC were tested by Sonebi et al. [291]. The investigated concrete mixtures were prepared with Portland cement, ground granulated blast slag and limestone filler as powder material, quartzite sand with continuously crushed granite aggregates (nominal particle size of 10 and 20 mm) were used as aggregates in combination with a copolymer-based superplasticiser: precise compositions mentioned in [291]. In total, eight columns were cast: two columns out of SCC and two with VC intended for housing purposes (strength of 35 MPa (C35)), and two columns with SCC and two with VC intended for civil engineering purposes (strength of 60 MPa (C60)). For the variation in situ strength of columns, the results obtained at the bottom are higher than those obtained at the top, except for the C60-VC mixture used for civil engineering purposes. The results also show that the strength variation within the column is less significant for C60 strength grade mixtures than for C35 mixtures, with slightly smaller variations obtained for SCC. The study also indicates that the in situ strength achieved in the columns varies between 80 and 100 % for all mixtures (SCC and VC), which is above the average value of 65 % reported in literature for such elements. In general, the in situ compressive strengths of SCC were closer to standard cube strength than those of VC. Moreover, the distribution of in situ properties was found to be more uniform in the case of SCC. Finally, the maximum load of the SCC column was slightly higher (4 %) but the SCC columns did not behave in the same manner as the VC columns in terms of initial stiffness and axial strain at maximum load, where the VC columns exhibited greater ductility.

Similar findings were reported by Zhu et al. [233], who tested realistic size columns (3.0 m in height) and beams (3.8 m in length), with different types of reinforcement depending on the chosen application: C35 strength class (housing) and C60 (civil engineering purposes). The investigated SCC and VC concrete mixtures were prepared with Portland cement (42.5 N strength class), ground granulated blast slag and limestone filler as powder material, locally available sand and crushed aggregates (nominal particle size of 10 and 20 mm) were used as aggregates: precise compositions mentioned in [291]. By means of precast cubes and withdrawn cores (compressive strength), pre-embedded inserts (pull-out tests) and rebound hammer values for near-surface properties, comparable conclusions regarding the behaviour of the columns and beams were drawn: (i) the strength-differences along the height of the columns were statistically significant for all mixtures. The strength of the used concrete, SCC and VC, increased from top to bottom, the difference between top and bottom being greater for C35 strength grading, (ii) the results from SCC mixtures were similar to those of the VC mixtures, although the properties of SCC were marginally more uniform.

Overall, it can be confirmed that SCC cast in situ could provide similar (or even slightly enhanced) properties compared to those obtained by means of properly compacted VC. However, the robustness of a SCC mixture, i.e. the stability of its properties when affecting the original mixture design by errors in weighing the constituents, is an important parameter that cannot be overlooked. According to Rigueira et al. [292], the tolerances known for VC have proved to be valid for SCC as well. Especially errors in weighing water (effective water and water transported by means of the aggregates) and fines contents are of capital importance for SCC, and are admissible up to ± 6 % interval. Results of a pilot survey conducted by Ferrara [293] show that SCC can be reliably and robustly produced in precast plant for manufacturing precast pre-stressed elements.

2.8 Conclusions

Several studies have addressed the hardened properties of SCC. For the sake of providing a general view of obtained results, an extensive database was developed and analysed in the framework of the RILEM technical committee on mechanical properties of SCC rather than reviewing individual studies.

With respect to compressive strength, the influence of specimen size and shape is well documented in the case of VC. However, for SCC the data are limited. Based on the database, a tendency for the development of higher conversion factors, (f_{ccyl}/f_{ccub}) up to 0.90, is noticed in the case of SCC. With respect to the influencing factors on the obtained compressive strength, W/C and cement strength class does seem to affect the obtained strengths as they do for VC.

The most common addition types to produce SCC in the studies reported in the database are limestone filler, fly ash, and silica fume. Higher 28-day compressive strengths are obtained with fly ash and natural pozzolans compared to limestone filler, while maintaining C/P and the cement content. The influence of C/P was analysed in case of the addition of limestone filler, fly ash, slag, and silica fume. At a constant W/C, a higher C/P generally leads to lower strength. Increased air content decreases the compressive strength of SCC with about 4 MPa per 1 % increase in air content. No significant effect of the type, origin and shape of the aggregate was noted, except for some general trends. As in the case of VC, the higher the compressive strength is, the more critical is the compressive strength of the rock.

Different conclusions can be found in the literature concerning the stress-strain relationship of SCC, due to different test set-ups. However, they all show some differences between SCC and VC. For limestone filler-based SCC, the peak strain at different ages and concrete strengths is higher than for VC. Besides the influence of limestone filler, other filler materials resulted in higher peak strains as well.

The strain softening behaviour of SCC and VC is comparable for the same compressive strength level. The toughness of limestone filler-based SCC is slightly higher than that of VC made without limestone filler. The largest difference was found for the ascending branch of the stress-strain diagram, whereas the surface under the descending branch was comparable. However, the difference in toughness between VC and SCC does not seem to be significant.

The direct tensile strength tends to increase with increasing cube compressive strength. Based on limited results, no significant effect of the paste volume or filler type on the correlation between direct tensile strength and compressive strength is found. However, the C/P seems to have some visible effect, yielding lower direct tensile strength values in case of C/P < 75 %, for a given strength.

An increasing trend of splitting tensile strength is also noticed for increasing compressive strength. Splitting tensile strength results of SCC using granite coarse aggregate lie in the upper half or above the ranges proposed by the EC2 and MC2010. This was not the case when limestone or gravel aggregates are used. Further analysis of available literature data shows that there is no significant effect of coarse aggregate size or paste volume on the correlation between splitting tensile strength and compressive strength. The influence of the C/P or filler type is also not significant according to the available data.

For the evaluation of the flexural tensile strength of SCC, only a limited amount of test results could be found in literature. A logically increasing trend of flexural strength is noticed for increasing cube compressive strength. As could be expected, higher values are found in case of 3-point bending tests, with most of the results lying above the mean value and even above the upper range proposed by EC2 and MC2010. In case of 4-point bending, the test results follow the mean value proposed by EC2 and MC2010, and all data are between the lower and upper range as proposed by the codes.

Due to the considerable contribution of aggregates to the overall stiffness of concrete, it is often assumed that SCC is characterised by a lower modulus of elasticity. Nevertheless, analysis of the data available in the RILEM database shows that the available results fit well into the range acceptable for design using the CEB-fib Model Code. The modulus of elasticity of SCC thus seems to be very similar to VC, with an important but similar scatter on the results for both types of concrete.

The denser microstructure of SCC in comparison with VC seems to be a disadvantage when exposed to fire, as it is the case for HPC. On average, the strength loss after heating of SCC is comparable to VC. The strength of SCC at fire temperature (hot strength) and the residual or post-cooling strength of SCC more or less follow the strength decrease as predicted by EC2. The strength decrease of SCC with temperature increase is less than what is observed for HPC.

The probability of spalling of SCC is higher compared to VC, although sometimes conflicting results are given in literature. The explosive fire spalling is mainly dependent on the stress in the concrete, the C/P, and the W/C.

In literature, several studies concerning the in situ properties of SCC can be found. Overall, it can be confirmed that SCC cast in situ can provide similar or even slightly better properties compared to VC.

References

1. Sonebi, M., Bartos, P.J.M., Zhu, W., Gibbs, J., Tamimi, A.: Brite-EuRam—Task 4-Properties of Hardened Self-Compacting Concrete—Final Report. p. 73 (2005)
2. Domone, P.L.: A review of the hardened mechanical properties of self-compacting concrete. Cement Concr. Compos. **29**, 1–12 (2007)
3. Holschmacher, K., Klug., Y.: A Database for the evaluation of hardened properties of SCC. LACER **7**, 123–134 (2002)
4. De Schutter, G., Bartos, P.J.M., Domone, P., Gibbs, J.: Self-Compacting Concrete, p. 293. Whittles Publishing, Caithness (2008)
5. Abdel-Jawad, Y., Dehn, F.: Self-healing of self-compacting concrete. In: 4th International RILEM Symposium on Self-Compacting Concrete, pp. 1023–1029 (2005)
6. Abidelah, A., Bouchaïr, A., Kerdal, D., Ayed, K.: Characterization of a self-compacting sand concrete using the quarry waste. Can. J. Civ. Eng. **36**, 1773–1783 (2009)
7. Adam, T., Ruehl, M., Gruebl, P.: The effect of different binders and chemical admixtures on the early hardening of high-strength self-consolidating concrete. In: 4th International RILEM Symposium on Self-Compacting Concrete, pp. 973–979 (2005)
8. Akalin, O., Ulas Akay, K., Sennaroglu, B.: Self-consolidating high-strength concrete optimization by mixture design method. ACI Mater. J. **107**(4), 357–364 (2010)
9. Akcay, B., Tasdemir, M.: Mechanical behaviour and fiber dispersion of hybrid steel fiber reinforced self-compacting concrete. Constr. Buil. Mater. **28**, 287–293 (2012)
10. Akram, T., Memon, S.A., Obaid, H.: Production of low cost self-compacting concrete using bagasse ash. Constr. Build. Mater. **23**, 703–712 (2009)
11. Almeida Filho, F.M., Barragán, B.E., Casas, J.R., El Debs, A.L.H.C.: Hardened properties of self-compacting concrete—a statistical approach. Constr. Build. Mater. **24**, 1608–1615 (2010)
12. Al-Tamimi, A.K., Sonebi, M.: Assessment of self-compacting concrete immersed in acidic solutions. J. Mater. Civ. Eng. **15**(4), 354–357 (2003)

13. Alyamaç, K.E., Ince, R.: A preliminary concrete mix design for SCC with marble powders. Constr. Build. Mater. **23**, 1201–1210 (2009)
14. Anagnostopoulos, N., Sideris, K.K., Georgiadis, A.: Mechanical characteristics of self-compacting concretes with different filler materials, exposed to elevated temperatures. Mater. Struct. **42**, 1393–1405 (2009)
15. Anagnostopoulos, N.S., Georgiadis, A.S., Sideris, K.K.: Carbonation of self-compacting concretes produced with different materials. In: 5th International RILEM Symposium on Self-Compacting Concrete, pp. 721–727 (2007)
16. Annerel, E., De Schutter, G.: Microstructure and aesthetic appearance of SCC. In: 5th International RILEM Symposium on Self-Compacting Concrete, pp. 381–386 (2007)
17. Arbelàez, J., Rigueira, V., Marti, V., Serna, R., Pinto, B.: Bond characteristics of prestressed strand in self-compacting concrete. In: 3rd International RILEM Symposium on Self-Compacting Concrete, pp. 684–691 (2003)
18. Assié, S.: Durabilité des Bétons Autoplaçants, PhD, Institut national des sciences appliquees de toulouse, p. 249 (2004)
19. Assié, S., Escadeillas, G.: Estimates of self-compacting concrete 'potential' durability. Constr. Build. Mater. **21**, 1909–1917 (2007)
20. Assié, S., Escadeillas, G., Waller, V., Marchese, G., Vachon, M.: Self-compacting and vibrated concrete compared by their physico-chemical durability properties. In: 4th International RILEM Symposium on Self-Compacting Concrete, pp 373–379 (2005)
21. Audenaert, K., De Schutter, G.: Chloride penetration in self compacting concrete. In: 3rd International RILEM Symposium on Self-Compacting Concrete, pp. 818–825 (2003)
22. Audenaert, K., De Schutter, G.: Modelling the carbonation process of self-compacting concrete. In: 5th International RILEM Symposium on Self-Compacting Concrete, pp. 689–694 (2007)
23. Ayding, A.C.: Self compactability of high volume hybrid fiber reinforced concrete. Constr. Build. Mater. **21**, 1149–1154 (2007)
24. Barbhuiya, S.: Effects of fly ash and dolomite powder on the properties of self-compacting concrete. Constr. Build. Mater. **25**, 3301–3305 (2011)
25. Bassuoni, M.T., Nehdi, M.L.: Durability of self-consolidating concrete to sulfate attack under combined cyclic environments and flexural loading. Cem. Concr. Res. **39**, 206–225 (2009)
26. BBRI: Onderzoeksresultaten Prenormatief onderzoek: Mechanische prestaties van zelfverdichtend beton: naar een mogelijke toepassing van Eurocode 2, Biënnale 2008–2010, Conventie NBN: CC CCN/PB/NBN-513, Eindverslag, (2011)
27. Beaupré, D., Lacombe, P., Khayat, K.H.: Laboratory investigation of rheological properties and scaling resistance of air entrained self-consolidating. Mater. Struct. **32**(3), 235–240 (1999)
28. Boel, V., Audenaert, K., De Schutter, G.: Behaviour of self-compacting concrete concerning frost action with deicing salts. In: 3rd International RILEM Symposium on Self-Compacting Concrete, pp. 837–843 (2003)
29. Boel, V., Audenaert, K., De Schutter, G.: Gas permeability and capillary porosity of self-compacting concrete. Mater. Struct. **41**, 1283–1290 (2008)
30. Boel, V., Audenaert, K., De Schutter, G., Heirman, G., Vandewalle, L.: Experimental durability evaluation of self-compacting concrete with limestone filler. In: 4th International RILEM Symposium on Self-Compacting Concrete, pp. 297–303 (2005)
31. Boel, V.: Microstructure of self-compacting concrete in relation to gas permeability and durability (in Dutch). PhD, Ghent University (2010)
32. Bokan Bosiljkov, V., Duh, D., Zarnic, R.: Salt frost scaling of SCC related to curing regime and air void system. In: 5th International RILEM Symposium on Self-Compacting Concrete, pp. 729–734 (2007)
33. Bokan Bosiljkov, V.: SCC mixes with poorly graded aggregate and high volume of limestone filler. Cem. Concr. Res. **33**, 1279–1286 (2003)

34. Boukendakdji, O., Kenai, S., Kadri, E.H., Rouis, F.: Effect of slag on the rheology of fresh self-compacted concrete. Constr. Build. Mater. **23**, 2593–2598 (2009)
35. Bouzoubaâ, N., Lachemi, M.: Self-compacting concrete incorporating high volumes of class F fly ash: preliminary results. Cem. Concr. Res. **31**(3), 413–420 (2001)
36. Braam, C.R.: Horizontale belasting door zelfverdichtende betonspecie. Cement **3**, 82–85 (2002)
37. Brewe, J.E., Myers, J.J.: High strength self-consolidating concrete girders subjected to elevated compressive fiber stresses, part 1: Prestress loss and camber behavior. PCI J. **3**, 59–77 (2010)
38. Brouwers, H.J., Radix, H.J.: Self-compacting concrete: theoretical and experimental study. Cem. Concr. Res. **35**, 2116–2136 (2005)
39. Brunner, M.: Durability of SCC with high water content. In: 4th International RILEM Symposium on Self-Compacting Concrete, pp. 333–340 (2005)
40. Bui, V.K., Akkaya, Y., Shah, S.P.: Rheological model for self-consolidating concrete. ACI Mater. J. **99**(6), 549–559 (2002)
41. Byun, K.-J., Kim, J.-K., Song, H.-W.: Self-Compacting Concrete in Korea, International Workshop on Self-Compacting Concrete Japan, pp. 23–33 23–26 August 1998
42. Castel, A., Vidal, T., François, R.: Bond and cracking properties of self-consolidating concrete. Constr. Build. Mater. **24**, 1222–1231 (2010)
43. Castel, A., Vidal, T., Viriyametanont, K., François, R.: Effect of reinforcing bar orientation and location on bond with self-consolidating concrete. ACI Mater. J. **4**, 559–567 (2006)
44. Chopin, D., Francy, O., Lebourgeois, S., Rougeau, P.: Creep and shrinkage of heat-cured self-compacting concrete (SCC). In: 3rd International RILEM Symposium on Self-Compacting Concrete, pp. 672–683 (2003)
45. Collepardi, M., Borsoi, A., Collepardi, S., Troli, R.: Strength, shrinkage and creep of SCC and flowing concrete. In: 4th International RILEM Symposium on Self-Compacting Concrete, pp. 911–919 (2005)
46. Collepardi, M., Collepardi, S., Ogoumah Olagot, J.J., Troli, R.: Laboratory-tests and field-experiences of high-performance SCCs. In: 3rd International RILEM Symposium on Self-Compacting Concrete, pp. 904–912 (2003)
47. Collepardi, M., Passuelo, A.: The best SCC: stable, durable, colorable. In: 4th International ACI/CANMET Conference on Quality of Concrete Structures and Recent Advances in Concrete Materials and Testing (2005)
48. Coppola, L., Cerulli, T., Salvioni, D.: Sustainable development and durability of self-compacting concretes. In: 11th International Conference on Fracture (2005)
49. Corinaldesi, V., Moriconi, G.: Characterization of self-compacting concretes prepared with different fibers and mineral additions. Cement Concr. Compos. **33**, 596–601 (2011)
50. Corinaldesi, V., Moriconi, G.: Durable fiber reinforced self-compacting concrete. Cem. Concr. Res. **34**, 249–254 (2004)
51. Corinaldesi, V., Moriconi, G.: The role of industrial by-products in self-compacting concrete. Constr. Build. Mater. **25**, 3181–3185 (2011)
52. Craeye, B., De Schutter, G., Desmet, B., Vantomme, J., Heirman, G.: Effect of mineral filler type on autogenous shrinkage of self-compacting concrete. Cem. Concr. Res. **40**, 908–913 (2010)
53. Craeye, B.: Early-age thermo-mechanical behaviour of concrete supercontainers for radwaste disposal, Doctoral thesis, p. 376 (2010)
54. Daoud, A., Lorrain, M.: Influence de la position des armatures sur l'adhérence des bétons autoplaçants: interprétation par analyse d'image. Mater. Struct. **36**, 231–237 (2003)
55. Daoud, A., Lorrain, M., Laborderie, C.: Anchorage and cracking behaviour of self-compacting concrete. In: 3rd International RILEM Symposium on Self-Compacting Concrete, pp. 692–702 (2003)
56. de Almeida Filho, F.M., El Debs, M.K., El Debs, A.L.H.C.: Bond-slip behavior of self-compacting concrete and vibrated concrete using pull-out and beam tests. Mater. Struct. **41**, 1073–1089 (2008)

57. De Schutter, G., Audenaert, K., Boel, V., Vandewalle, L., Heirman, G., Dupont, D.: Transport properties in self-compacting concrete and relation with durability: overview of a belgian research project. In: 3rd International RILEM Symposium on Self-Compacting Concrete, pp. 799–807 (2003)
58. Dehousse, E., Marion, A.-M.: Early age compressive strength prediction of self-compacting concrete: first results. In: 3rd International RILEM Symposium on Self-Compacting Concrete, pp. 586–595 (2003)
59. Desmet, B., Atitung, K.C., Sanchez, M.A.A., Vantomme, J., Feys, D.: Monitoring the early-age hydration of self-compacting concrete using ultrasonic p-wave transmission and isothermal calorimetry. Mater. Struct. **44**, 1537–1558 (2011)
60. Desnerck, P.: Compression, bond and shear behaviour of powder-type self-compacting concrete. PhD Thesis (2011)
61. Desnerck, P., De Schutter, G., Taerwe, L.: Experimental determination of bond strength of reinforcing bars in self-compacting concrete. In: 5th International RILEM Symposium on Self-Compacting Concrete, pp. 659–664 (2007)
62. Dinakar, P., Babu, K.G., Santhanam, M.: Durability properties of high volume fly ash self-compacting concretes. Cement Concr. Compos. **30**, 880–886 (2008)
63. Dinakar, P., Babu, K.G., Santhanam, M.: Mechanical properties of high volume fly ash self compacting concrete. In: 5th International RILEM Symposium on Self-Compacting Concrete, pp. 651–657 (2007)
64. Ding, Y., Zhang, Y., Thomas, A.: The investigation on strength and flexural toughness of fiber cocktail reinforced self-compacting high performance concrete. Constr. Build. Mater. **23**, 448–452 (2009)
65. Druta, C.: Tensile Strength and bonding characteristics of self-compacting concrete. MSc Thesis, Louisiana State University-USA, p. 115 (2003)
66. El Chabib, H., Nehdi, M., El Naggar, M.-H.: Behavior of SCC confined in short GFRP tubes. Cement Concr. Compos. **27**, 55–64 (2005)
67. Emdadi, A., Libre, N.A., Mehdipour, I., Vahdani, M., Dara, S.: SCC mixtures with different aggregate gradation and limestone powder, 5th International RILEM Symposium on Self-Compacting Concrete, pp. 155–162 (2007)
68. Emdadi, A., Libre, N.A., Mohebbi, A.R., Yekta, S., Shekarchi, M.: Self-compacting concrete for developing countries using raw materials. In: 5th International RILEM Symposium on Self-Compacting Concrete, pp. 181–188 (2007)
69. Fantilli, A.P., Vallini, P., Chiaia, B., Chiaia, B.: Ductility of fiber-reinforced self-consolidating concrete under multi-axial compression. Cement Concr. Compos. **33**, 520–527 (2011)
70. Fares, H., Noumowe, A., Remond, S.: Self-consolidating concrete subjected to high temperature—mechanical and physicochemical properties. Cem. Concr. Res. **39**, 1230–1238 (2009)
71. Fares, H., Remond, S., Noumowe, A., Cousture, A.: High temperature behaviour of self-consolidating concrete microstructure and physicochemical properties. Cem. Concr. Res. **40**, 488–496 (2010)
72. Fava, C., Bergol, L., Fornasier, G., Giangrasso, F., Rocco, C.: Fracture behaviour of self compacting concrete. In: 3rd International RILEM Symposium on Self-Compacting Concrete, pp. 628–636 (2003)
73. Felekoglu, B.: A comparative study on the performance of sands rich and poor in fines in self-compacting concrete. Constr. Build. Mater. **22**, 646–654 (2008)
74. Felekoglu, B., Sarikahya, H.: Effect of chemical structure of polycarboxylate-based superplasticizers on workability retention of self-compacting concrete. Constr. Build. Mater. **22**, 1972–1980 (2008)
75. Felekoglu, B., Türkel, S., Baradan, B.: Effect of water/cement ratio on the fresh and hardened properties of self-compacting concrete. Build. Environ. **42**, 1795–1802 (2007)

76. Felekoglu, B., Yardimci, M.Y., Baradan, B.: A comparative study on the use of mineral and chemical types of viscosity enhancers in self-compacting concrete. In: 3rd International RILEM Symposium on Self-Compacting Concrete, pp. 446–456 (2003)
77. Floyd, R.W., Ruiz, E. D., Do, N. H., Staton, B. W., Hale, W. M.: Development lengths of high-strength self-consolidating concrete beams, PCI J. **52**(4), 44–57 (2011)
78. Georgiadis, A.S., Anagnostopoulos, N.S., Sideris, K.K.: Mechanical characteristics of self-compacting concrete produced with different filler materials. In: 5th International RILEM Symposium on Self-Compacting Concrete, pp. 611–618 (2007)
79. Georgiadis, A.S., Sideris, K.K., Anagnostopoulos, N.S.: Properties of SCC produced with limestone filler or viscosity modifying admixture viscosity modifying admixture. J. Mater. Civ. Eng. **22**(4), 352–360 (2010)
80. Gesoglu, M., Güneyisi, E.: Properties of self-compacting concretes made with binary, ternary, and quaternary cementitious blends of fly ash, blast furnace slag, and silica fume. Constr. Build. Mater. **23**, 1847–1854 (2009)
81. Ghoddousi, P., Javid, A.A.S.: Effect of reinforcement on plastic shrinkage and settlement of self-consolidating concrete as repair material. Mater. Struct. **45**, 41–52 (2012)
82. Giordano, L., Mancini, G.: Fatigue behaviour of bridge deck repaired with self compacting concrete. J. Adv. Concr. Technol. **7**(3), 415–424 (2009)
83. Girgis, A.F.M., Tuan, C.Y.: Bond strength of pretensioned bridge girders cast with self-consolidating concrete, PCI J, **50**(6), 72–87 (2005)
84. Gram, H.-E., Pentti, P., Properties of SCC—especially early age and long term shrinkage and salt frost resistance. In: 1st International RILEM Symposium on Self-Compacting Concrete, pp. 211–225 (1999)
85. Grünewald, S., Walraven, J.: Stabiliseerders in zelfverdichtend beton (in Dutch). Cement **1**, 87–91 (2006)
86. Grünewald, S., Walraven, J.C.: Parameter-study on the influence of steel fibers and coarse aggregate content on the fresh properties of self-compacting concrete. Cem. Concr. Res. **31**, 1793–1798 (2001)
87. Grünewald, S., Walraven, J.C.: The effect of viscosity agents on the characteristics of self-compacting concrete. In: 4th International RILEM Symposium on Self-Compacting Concrete, pp. 9–15 (2005)
88. Güneyisi, E., Gesoglu, M., Özbay, E.: Evaluating and forecasting the initial and final setting times of self-compacting concretes containing mineral admixtures. Mater. Struct. **42**, 469–484 (2009)
89. Güneyisi, E., Gesoglu, M., Özbay, E.: Strength and drying shrinkage properties of self-compacting concretes incorporating multi-system blended mineral admixtures. Constr. Build. Mater. **24**, 1878–1887 (2010)
90. Güneyisi, E., Gesoğlu, M.: Properties of self-compacting Portland Pozzolana and limestone blended cement concretes containing different replacement levels of slag. Mater. Struct. **44**, 1399–1410 (2011)
91. Hammer, T.A.: The influence of some mix design parameters on drying shrinkage of SCC. In: 5th International RILEM Symposium on Self-Compacting Concrete, pp. 559–564 (2007)
92. Hassan, A.A.A., Hossain, K.M.A., Lachemi, M.: Behavior of full-scale self-consolidating concrete beams in shear. Cement Concr. Compos. **30**, 588–596 (2008)
93. Hassan, A.A.A., Hossain, K.M.A., Lachemi, M.: Bond strength of deformed bars in large reinforced concrete members cast with industrial self-consolidating concrete mixture. Constr. Build. Mater. **24**, 520–530 (2010)
94. Hassan, A.A.A., Hossain, K.M.A., Lachemi, M.: Corrosion resistance of self-consolidating concrete in full-scale reinforced beams. Cement Concr. Compos. **31**, 29–38 (2009)
95. Hauke, B., Self-compacting concrete for precast concrete products in Germany. In: 2nd International RILEM Symposium on Self-Compacting Concrete, pp. 33–642 (2001)
96. Hegger, J., Görtz, S., Will, N., Drössler, C., Tigges, C.: Tragfähigkeit von Bauteilen aus selbstverdichtendem Beton. Beton- und Stahlbetonbau **96**(11), 691–697 (2001)

97. Heirman, G., De Geyter, N.: De Invloed van Vulstoffen op de Eigenschappen van Zelfverdichtend Beton in Vloeibare en Verharde Toestand (in Dutch), Master thesis—KULeuven (2002)
98. Heirman, G., Hendrickx, R., Vandewalle, L., Van Gemert, D., Vantomme, J., Feys, D., De Schutter, G., Desmet, B.: Integration approach of the Couette inverse problem of powder type self-compacting concrete in a wide-gap concentric cylinder rheometer Part II: influence of mineral additions and chemical admixtures on the shear thickening flow behaviour. Cem. Concr. Res. **39**, 171–181 (2009)
99. Heirman, G.: Modelling and quantification of the effect of mineral additions on the rheology of fresh powder type self-compacting concrete, PhD, ISBN 978-94-6018-372-0, 202-206 (2011)
100. Heirman, G., Vandewalle, L.: Shrinkage, creep and frost resistance of self-compacting concrete. In: International CANMET/ACI/JCI Symposium on Sustainable Development of Cement, Concrete and Concrete Structures (2005)
101. Heirman, G., Vandewalle, L.: The influence of fillers on the properties of self-compacting concrete in fresh and hardened state. In: 3rd International RILEM Symposium on Self-Compacting Concrete, pp. 606–618 (2003)
102. Ho, D.W.S., Sheinn, A.M.M., Tam, C.T.: The sandwich concept of construction with SCC. Cem. Concr. Res. **31**, 1377–1381 (2001)
103. Hodgson III, D., Schindler, A.K., Brown, D.A., Gardiner, M.S.: Self-consolidating concrete for use in drilled shaft applications. J. Mater. Civ. Eng. **17**(3), 363–369 (2005)
104. Hoffmann, C., Leemann, A.: Homogeneity of structures made with self-compacting concrete and conventional concrete. In: 3rd International RILEM Symposium on Self-Compacting Concrete, pp. 619–627 (2003)
105. Hossain, K.M.A., Lachemi, M.: Bond behavior of self-consolidating concrete with mineral and chemical admixtures. J. Mater. Civ. Eng. **20**(9), 608–616 (2008)
106. Hossain, K.M.A., Lachemi, M.: Fresh mechanical, and durability characteristics of self-consolidating concrete incorporating volcanic ash. J. Mater. Civ. Eng. **22**(7), 651–657 (2010)
107. Hunger, M., Entrop, A.G., Mandilaras, I., Brouwers, H.J.H., Founti, M.: The behavior of self-compacting concrete containing micro-encapsulated phase change materials. Cement Concr. Compos. **31**, 731–743 (2009)
108. Hwang, S.-D., Khayat, K.H.: Comparison of in situ properties of wall elements cast using self-consolidating concrete. Mater. Struct. **45**, 123–141 (2012)
109. Hwang, S.-D., Khayat, K.H.: Effect of mix design on restrained shrinkage of self-consolidating concrete. Mater. Struct. **43**, 367–380 (2010)
110. Jau, W.-C., Yang, C.-T.: Development of a modified concrete rheometer to measure the rheological behavior of conventional and self-consolidating concretes. Cement Concr. Compos. **32**, 450–460 (2010)
111. Júlio, E., Dias, N., Lourenço, J., Silva, J.: Feret coefficients for white self-compacting concrete. Mater. Struct. **39**, 585–591 (2006)
112. Karatas, M., Turk, K.: Investigation of bond between lap-spliced steel bar and self-compacting concrete: the role of silica fume. Can. J. Civ. Eng. **37**, 420–428 (2010)
113. Khatib, J.M.: Performance of self-compacting concrete containing fly ash. Constr. Build. Mater. **22**, 1963–1971 (2008)
114. Khayat, K.H., Bickley, J., Lessard, M.: Performance of self-consolidating concrete for casting basement and foundation walls. ACI Mater. J. **97**(8), 374–380 (2000)
115. Khayat, K.H., Ghezal, A., Hadriche, M.S.: Factorial design models for proportioning self-consolidating concrete. Mater. Struct. **32**, 679–686 (1999)
116. Khayat, K.H., Manai, K., Trudel, A.: In-Situ mechanical properties of wall elements cast using self-consolidating concrete. ACI Mater. J. **94**(6), 491–500 (1997)
117. Khayat, K.H.: Use of viscosity-modifying admixtures to reduce top-bar effect of anchored bars cast with fluid concrete. ACI Mater. J. **95**(2), 158–167 (1998)

118. Kheder, G.F., Al Jadiri, R.S.: New method for proportioning self-consolidating concrete based on compressive strength requirements. ACI Mater. J. pp. 490–497 (2010)
119. Kim, J.-K., Han, S.H., Park, Y.D., Noh, J.H.: Material properties of self-flowing concrete. J. Mater. Civ. Eng. **10**(4), 244–249 (1998)
120. Kim, Y.H., Hueste, M.B.D., Trejo, D., Cline, D.B.H.: Shear characteristics and design for high-strength self-consolidating concrete. J. Struct. Eng. **136**(8), 989–1000 (2010)
121. Kim, Y.J., Choi, Y.W., Lachemi, M.: Characteristics of self-consolidating concrete using two types of lightweight coarse aggregates. Constr. Build. Mater. **24**, 11–16 (2010)
122. König, G., Holschemacher, K., Dehn, F., Weisse, D.: Verbundverhalten von Betonstahl in selbstverdichtendem Beton. Beton- und Stahlbetonbau **96**(3), 124–132 (2001)
123. König, G., Holschemacher, K., Dehn, F., Weiβe, D.: Bond of reinforcement in self-compacting concrete (scc) under monotonic and cyclic loading. In: 3rd International RILEM Symposium on Self-Compacting Concrete, pp. 939–947 (2003)
124. König, G., Holschemacher, K., Dehn, F., Weiβe, D.: Self-compacting concrete-time development of material properties and bond behaviour. In: 2nd International RILEM Symposium on Self-Compacting Concrete, pp. 507–516 (2001)
125. Kou, S.C., Poon, C.S.: Properties of self-compacting concrete prepared with coarse and fine recycled concrete aggregates. Cement Concr. Compos. **31**, 622–627 (2009)
126. Kraus, R.N., Naik, T.R., Ramme, B.W., Kumar, R.: Use of foundry silica-dust in manufacturing economical self-consolidating concrete. Constr. Build. Mater. **23**, 3439–3442 (2009)
127. Kumar, R., Kumar, R., Kumar, N.: In situ performance of self-compacting concrete in T-beams. J. Mater. Civ. Eng. **21**(3), 103–109 (2009)
128. Kwan, A.K.H., Ng, I.Y.T.: Improving performance and robustness of SCC by adding supplementary cementitious materials. Constr. Build. Mater. **24**, 2260–2266 (2010)
129. Lachemi, M., Hossain, K.M.A., Lambros, V., Bouzoubaa, N.: Development of cost-effective self-consolidating concrete incorporating fly ash Slag Cement, or Viscosity-Modifying Admixtures. ACI Mater J **100**(5), 419–425 (2003)
130. Lachemi, M., Hossain, K.M.A., Lambros, V., Nkinamubanzi, P.C., Bouzoubaa, N.: Self-consolidating concrete incorporating new viscosity modifying admixtures. Cem. Concr. Res. **34**, 917–926 (2004)
131. Lachemi, M., Hossain, K.M.A., Lambros, V.: Shear resistance of self-consolidating concrete beams—experimental investigations. Can. J. Civ. Eng. **32**, 1103–1113 (2005)
132. Leemann, A., Hoffmann, C.: Properties of self-compacting and conventional concrete—differences and similarities. Mag. Concr. Res. **57**(6), 315–319 (2005)
133. Leung, H.Y., Nadeem, A., Tse, G.K.W.: Permeability of OPC-FA-SF self-compacting concrete. Struct. Concr. **10**(1), 15–19 (2009)
134. Liu, M.: Incorporating ground glass in self-compacting concrete. Constr. Build. Mater. **25**, 919–925 (2011)
135. Liu, M.: Self-compacting concrete with different levels of pulverized fuel ash. Constr. Build. Mater. **24**, 1245–1252 (2010)
136. Loser, R., Leemann, A.: Effects of curing time and drying behaviour of scc in case of restrained shrinkage deformations. In: 5th International RILEM Symposium on Self-Compacting Concrete, pp. 539–544 (2007)
137. Loser, R., Leemann, A.: Shrinkage and restrained shrinkage cracking of self-compacting concrete compared to conventionally vibrated concrete. Mater. Struct. **42**(1), 71–82 (2009)
138. Lowke, D.: Effect of powder content and viscosity agents on creep and shrinkage of self-compacting concretes. BFT **08**, 50–60 (2007)
139. Lowke, D., Schmidt, K., Schiebl, P., Heinz, D.: Dauerhaftigkeitspotentiale selbstverdichtender Betone. Beton- und Stahlbetonbau **103**(5), 324–333 (2008)
140. Ma, J., Dietz, J., Dehn, F.: Ultra high performance self compacting concrete. In: 3rd International RILEM Symposium on Self-Compacting Concrete, pp. 136–142 (2003)
141. Ma, J., Dietz, J.: Ultra high performance self compacting concrete. Lacer **7**, 33–42 (2002)

142. Maeda, M.K., Yamada, K., Uchida, A.: Evaluation on the practicability of SCC. In: 1st International RILEM Symposium on Self-Compacting Concrete, pp. 617–628 (1999)
143. Marquardt, I., Vala, J.: Optimization of self-compacting concrete mixes. In: 2nd International RILEM Symposium on Self-Compacting Concrete, pp. 295–302 (2001)
144. Melo, K.A., Carneiro, A.M.P.: Effect of Metakaolin's finesses and content in self-consolidating concrete. Constr. Build. Mater. **24**, 1529–1535 (2010)
145. Memon, S.A., Shaikh, M.A., Akbar, H.: Utilization of rice husk ash as viscosity modifying agent in self compacting concrete. Constr. Build. Mater. **25**, 1044–1048 (2011)
146. Mnahoncakova, E., Pavlıkova, M., Grzeszczyk, S., Rovnanıkova, P., Cerny, R.: Hydric, thermal and mechanical properties of self-compacting concrete containing different fillers. Constr. Build. Mater. **22**, 1594–1600 (2008)
147. Mortsell, E., Rodum, E.: Mechanical and durability aspects of SCC for road structures. In: 2nd International RILEM Symposium on Self-Compacting Concrete, pp. 459–467 (2001)
148. Nachbaur, L., Waller, V., Haddad, O., Aggoun, S.: Estimating SCC early ages strength through simple tests. In: 4th International RILEM Symposium on Self-Compacting Concrete, pp. 757–763 (2005)
149. Nachbaur, L., Waller, V., Haddad, O., Vachon, M.: Extension of concrete equivalent mortar (CEM) method to self-compacting concrete. In: 4th International RILEM Symposium on Self-Compacting Concrete, pp. 1235–1241 (2005)
150. Naito, C.J., Parent, G., Brunn, G.: Performance of bulb-Tee girders made with self-consolidating concrete. PCI J. **51**(6), 72–85 (2006)
151. Nedhi, M., El Chabib, H., Hesham El Naggar, M.: Predicting performance of self-compacting concrete mixtures using artificial neural networks. ACI Mater. J. **98**(5), 394–401 (2001)
152. Nehdi, M., Pardhan, M., Koshowski, S.: Durability of self-consolidating concrete incorporating high-volume replacement composite cements. Cem. Concr. Res. **34**, 2103–2112 (2004)
153. Nguyen-Minh, L., Rovnak, M.: Shear resistance of GFRP-reinforced beams. Mag. Concr. Res. **63**(3), 215–233 (2011)
154. Nunes, S., Figueiras, H., Coutinho, J.S., Figueiras, J.: SCC and conventional concrete on site: property assessment. Ibracon Struct. Mater. J. **2**(1), 25–36 (2009)
155. Opçu, I.B., Bilir, T., Uygunoglu, T.: Effect of waste marble dust content as filler on properties of self-compacting concrete. Constr. Build. Mater. **23**, 1947–1953 (2009)
156. Pandurangan, K., Kothandaraman, S., Sreedaran, D.: A study on the bond strength of tension lap splices in self-compacting concrete. Mater. Struct. **43**, 1113–1121 (2010)
157. Panesar, D.K., Shindman, B.: Elastic properties of self-consolidating concrete. Constr. Build. Mater. **25**, 3334–3344 (2011)
158. Parra, C., Valcuende, M., Gómez, F.: Splitting tensile strength and modulus of elasticity of self-compacting concrete. Constr. Build. Mater. **25**, 201–207 (2011)
159. Paultre, P., Khayat, K.H., Cusson, D., Tremblay, S.: Structural performance of self-consolidating concrete used in confined concrete columns. ACI Struct. J. **102**(4), 560–568 (2005)
160. Persson, B.: Chloride migration coefficient of self-compacting concrete. Mater. Struct. **37**(2), 82–91 (2004)
161. Persson, B.: Creep, shrinkage and elastic modulus of self-compacting concrete. In: 1st International RILEM Symposium on Self-Compacting Concrete, pp. 239–250 (1999)
162. Persson, B.: Fire Resistance of Self-Compacting Concrete. SCC Mater. Struct. **37**(11), 575–584 (2004)
163. Persson, B.: Internal frost resistance and salt frost scaling of self-compacting concrete. Cem. Concr. Res. **33**, 373–379 (2003)
164. Persson, B.: Sulphate resistance of self-compacting concrete. Cem. Concr. Res. **33**, 1933–1938 (2003)
165. Peter, J.A., Lakshmanan, N., Manoharan, P.D.: Investigations on the static behavior of self-compacting concrete under-reamed piles. J. Mater. Civ. Eng. **18**(3), 408–414 (2006)

166. Piérard, J., Dieryck, V., Desmyter, J.: Autogenous shrinkage of self-compacting concrete. In: 4th International RILEM Symposium on Self-Compacting Concrete, pp. 1013–1021 (2005)
167. [167] Pineaud, A., Cabrillac, R., Rémond, S., Pimienta, P., Rivillon, P.: Mechanical properties of self-compacting concrete—influence of composition parameters. In: 4th International RILEM Symposium on Self-Compacting Concrete, pp. 863–868 (2005)
168. Pons, G., Mouret, M., Alcantara, M., Granju, J.L.: Mechanical behaviour of self-compacting concrete with hybrid fiber reinforcement. Mater. Struct. **40**, 201–210 (2007)
169. Pons, G., Proust, E., Assié, S.: Creep and shrinkage of self-compacting concrete: a different behaviour compared with vibrated concrete? In: 3rd International RILEM Symposium on Self-Compacting Concrete, pp. 645–654 (2003)
170. Poon, C.S., Ho, D.W.S.: A feasibility study on the utilization of r-FA in SCC. Cem. Concr. Res. **34**, 2337–2339 (2004)
171. Poppe, A.-M., De Schutter, G.: Creep and shrinkage of self-compacting concrete. In: 1st International Symposium on Design, Performance and Use of Self-Consolidating Concrete, pp. 329–336 (2005)
172. Poppe, A.-M., De Schutter, G.: Effect of limestone filler on the cement hydration in self-compacting concrete. In: 3rd International RILEM Symposium on Self-Compacting Concrete, pp. 558–566 (2003)
173. Pozolo, A., Andrawes, B.: Analytical prediction of transfer length in prestressed self-consolidating concrete girders using pull-out test results. Constr. Build. Mater. **25**, 1026–1036 (2011)
174. Radix, H.J., Brouwers, H.J.: Zelfverdichtend beton volgens de Chinese methode—Deel 2: Proeven op mortel en betonspecie. Cement **6**, 104–108 (2004)
175. Radix, H.J., Brouwers, H.J.: Zelfverdichtend beton volgens de Chinese methode—Deel 3: Proeven op verhard beton. Cement **7**, 76–79 (2004)
176. Prasad, Raghu: B.K., Eskandari, H., Venkatarama Reddy, B.V.: Prediction of compressive strength of SCC and HPC with high volume fly ash using ANN. Constr. Build. Mater. **23**, 117–128 (2009)
177. Reinhardt, H.W., Jooss, M.: Permeability and diffusivity of self-compacting concrete as function of temperature. In: 3rd International RILEM Symposium on Self-Compacting Concrete, pp. 808–817 (2003)
178. Reinhardt, H.W., Wüstholz, T.: Tensile deformation behaviour of self-compacting concrete under sustained loading. In: 5th International RILEM Symposium on Self-Compacting Concrete, pp. 591–598 (2007)
179. Reza Esfahani, M., Lachemi, M., Reza Kianoush, M.: Top-bar effect of steel bars in self-consolidating concrete (SCC). Cement Concr. Compos. **30**, 52–60 (2008)
180. Reza Esfahani, M., Reza Kianoush, M., Lachemi, M.: Bond strength of glass fiber reinforced polymer reinforcing bars in normal and self-consolidating concrete. Can. J. Civ. Eng. **32**, 553–560 (2005)
181. Rols, S., Ambroise, J., Péra, J.: Effects of different viscosity agents on the properties of self-leveling concrete. Cem. Concr. Res. **29**, 261–266 (1999)
182. Roussel, N., Staquet, S., D'Aloia Schwarzentruber, L., Le Roy, R., Toutlemonde, F.: SCC casting prediction for the realization of prototype VHPC-precambered composite beams. Mater. Struct. **40**(9), 877–887 (2007)
183. Rozière, E., Granger, S., Turcry, Ph, Loukili, A.: Influence of paste volume on shrinkage cracking and fracture properties of self-compacting concrete. Cement Concr. Compos. **29**, 626–636 (2007)
184. Rozière, E., Turcry, P., Loukili, A., Loukili, A.: Influence of paste volume, addition content and addition type on shrinkage cracking of self-compacting concrete. In: 4th International RILEM Symposium on Self-Compacting Concrete, pp. 945–951 (2005)
185. Safawi, M.I., Iwaki, I., Miura, T.: The segregation tendency in the vibration of high fluidity concrete. Cem. Concr. Res. **34**, 219–226 (2004)

186. Safiuddin, M., West, J.S., Soudki, K.A.: Hardened properties of self-consolidating high performance concrete including rice husk ash. Cement Concr. Compos. **32**, 708–717 (2010)
187. Sahmaran, M., Keskin, S.B., Ozerkan, G., Yaman, I.O.: Self-healing of mechanically-loaded self-consolidating concretes with high volumes of fly ash. Cement Concr. Compos. **30**, 872–879 (2008)
188. Sahmaran, M., Yaman, I.O.: Hybrid fiber reinforced self-compacting concrete with a high-volume coarse fly ash. Constr. Build. Mater. **21**, 150–156 (2007)
189. Sahmaran, M., Yaman, I.Ö., Tokyay, M.: Transport and mechanical properties of self-consolidating concrete with high volume fly ash. Cement Concr. Compos. **31**, 99–106 (2009)
190. Sahmaran, M., Yaman, O., Tokyay, M.: Development of high-volume low-lime and high-lime fly-ash-incorporated self-consolidating concrete. Mag. Concr. Res. **59**(2), 97–106 (2007)
191. Sahmaran, M., Yurtseven, A., Yaman, I.O.: Workability of hybrid fiber reinforced self-compacting concrete. Build. Environ. **40**, 1672–1677 (2005)
192. Schindler, A.K., Barnes, R.W., Roberts, J.B., Rodriguez, S.: Properties of self-consolidating concrete for prestressed members. ACI Mater. J. **104**(1), 53–61 (2007)
193. Siad, H., Mesbah, H.A., Khelafi, H., Kamali-Bernard, S., Mouli, M.: Effect of mineral admixture on resistance to sulphuric and hydrochloric acid attacks in self-compacting concrete. Can. J. Civ. Eng. **37**, 441–449 (2010)
194. Siddique, R.: Properties of self-compacting concrete containing class F fly ash. Mater. Des. **32**, 1501–1507 (2011)
195. Sideris, K.K.: Mechanical characteristics of self-consolidating concretes exposed to elevated temperatures. J. Mater. Civ. Eng. **19**(8), 648–654 (2007)
196. Sonebi, M., Bartos, P.J.M.: Filling ability and plastic settlement of self-compacting concrete. Mater. Struct. **35**(5), 462–469 (2002)
197. Sonebi, M., Bartos, P.J.M., Zhu, W., Gibbs, J., Tamimi, A.: SCC: Properties of Hardened Concrete (Task 4), Brite EuRam Proposal No. BE96-3801, 73 (2000)
198. Sonebi, M., Cevik, A.: Genetic programming based formulation for fresh and hardened properties of self-compacting concrete containing pulverised fuel ash. Constr. Build. Mater. **23**, 2614–2622 (2009)
199. Sonebi, M., Cevik, A.: Prediction of fresh and hardened properties of self-consolidating concrete using neurofuzzy approach. J. Mater. Civ. Eng. **21**(11), 672–679 (2009)
200. Sonebi, M.: Medium strength self-compacting concrete containing fly ash: Modelling using factorial experimental plans. Cem. Concr. Res. **34**, 1199–1208 (2004)
201. Sonebi, M., Tamimi, A.K., Bartos, P.J.M.: Performance and cracking behavior of reinforced beams cast with self-consolidating concrete. ACI Mater. J. **100**(6), 492–500 (2003)
202. Söylev, T.A., François, R.: Effect of bar-placement conditions on steel-concrete bond. Mater. Struct. **39**, 211–220 (2006)
203. Soylev, T.A., François, R.: Quality of steel-concrete interface and corrosion of reinforcing steel. Cem. Concr. Res. **33**, 1407–1415 (2003)
204. Sua, N., Hsub, K.C., Chaic, H.W.: A simple mix design method for self-compacting concrete. Cem. Concr. Res. **31**, 1799–1807 (2001)
205. Sukumar, B., Nagamani, K., Raghavan, R.S.: Evaluation of strength at early ages of self-compacting concrete with high volume fly ash. Constr. Build. Mater. **22**, 1394–1401 (2008)
206. Tam, C.T., Sheinn, A.M.M., Teng, S. Swaddiwudhipong, S.H.N., Teng, S.H.N.: Comparing the volume change between scc and conventional slump concrete. In: 4th International RILEM Symposium on Self-Compacting Concrete, pp. 997–1002 (2005)
207. Tao, J., Yuan, Y., Taerwe, L.: Compressive strength of self-compacting concrete during high-temperature exposure. J. Mater. Civ. Eng. **22**(10), 1005–1011 (2010)
208. Terpstra, J.: Stabilizing low-viscous self-compacting concrete. In: 4th International RILEM Symposium on Self-Compacting Concrete, pp. 47–53 (2005)

209. Terpstra, J., Stabilizing self-levelling concrete with polysaccharide additives. In: 1st International Symposium on Design, Performance and Use of Self-Consolidating Concrete, pp. 207–213 (2005)
210. Topçu, I.B., Uygunoglu, T.: Effect of aggregate type on properties of hardened self-consolidating lightweight concrete (SCLC). Constr. Build. Mater. **24**, 1286–1295 (2010)
211. Turcry, P., Loukili, A.: A study of plastic shrinkage of self-compacting concrete. In: 3rd International RILEM Symposium on Self-Compacting Concrete, 576–585 (2003)
212. Turcry, P., Loukili, A.: Evaluation of plastic shrinkage cracking of self-consolidating concrete. ACI Mater. J. **103**(4), 272–279 (2006)
213. Turcry, P., Loukili, A., Haidar, K., Pijaudier-Cabot, G., Belarbi, A.: Cracking tendency of self-compacting concrete subjected to restrained shrinkage: experimental study and modeling. J. Mater. Civ. Eng. **18**(1), 46–54 (2006)
214. Turk, K., Karatas, M., Ulucan, Z.C.: Effect of the use of different types and dosages of mineral additions on the bond strength of lap-spliced bars in self-compacting concrete. Mater. Struct. **43**, 557–570 (2010)
215. Türkel, S., Kandemir, A.: Fresh and hardened properties of scc made with different aggregate and mineral admixtures. J. Mater. Civ. Eng. **22**(10), 1025–1032 (2010)
216. Türkmen, I., Kantarci, A.: Effects of expanded perlite aggregate and different curing conditions on the physical and mechanical properties of self-compacting concrete. Build. Environ. **42**, 2378–2383 (2007)
217. Uygunoglu, T., Topçu, I.B.: Thermal expansion of self-consolidating normal and lightweight aggregate concrete at elevated temperature. Constr. Build. Mater. **23**, 3063–3069 (2009)
218. Uysal, M., Yilmaz, K.: Effect of mineral admixtures on properties of self-compacting concrete. Cement Concr. Compos. **33**, 771–776 (2011)
219. Valcuende, M., Parra, C.: Bond behaviour of reinforcement in self-compacting concretes. Constr. Build. Mater. **23**, 162–170 (2009)
220. Valdez, P., Barragán, B., Girbes, I., Shuttleworth, N., Cockburn, A.: Use of waste from the marble industry as filler for the production of self-compacting concretes. Materiales de Construcción **61**(301), 61–76 (2011)
221. Vejmelková, E., Keppert, M., Grzeszczyk, S., Skalinski, B., Cerny, R.: Properties of self-compacting concrete mixtures containing metakaolin and blast furnace slag. Constr. Build. Mater. **25**, 1325–1331 (2011)
222. Vieira, M., Bettencourt, A.: Deformability of Hardened SCC. In: 3rd International RILEM Symposium on Self-Compacting Concrete, pp. 637–644 (2003)
223. Wang, H.-Y., Huang, W.-L.: Durability of self-consolidating concrete using waste LCD glass. Constr. Build. Mater. **24**, 1008–1013 (2010)
224. Weiße, D., Holschemacher, K.: Sandreiche Selbstverdichtende Betone. Beton **3**, 122–127 (2004)
225. Whitcomb, B.L., Kiousis, P.D.: Development of self-consolidating concrete for thin wall applications including validation. J. Mater. Civ. Eng. **21**(10), 587–593 (2009)
226. Xie, Y., Liu, B., Yin, J., Zhou, S.: Optimum mix parameters of high-strength self-compacting concrete with ultrapulverized fly ash. Cem. Concr. Res. **32**, 477–480 (2002)
227. Yadegaran, I., Mahoutian, M., Shekarchi, M., Libre, N.A.: Effect of polypropylene fibers on shrinkage of self-compacting concrete. In: 5th International RILEM Symposium on Self-Compacting Concrete, pp. 707–713 (2007)
228. Yazici, H.: The effect of silica fume and high-volume Class C fly ash on mechanical properties, chloride penetration and freeze–thaw resistance of self-compacting concrete. Constr. Build. Mater. **22**, 456–462 (2008)
229. Yin, J., Yu, Z., Xie, Y.: Research on the influence of different curing condition on the peformance of the self-compacting high-performance concrete. In: 1st International Symposium on Design, Performance and Use of Self-Consolidating Concrete, pp. 439–447 (2005)

230. Yurtdas, I., Burlion, N., Shao, J.-F., Li, A.: Evolution of the mechanical behaviour of a high performance self-compacting concrete under drying. Cement Concr. Compos. **33**, 380–388 (2011)
231. Zhu, W., Bartos, P.: Microstructure and properties of interfacial transition zone in SCC. In: 1st International Symposium on Design, Performance and Use of Self-Consolidating Concrete, pp. 319–327 (2005)
232. Zhu, W., Bartos, P.: Permeation properties of self-compacting concrete. Cem. Concr. Res. **33**, 921–926 (2003)
233. Zhu, W., Gibbs, J.C., Bartos, P. J.M.: Uniformity of in situ properties of SCC in full-scale structural elements. Cement Concr. Compos. **23**(1), 57–64 (2001)
234. Zhu, W., Gibbs, J.C.: Use of different limestone and chalk powders in self-compacting concrete. Cem. Concr. Res. **35**, 1457–1462 (2005)
235. Zhu, W., Quinn, J., Bartos, P.: Transport properties and durability of self-compacting concrete. In: 2nd International RILEM Symposium on Self-Compacting Concrete, pp. 451–458 (2001)
236. Zhu, W., Sonebi, M., Bartos, P.J.M.: Bond and interfacial properties of reinforcement in self-compacting concrete. Mater. Struct. **37**, 442–448 (2004)
237. Powers, T., Brownyard, L.: Studies of the physical properties of hardened Portland cement paste (nine parts). J. Am. Concr. Inst. **43**, 101–132 (1946)
238. Zsigovics, I.: Effect of limestone powder on the consistency and compressive strength of SCC. In: 4th International RILEM Symposium on Self-Compacting Concrete, pp. 173–180 (2005)
239. European Committee for Standardization: EN 206-9: Concrete—Part 9: Additional Rules for Self-Compacting Concrete (SCC), p. 27 (2010)
240. European Committee for Standardization: EN 197-1: Cement. Composition, specifications and conformity criteria for low heat common cements, p. 52 (2002)
241. ASTM C150/C150 M: Standard Specifications for Portland Cement, p. 9 (2012)
242. European Committee for Standardization: EN 206-1: Concrete—Part 1: Specification, Performance, Production and Conformity, p. 72 (2000)
243. Khayat, K., Mitchell, D.: NCHRP report 628: Self-Consolidating Concrete for Precast, Prestressed Concrete Bridge Elements, National Cooperative Highway Research Program, p. 99 (2009)
244. Vandegrift, D. Jr., Schindler, A.K.: The effect of test cylinder size on the compressive strength of sulfur capped concrete specimens. Highway Research Center and Department of Civil Engineering at Auburn University, p. 83 (2006)
245. Elwell, D.J., Fu, G.: Compression testing of concrete: cylinders vs. cubes, Report FHWA/NY/SR-95/119, special report 119. Transportation Research and Development Bureau New York State Department of Transportation (1995)
246. EFNARC: The European Guidelines for Self-Compacting Concrete, p. 63 (2005)
247. European Committee for Standardization: EN 12390-3: Testing hardened concrete. Compressive strength of test specimen, p. 22 (2009)
248. ASTM C39: Standard Test Method for Compressive Strength of Cylindrical Concrete Specimens, p. 7 (2012)
249. American Association of State and Highway Transportation Officials, AASHTO T 22-06: Standard Method of Test for Compressive Strength of Cylindrical Concrete Specimens, p. 11 (2006)
250. Neville, A.M.: A general relation for strengths of concrete specimens of different shapes and sizes. J. Am. Concr. Inst. **63**(10), 1109–1195 (1966)
251. Belgian Institute for Normalisation (BIN), NBN B15-220: Concrete testing—Compressive strength: Addendum 1 (In Dutch) (1973)
252. UNESCO: Reinforced Concrete : An International Manual, p. 20. Butterworths, Sydney (1971)
253. American Concrete Institute, ACI 318-08: Building Code Requirements for Structural Concrete and Commentary (2008)

254. Klug, Y., Holschemacher, K.: Comparison of the Hardened Properties of Self-Compacting and Normal Vibrated Concrete. In: 3rd International RILEM Symposium on Self-Compacting Concrete (Reykjavik 2003), pp. 596–605 (2003)
255. Van Itterbeeck, P., Cauberg, N., Parmentier, B., Vandewalle, L., Lesage, K.: Eurocode 2 predictions vs. mechanical properties of self compacting concrete. In: CD Proceedings of Design, Production and Placement of Self-Consolidating Concrete (SCC2010), Montreal, Canada (2010)
256. Boel, V., Craeye, B., Desnerck, P., Van Der Vurst, F., De Schutter, G.: Influence of specimen shape and dimension on the compressive strenght of self-compacting concrete. In: Proceedings of SCC2013, Paris, France (2013)
257. European Committee for Standardization: EN 1992-1-1, Eurocode 2: Design of concrete structures—Part 1-1: General Rules and Rules for Buildings, p. 225 (2004)
258. RILEM, Report 038: Durability of Self-Compacting Concrete—State-of-the-Art Report of RILEM Technical Committee 205-DSC, p. 208 (2007)
259. Schiessl, A, Zilch, K.: The effects of the modified composition of SCC on shear and bond behaviour. In: Proceedings of the second RILEM international symposium on self-compacting concrete, Japan, COMS Engineering Corporation, pp. 501–506 (2001)
260. Hewlett, P.: Lea's Chemistry of Cement and Concrete. Arnold Publishers, London (1998)
261. Feret, R.: Sur la compacité des mortiers hydrauliques, Annales des Ponts et Chaussées. Série **7**(4), 5–164 (1892)
262. ASTM C109: Standard Test Method for Compressive Strength of Hydraulic Cement Mortars, p. 10 (2012)
263. Domone, P.: Self-compacting concrete: An analysis of 11 years of case studies. Cement Concr. Compos. **28**, 197–208 (2006)
264. Dewar, J.D.: Computer Modeling of Concrete Mixtures, ISBN 0-419-23020-3 (1999)
265. Poppe, A.-M.: Influence of fillers on hydration and properties of self-compacting concrete (in Dutch). PhD, Ghent University (2004)
266. Lambotte, H.: Rheologie (in Dutch). Ghent University (1979)
267. RILEM, TC 148-SSC: Test method for measurement of the strain-softening behaviour of concrete under uniaxial compression. Mater. Struct. **33**, 347–351 (2000)
268. CEB-FIP: Bulletin 55: Model Code 2010—First complete draft, CEB-FIP, vol. 1, p. 292 (2010)
269. Comite Euro-International du Beton: CEB-Fib model code 1990: Design Code. Thomas Telford, London (1993)
270. Comite Euro-International du Beton: CEB-Fib model code 1978: Design Code. Thomas Telford, London (1978)
271. Annerel, E.: Assessment of the residual strength of concrete structures after fire exposure. PhD thesis, Ghent University (2010)
272. Liu, X.: Microstructural behaviour of self-compacting concrete during hydration and after exposure to high temperatures. PhD thesis, Ghent University (2006)
273. Uysal, M., Yilmaz, K., Ipek, M.: Properties and behavior of self-compacting concrete produced with gbfs and fa additives subjected to high temperatures. Constr. Build. Mater. **28**, 321–326 (2012)
274. Sashar, R., Khoury, G.A.: Material and environmental factors influencing the compressive strength of unsealed cement paste and concrete at high temperatures. Mag. Concr. Res. **45**, 51–61 (1993)
275. Bamonte, P., Felicetti, R., Gambarova, P.G.: Mechanical properties of self-compacting concrete at high temperature and after cooling. In: The Third North American Conference on the Design and Use of Self-Consolidating Concrete (2008)
276. Craeye, B., De Schutter, G., Wacquier, W., Van Humbeeck, H., Van Cotthem, A., Areias, L.: Closure of the concrete supercontainer in hot cell under thermal load. Nucl. Eng. Des. **241**, 1352–1359 (2011)

277. Khaliq, W., Kodur, V.: Thermal and mechanical properties of fiber reinforced high performance self-consolidating concrete at elevated temperatures. Cem. Concr. Res. **41**, 1112–1122 (2011)
278. Uysal, M.: Self-compacting concrete incorporating filler additives: performance at high temperatures. Constr. Build. Mater. **26**, 701–706 (2012)
279. Ye, G., Liu, X., De Schutter, G., Taerwe, L., Vandevelde, P.: Phase distribution and microstructural changes of self-compacting cement paste at elevated temperature. Cem. Concr. Res. **37**, 978–987 (2007)
280. Bhal, N.S., Jain, J.P.: Compressive strength of concrete subjected to elevated temperatures. Indian Concr. J. **73**(8), 503–507 (1999)
281. Chan, Y.N., Peng, G.F., Anson, M.: Residual strength and pore structure of high strength concrete and normal strength concrete after exposure to high temperatures. Cement Concr. Compos. **21**(1), 23–27 (1999)
282. Khoury, G.A.: Compressive strength of concrete at high temperatures: reassessment. Mag. Concr. Res. **44**(161), 291–309 (1994)
283. Malhorta, H.L.: Spalling of Concrete in Fire, Construction Industry Research and Information Association, Technical Report, No. 118 (1984)
284. Jansson, R.: Liquid/steam pressure measurement inside concrete exposed to fire. In: Proceedings of the 4th International Workshop on Structures in Fire, Aveiro, Portugal (2006)
285. Norling-Mjörnell, K.: Uttorkning av bygfukkt i självkompakterande betong, Report White Arkitekter AB (2003)
286. Domone, P.L.: Self-compacting concrete: an analysis of 11 years of case studies. Cement Concr. Compos. **28**, 197–208 (2006)
287. Sonebi, M., Rooney, M., Bartos, P.: In-situ compressive strength of self-compacting concrete. Concrete **36**(3), 48–49 (2002)
288. Khayat, K.H., Trembley, S., Paultre, P.: Structural response of self-compacting concrete columns. In: Proceeding of the First International RILEM Symposium on Self-Compacting Concrete, Stockholm, pp. 291–306, (1999)
289. Craeye, B., De Schutter, G., Van Humbeeck, H., Van Cotthem, A.: Early age behaviour of concrete supercontainers for radioactive waste disposal. Nucl. Eng. Des. **239**, 23–35 (2009)
290. Piérard, J., Pollet, V.: Testing of cores and precast samples out of cast column in SCC and VC. Belgian Building Research Institute, Report DE611XB050, Belgium, p. 23 (2007)
291. Sonebi, M., Bartos, P.: Performance of reinforced columns cast with self-compacting concrete. In: Proceedings of the 5th CANMET/ACI International Conference on Recent Advances in Concrete Technology, SP 200-25, Singapore, pp. 415–431, (2001)
292. Rigueira, J.W., Garcia-Taengua, E., Serna-Ros, P.: Self-consolidating concrete robustness in continuous production regarding fresh and hardened state properties, ACI Mater. J. **106**(3), 301–307 (2009)
293. Ferrara, L.: Statistical properties of steam-cured-plant produced SCC for prestressed precast applications. In: Proceedings of the 2nd International Symposium on Design, Performance and Use of Self Consolidating Concrete, Bejing, China, pp. 483–494, (2009)

Chapter 3
Creep and Shrinkage of SCC

Andreas Leemann and Pietro Lura

3.1 Introduction

Strain in concrete can occur due to different reasons. When a stress is applied, concrete shows an immediate, reversible strain (elastic deformation), and a further deformation that increases with time (creep). After stress removal, part of this deformation is recovered immediately due to the elastic properties of the material, and a minor part is recovered in time (reversible creep). Another other part of the deformation, however, is not recovered (irreversible creep). In addition to creep, strain in concrete develops as well because of a volume decrease caused by shrinkage. Depending on the stress-strain situation at specific points in a structure, strain caused by shrinkage and creep may add up or creep may lead to relaxation as it can reduce stress caused by shrinkage strain. This interaction of stress and strain is of vital importance for concrete structures as it influences cracking, deflection and prestress loss. As a result, creep and shrinkage have been taken into account by numerous standards, and they are a focus point of research. With the appearance of self-compacting concrete (SCC) as a valuable alternative for conventionally vibrated concrete (VC), some of the established stress-strain-relations have to be questioned or at least reconfirmed. Looking at the scene in a broad sense, the most prominent changes in mix design from VC to SCC are the higher paste volume, the substantial use of mineral additions, and the high dosage of superplasticiser, often in combination with a viscosity-modifying agent (VMA). The changes in paste volume and binder composition influence the viscoelastic properties of the concrete.

A. Leemann (✉) · P. Lura
EMPA, Dübendorf, Switzerland
e-mail: Andreas.Leemann@empa.ch

K. H. Khayat and G. De Schutter (eds.), *Mechanical Properties of Self-Compacting Concrete*, RILEM State-of-the-Art Reports 14,
DOI: 10.1007/978-3-319-03245-0_3,

3.2 Creep

Creep of concrete is a viscous deformation, i.e. an increase of strain as a function of time in concrete under stress. When stress is applied, the deformation registered during the application of the stress is regarded as elastic and the following increase in strain as creep. The magnitude of creep strain can be substantially larger than that of the elastic deformation, depending on the elastic modulus of the concrete and the magnitude of the applied stress. When the applied load is smaller than 40 % of the compressive strength, a linear relationship between stress and creep strain can be assumed [1].

Creep of concrete is generally measured on cylinders or prisms that are loaded at an age of 2–28 days. The applied load usually corresponds to 20–35 % of the compressive strength of the concrete at the time of loading and sometimes is increased stepwise with age. The specimens can either be sealed or exposed to a specific relative humidity, usually in the range of 50–70 %. In the first exposure condition (sealed), basic creep is determined, while total creep is measured in the second one (unsealed). Drying creep can be calculated by subtracting basic creep from total creep.

In order to obtain creep strain ε_c from the time of loading t′ until time t, the elastic deformation ε_{el} at the time of loading and the total shrinkage strain ε_{st} have to be subtracted from the total deformation ε_{ct}:

$$\varepsilon_c(t,t') = \varepsilon_{ct}(t,t') - \varepsilon_{st}(t,t') - \varepsilon_{el}(t') \tag{3.1}$$

A comparison of creep strain obtained in different studies is difficult, as various parameters like specimen size, ratio between compressive strength and applied load, age at time of loading, stiffness of test equipment or relative humidity can vary. As an example, when compressive strength varies within a series, and the applied load corresponds to a certain percentage of compressive strength at the time of loading, creep strain of the different mixtures cannot be compared directly.

In order to compare creep of different mixtures, the creep coefficient φ or the specific creep J (often referred to as "creep compliance") can be used. The creep coefficient, φ, can be calculated as:

$$\varphi(t,t') = \frac{\varepsilon_c(t,t')}{\varepsilon_{el}(t')} \tag{3.2}$$

and the specific creep or creep compliance, J, with σ being the applied load as:

$$J(t,t') = \frac{\varepsilon_c(t,t')}{\sigma} \tag{3.3}$$

3.2.1 Influence of Paste Volume

Heirman et al. [2] have determined creep coefficients for a variety of SCC mixtures and compared them with the creep coefficient of VC (Fig. 3.1). Creep coefficients of SCC are shown to be higher compared to VC. However, it has to be mentioned, that the VC was produced with OPC and SCC with OPC and a substantial amount of limestone powder. The two SCC mixtures with the highest paste volume and W/P by volume showed the highest creep coefficients. When an identical W/CM and an identical grain-size distribution of the aggregates are used for the VC and SCC, a higher paste volume of SCC (150 l/m^3) results in higher creep strain and higher creep coefficients of SCC compared to VC [3].

On the other hand, Turcry et al. [4] tested three pairs of VC/SCC mixtures with each pair having a similar compressive strength and found no difference of creep strain between the VC and SCC mixtures. However, this could be an effect of binder composition where between 25 and 33 % of the binder mass had limestone powder for the SCC mixtures. The paste volume of SCC was 55–70 l/m^3 higher than the one of the VC. A load corresponding to 20 % of the compressive strength was applied at the age of 7 days.

When creep strain and creep coefficients of VC and SCC of the same strength class, strength development and binder composition are compared, the differences of creep coefficients are smaller than the ones of creep strain (Figs. 3.2 and 3.3, [5]). Although the results of different studies are not consistent, the creep coefficient and the specific creep seem to be generally slightly higher (5–10 %) for SCC compared to VC with the same binder composition [6, 7].

3.2.2 Influence of Binder Composition

The replacement of cement with limestone powder results in an increase of basic and drying creep when W/P by vol. and volume of paste are kept constant (Fig. 3.4, [8]). However, when the specific creep is calculated, there are only insignificant differences between the four different SCC mixtures, as compressive strength decreases with increasing limestone content. This observation agrees with [9]. In spite of varying grain size distribution of the aggregates, W/CM, paste volume and binder composition, VC and SCC show the same relation between creep coefficient and compressive strength [10].

Evaluating the influence of binder composition on creep, the strength development after loading has to be considered. SCC produced with OPC or a combination of OPC with limestone powder has less strength increase after loading than SCC with fly ash or GGBFS. Therefore, the creep coefficient of the latter may increase less compared to SCC produced with OPC (Fig. 3.5). The same observations are made in [2, 3] with regard to GGBFS cement.

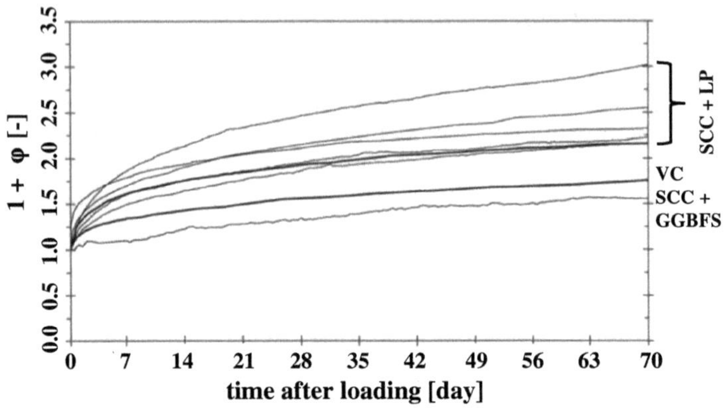

Fig. 3.1 Creep coefficients of VC produced with OPC and various SCC mixtures produced with OPC and limestone powder (LP) and one mixture produced with OPC and GGBFS (paste volume: VC = 283 l/m^3, SCC = 371–412 l/m^3/compressive strength: VC = 47.6 MPa, SCC = 39.9–73.3 MPa/adopted from [2])

Fig. 3.2 Creep strain of VC and SCC produced with OPC versus time (applied load = 10 MPa/numbers in parentheses = compressive strength at 28d/paste volume of SCC is 80 l/m^3 higher than the one of VC/data from [5])

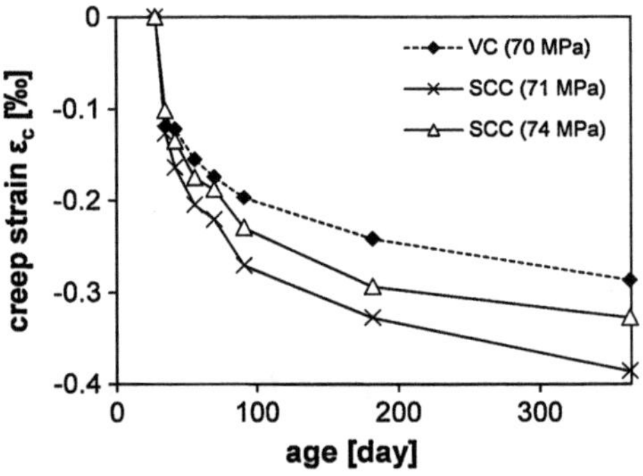

Fig. 3.3 Creep coefficient of VC and SCC produced with OPC versus time (applied load = 10 MPa/numbers in parentheses = compressive strength at 28d/paste volume of SCC is 80 l/m^3 higher than the one of VC/data from [5])

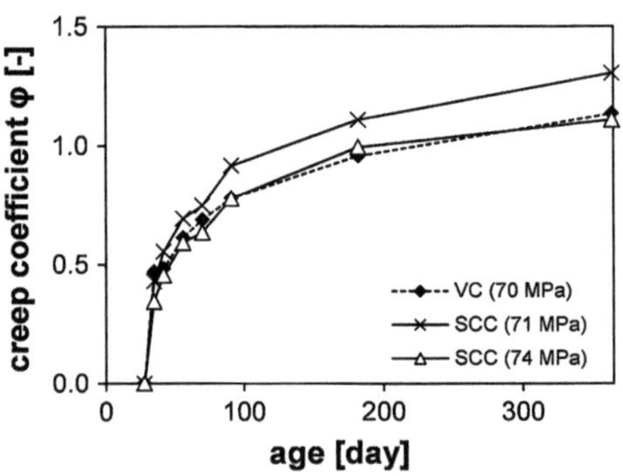

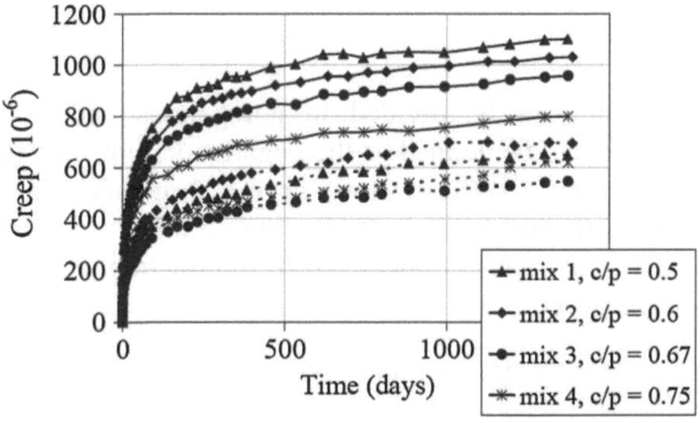

Fig. 3.4 Basic (*dotted lines*) and drying creep of SCC with varying limestone powder content, constant water-powder-ratio (0.28) and constant paste volume (cement content: 300–450 kg/m^3, limestone powder: 300–150 kg/m^3/c = cement, p = cement + limestone powder/loading at the age of 28d with one-third of the compressive strength/compressive strength 59.0–74.3 MPa/ values in μm per m/[8])

Fig. 3.5 Creep coefficient of SCC produced with OPC and 30 mass % limestone powder and fly ash respectively (numbers in parentheses refer to compressive strength at 28 days/data from [6])

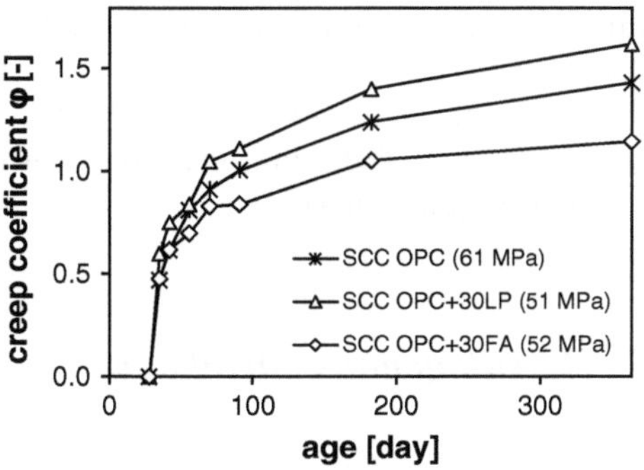

Creep measurements of SCC in tension are very rare [11]. No studies comparing creep of SCC under tensile and compressive stress have been found. The relation between creep in tension and in compression of VC is not well established either. While Gutsch [12] observed similar creep coefficients of VC in tension and compression, differences can occur in HPC [13]. Atrushi [13] observed that initial creep compliance in tension is lower but increases more with time, leading to an intersection of creep compliance in tension and compression few days after loading. On the other hand, Li [14] measured specific tensile creep of VC being approximately four times higher than specific compressive creep. In one case, the stress–strain relation in tension was linear up to 50 % of the ultimate tensile strength [15] and in this regard seems to be comparable to creep in compression. The creep coefficient of SCC under tensile stress can be approximately calculated by combining the tensile stress measured under restrained conditions (ring test according to ASTM C 1581) with the E-modulus [16]. With this approach an increase in tensile creep coefficient was calculated with increasing W/CM of SCC.

3.3 Shrinkage

Shrinkage of concrete can be divided in autogenous and drying shrinkage. Autogenous shrinkage can be defined as the negative volume change under sealed and isothermal conditions. Concrete shows drying shrinkage when it loses moisture due to exposure to unsaturated air. Drying shrinkage ε_{sd} from the time of demoulding t_0 until time t can be calculated by subtracting autogenous shrinkage ε_{sa} from total shrinkage ε_{st}:

$$\varepsilon_{sd}(t_0, t) = \varepsilon_{st}(t_0, t) - \varepsilon_{sa}(t_0, t) \tag{3.4}$$

However, it has to be pointed out that this is not fully correct as both drying shrinkage and autogenous shrinkage are ultimately caused by the same driving force; the lowering of the internal relative humidity of the concrete. In the case of autogenous shrinkage, this lowering is due to self-desiccation, while in the case of drying shrinkage, the internal relative humidity decreases due to moisture loss to the environment. If these processes occur simultaneously, it is not possible to account for the single mechanism taking place, and accordingly the application of Eq. 3.4 is not correct. In practice, total shrinkage is often the only measured deformation. However, as autogenous shrinkage is usually significantly lower than drying shrinkage in normal strength SCC, total shrinkage is dominated by drying shrinkage [17]. Therefore, the latter can be used for comparing the influence of mix design parameters.

3.3.1 Autogenous shrinkage

Autogenous shrinkage of concrete is determined as strain measurements that are usually started at one day, or earlier, when concrete strength is sufficient to demould the samples. The cylinders or prisms can be sealed with epoxy resin or adhesive aluminium tape in order to prevent loss of moisture. Alternatively, special measuring techniques allow measuring length changes of concrete from setting time, without the need of demoulding the specimens [18, 19].

3.3.1.1 Influence of Paste Volume

Cement content and W/P by vol. are the most influential parameters governing autogenous shrinkage; a decrease in W/P by vol. and an increase in cement content increase strain. Therefore, autogenous shrinkage of SCC increases with increasing paste volume at constant W/P by vol. [17, 20]. As a result, VC shrinks less than SCC when binder composition and strength class are the same.

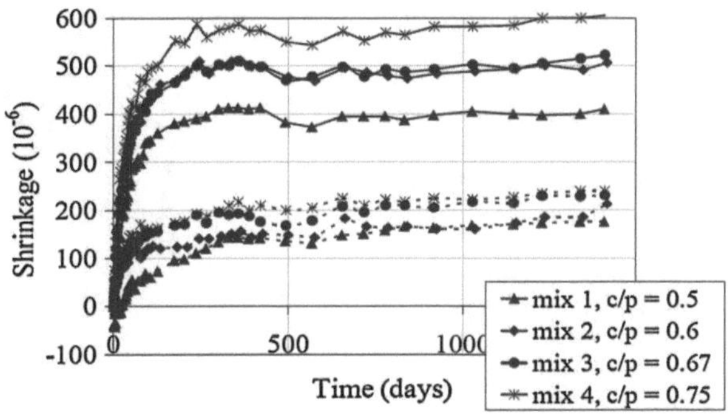

Fig. 3.6 Autogenous (dotted lines) and total shrinkage of SCC with varying limestone powder content, constant W/P by vol. (0.28) and constant volume of paste (cement content: 300–450 kg/m^3, limestone powder: 300–150 kg/m^3/values in μm per m/according to [8])

3.3.2 Influence of Binder Composition

When the amount of limestone powder is increased at constant W/P by vol. and constant paste volume, autogenous shrinkage decreases (Fig. 3.6/[8, 17, 21]). Increasing only the content of limestone powder but keeping cement and water content constant has only a minor effect on autogenous shrinkage [7, 17]. Comparing the effect of fly ash, limestone and quartz powder on autogenous shrinkage, no significant differences were observed [19]. Increasing the fineness of limestone filler leads to an increase of the autogenous shrinkage in the first day [22].

A high-performance concrete (HPC) with a replacement level of 25 % cement with fly ash shows approximately the same level of autogenous shrinkage as a concrete with the same paste volume but with only OPC as binder. However, autogenous shrinkage is considerably decreased when the replacement level of cement with fly ash is increased to 50 % [23]. This relation should be applicable for SCC as well. The use of GGBFS cement may lead to an initial expansion of SCC [24]. Generally, swelling peaks can appear once time zero is passed (up to 500 μm/m) with higher values for higher W/CM when autogenous shrinkage is measured during the first 24 h [19]. The water absorption on the filler surface and resulting disjoining pressure could be a possible explanation.

Shrinkage reducing admixtures (SRA) can decrease part of the resulting strain caused by autogenous shrinkage (see Sect. 3.3.2.3). Another strategy to reduced autogenous shrinkage is the application of internal curing [25].

3.3.3 Drying Shrinkage

Drying shrinkage is determined with linear strain measurements that are usually started at one day when concrete strength is sufficient to demould the samples. In most cases, the prisms or cylinders are exposed to a relative humidity in the range of 50–70 %. As drying shrinkage is usually significantly larger than autogenous shrinkage in normal strength SCC [17], total shrinkage can be used for comparing the influence of mix design parameters.

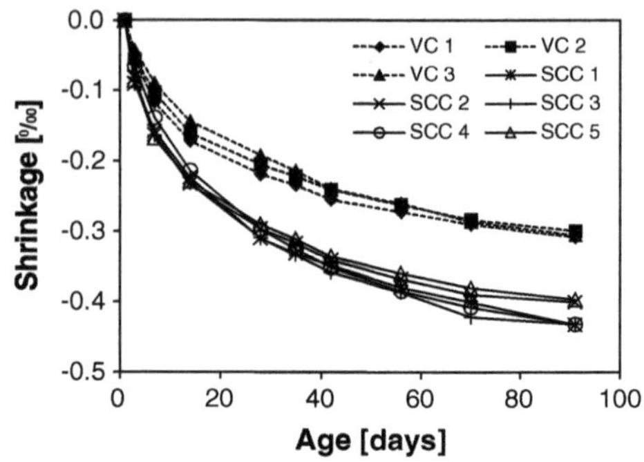

Fig. 3.7 Total shrinkage of VC (paste volume: 254 l/m^3/ W/CM = 0.40–0.60/ compressive strength: 42.3–69.7 MPa) and SCC (paste volume: 316–349 l/m^3/ W/CM = 0.35–0.46/ compressive strength: 59.0–74.0 MPa) both produced with OPC [7]

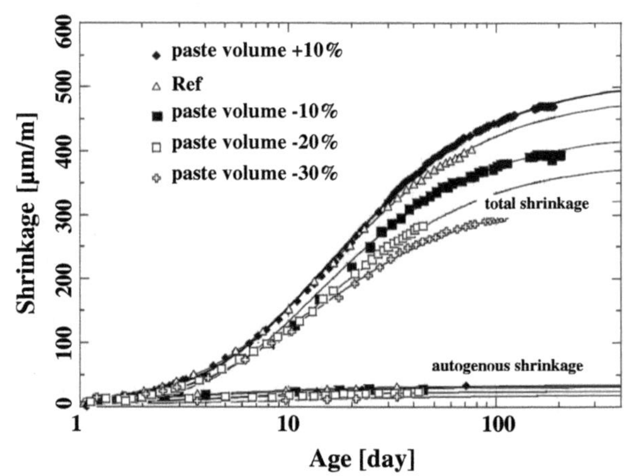

Fig. 3.8 Shrinkage of SCC with varying paste volume (paste volume: 291–457 l/m^3) and constant W/P by vol. of 0.32/data adopted from [17])

3.3.3.1 Influence of Paste Volume

Compared to VC, total shrinkage of SCC is increased due to its higher volume of paste (Fig. 3.7/[7, 26]). Comparing SCC with constant W/P by vol. but varying paste volume reveals that the relation between shrinkage and paste volume is approximately linear and can be regarded as the dominating parameter in drying shrinkage (Fig. 3.8/[17]). When the volume of paste is kept constant, but the water content is increased, total shrinkage increases as well [21, 27]. Consequently, when the increase of paste volume goes together with an increase in water content, shrinkage is increased considerably [17]. However, the use of mineral additions in SCC generally leads to a decrease in shrinkage (see following paragraph).

3.3.3.2 Influence of Binder Composition

The replacement of cement with siliceous mineral additions generally leads to a reduction of shrinkage at constant W/P by vol. [28, 29]. The only exception is silica fume that may lead to a slight increase in shrinkage (Fig. 3.9/[29]).

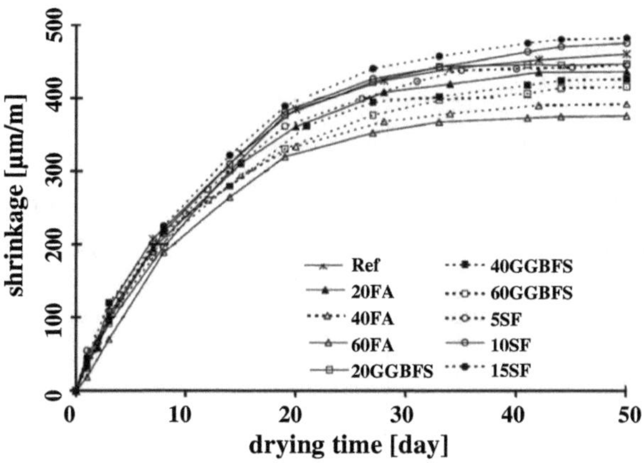

Fig. 3.9 Total shrinkage of SCC with different mineral additions after curing for 24 h (constant water content of 198 l/m^3 and constant mass of binder of 450 kg/m^3/compressive strength of reference mixture = 73.6/FA = fly ash, compressive strength = 42.5–68.0 MPa/S = GGBFS, compressive strength = 65.7–72.6 MPa/SF = silica fume, compressive strength = 71.2–76.1 MPa/data adopted from [28])

A replacement of cement with limestone powder at constant W/P by vol. generally leads to a decrease in total shrinkage (Fig. 3.6/[8, 30, 31]). An increased fineness of the cement increases shrinkage as a result of pore refinement [32].

When SCC mixtures produced with different binders are compared, the absolute values of shrinkage strain and mass loss do not show the same relation [3, 32]. Generally, binders with mineral additions show a higher mass loss during the first days even when shrinkage strain is the same or lower than the one of mixtures produced with plain OPC [32]. However, when the values of shrinkage strain and mass loss at an age of 91 days are used as a reference, the relative changes of these two parameters are the same and appear to be unaffected by binder type or volume of paste (Fig. 3.10). This suggests that the mechanisms leading to shrinkage are the same for different types of concrete, even when the absolute values differ.

3.3.3.3 Influence of Organic Admixtures

The use of shrinkage-reducing agents (SRA) can result in a significant decrease of drying shrinkage of SCC by reducing surface tension [7, 33–35]. As both VMA and SRA increase the viscosity of the pore fluid [36], they may help slowing down evaporation and alter the rate of drying shrinkage. However, no influence was observed in drying shrinkage of high strength SCC containing VMA [20].

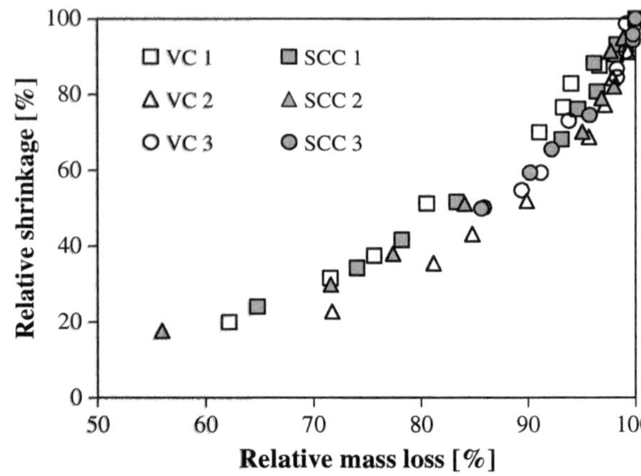

Fig. 3.10 Mass loss relative to the values at 91 days as a function of shrinkage strain relative to the values at 91 days of VC and SCC with 150 l/m^3 difference in paste volume (cement types: VC I/SCC I = CEM I 42.5 N, VC II/SCC II = CEM II/B-M (V-LL), VC III/SCC III = CEM III/B 32.5 R/[3])

3.3.4 Shrinkage Induced Cracking

3.3.4.1 Plastic State

Very few studies are available where either plastic shrinkage or plastic shrinkage cracking of VC and SCC are compared. Plastic shrinkage of SCC has been found to be higher than for VC of the same W/CM [37]. Turcry [38] found that both plastic shrinkage and cracking of SCC and VC of the same strength class were comparable at high evaporation rate. However, the lack of bleeding of SCC produced a higher plastic shrinkage at low evaporation rate, possibly increasing the susceptibility to plastic shrinkage cracking. On the contrary, Holt [39] found that horizontal plastic shrinkage of VC started earlier and was higher than that of SCC with different filler types (Fig. 3.11).

In the same study, increasing the limestone powder content by 50 or 100 % (and reducing the aggregates accordingly) in one of the SCC mixtures did not influence the horizontal shrinkage. The different mixtures had comparable compressive strength at 28 days, but paste volume and W/CM were different. Comparing different VC and SCC mixtures the latter showed a higher plastic shrinkage [40]. Plastic shrinkage of SCC was decreased with increased bleeding, the use of relatively large aggregate sizes and rapid strength development. In an experimental set-up based on ASTM C1579-06, the risk of plastic shrinkage cracking significantly decreases at a w/b below 0.44 [41]. Besides the w/b, the fineness of the binder is the main parameter governing cracking risk: the risk increases with increasing fineness. In any case, special care should be taken to prevent plastic shrinkage cracking of SCC.

Hammer [42] observed that due to reduced W/P by vol., finer cements, and higher volume of filler, SCC should be more susceptible to plastic shrinkage cracking than VC and accordingly special care should be taken to reduce evaporation in the first hours after casting. However, these considerations seem to be

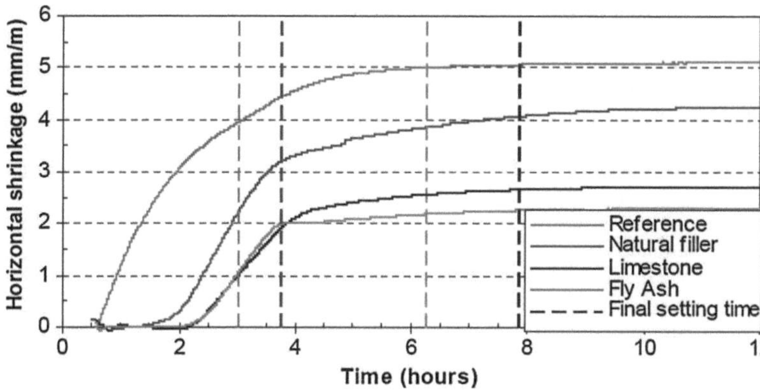

Fig. 3.11 Horizontal shrinkage at early ages of SCC with different mineral additions [39]. The vertical lines show the time of final setting

partially contradicted by experiments on SCC with limestone filler with different fineness, where it was observed that a high fineness reduced both the evaporation rate and the risk of cracking [22]. Obviously, shrinkage-induced cracking of SCC in the plastic state is highly dependable on the materials used, the mix design and the experimental set-up.

3.3.4.2 Hardened State

Cracking of concrete due to shrinkage occurs when shrinkage stress caused by restraint exceeds the tensile strength. Shrinkage stress σ_s is dependent on the total shrinkage ε_{st}, degree of restraint k and E-modulus reduced by creep-induced relaxation, E_{red} [7]:

$$\sigma_s = k \cdot E_{red} \cdot \varepsilon_{st} \tag{3.5}$$

The tensile stress of SCC developing in the ring test and in a shrinkage frame with passive restraint increases with paste volume when W/P by vol. is constant and tensile strength is similar [3, 17]. As a result, the time of cracking follows the same pattern [3, 17]. The correlation between the increase of tensile stress in the ring test and the total shrinkage of VC and SCC with similar tensile strength is confirmed by [4]. When the relaxation between VC and SCC having a considerable higher paste volume are compared, SCC shows a higher degree of relaxation (Fig. 3.12). In this case [43], this is a result of the generally higher creep of SCC as E-modulus of VC and SCC were comparable (see Sect. 3.2.1).

The replacement of cement with either limestone powder or fly ash can lead to a decrease of shrinkage and an increase in the time of cracking in a passive restrained shrinkage frame [44]. Only a slight increase of the time to cracking was observed adding limestone powder or fly ash in [7]. This apparent contradiction is

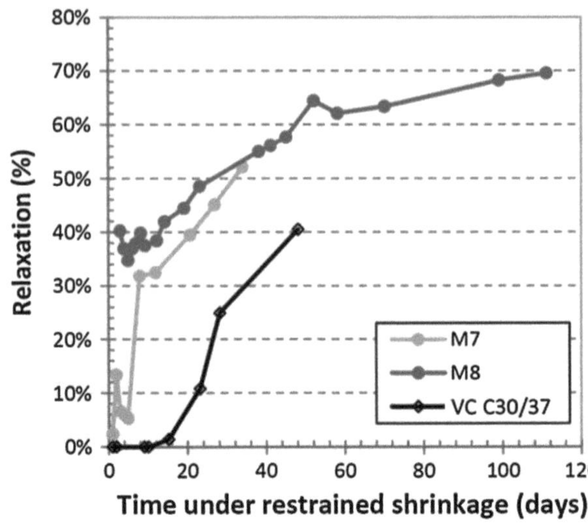

Fig. 3.12 Relaxation of SCC (M7, M8/W/P by vol. = 0.37, paste volume = 317–330 l/m^3) compared to VC ("TVC"/W/P by vol. = 0.55, paste volume = 265 l/m^3) in the ring test calculated as the difference between the elastic stress and the measured stress [43]

most likely caused by different curing regimes. In the first case, the time of curing was 7 days, and in the second case it was 2 days. The substitution of cement with GGBFS reduces shrinkage and with it the stress developing in the ring test [43]. At similar rate of shrinkage, the cracking risk increases when the tensile strength is relatively low [43]. However, when the degree of restraint is high and the concrete dries relatively fast under the chosen relative humidity, stress development and risk of cracking are mainly dependent on the shrinkage rate and on tensile strength development (Figs. 3.13 and 3.14). When drying is relatively slow and concrete with low W/CM is used, autogenous shrinkage can contribute considerably to the cracking risk during the first days.

In a pair of SCC mixtures with identical paste volume and composition but different water contents cast into a passive restrained shrinkage frame, the SCC with the lower W/CM cracks first [44]. This behaviour can be traced back to the higher E-modulus and resulting higher stress rate directly after the 7-day curing period. A similar behaviour has been observed in [16].

The sand-to-gravel ratio influences the time to cracking, which increases with increasing gravel content of SCC with identical paste volume, binder composition, and W/P by vol. [33]. However, no difference in the time to cracking was observed in [7]. Cracking occurs later when rounded aggregates are used instead of crushed aggregate [33].

The use of a SRA causes a decrease of total shrinkage and as such prolongs time to cracking [7, 33]. The combination of SRA with fibers leads to a further decrease of the cracking risk [31]. The use of polycarboxylate-based superplasticiser seems to result in a higher cracking risk than the use of polynaphtalene sulfonate-based superplasticiser [16]. Decreasing the rate of drying shrinkage by the use of a VMA may increase stress relaxation by creep and may result in a lower cracking risk.

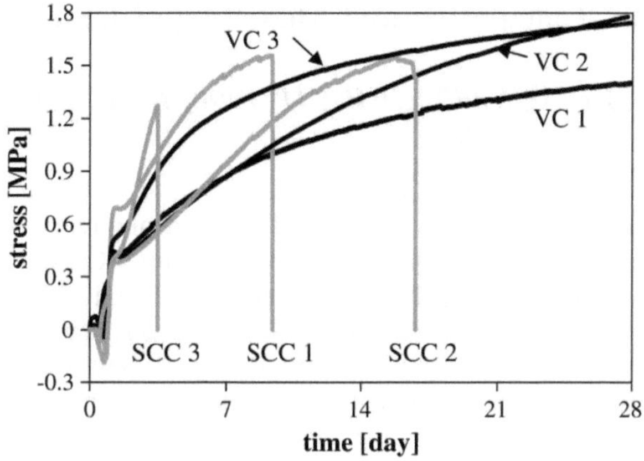

Fig. 3.13 Stress development in a passive-restrained shrinkage frame of SCC and VC of identical W/P by vol. (0.40), aggregate content and aggregate grain size distribution but 150 l/m^3 difference in paste volume (cement types: VC 1/SCC 1 = CEM I 42.5 N, VC 2/SCC 2 = CEM II/B-M (V-LL), VC 3/SCC 3 = CEM III/B 32.5 R/adapted from [3])

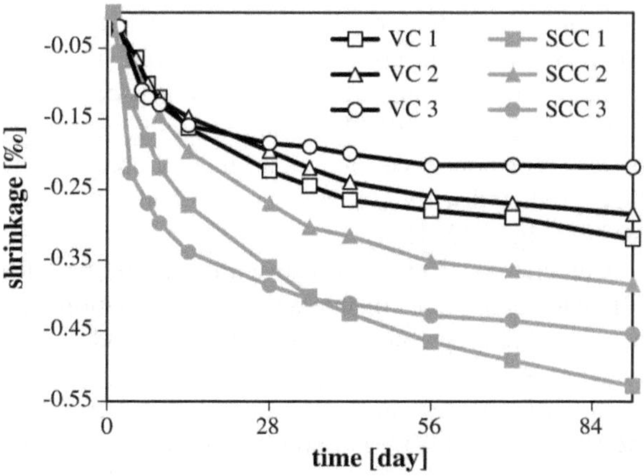

Fig. 3.14 Total shrinkage of SCC and VC of identical W/P by vol. (0.40), aggregate content and aggregate grain size distribution but 150 l/m^3 difference in paste volume (cement types: VC 1/SCC 1 = CEM I 42.5 N, VC 2/SCC 2 = CEM II/B-M (V-LL), VC 3/SCC 3 = CEM III/B 32.5 R/adapted from [3])

Curing has a significant influence on restrained shrinkage. The time to cracking after the end of curing is shortened by longer curing, because creep is reduced and the E-Modulus increased at long curing times [7, 33]. As a result, the low degree of relaxation leads to relatively fast cracking. Due to differences in creep, SCC can crack at the same time or even later than VC despite the higher shrinkage rate [7]; the higher creep of SCC can lead to a higher degree of stress relaxation and an increased time of cracking. However, when drying occurs fast, the influence of stress relaxation due to creep is significantly decreased and time to cracking is mainly dependent on the extent of shrinkage.

3.4 Calculation and Modelling of Creep and Shrinkage

3.4.1 ACI 209, CEB-FIP 1990, EHE, B3, GL2000, and EC2 Models

Several authors have compared experimental results of SCC with calculated creep and shrinkage using models developed for VC. An extensive review on shrinkage was done by [45]. The correspondence of ACI 209 [46], CEB-FIP 1990 [47], EHE [48], B3 [49], and GL2000 [50] with experimental data from various studies was evaluated. ACI 209R-92 and B3 agree best with the experimental data [45]. However, the agreement between models and experimental results is highly variable depending on the model used and on concrete mix design used in the experimental matrix as the following comparisons show. The relatively good accuracy of ACI 209 for shrinkage and its applicability for creep of limestone powder type SCC is confirmed by [8]. CEB-FIP shows a good correlation with creep of limestone powder type SCC, but in order to improve the accuracy of shrinkage predictions, modifications are necessary [2, 8]. The application of the Eurocode 2 [51] and BPEL leads to an underestimation of shrinkage but shows a better correspondence with experimental creep results [20]. Shrinkage of VC and SCC predicted with ACI 209 shows a better fit with the experimental results than MC90 (Fig. 3.15, [52]).

Calculation of tensile creep with a model based on the Eurocode 2 and CEB-FIP can lead to a considerable overestimation [11]. An extended comparison between creep predicted with different models and experimentally determined creep of SCC for precast, prestressed applications showed an accurate prediction using CEB-FIB 90, an overestimation with GL 2000 and an underestimation with ACI 209, AASHTO 2004 and AASHTO 2007 (Fig. 3.16, [53]).

In an attempt to model autogenous shrinkage of precast, prestressed SCC, the use of the CEB-FIP and the model proposed by Jonasson and Hedlund [54] lead to an underestimation at various ages [55]. On the other hand, a modification of the model proposed by Tazawa and Miyazawa [56] leads to less scattering and a reasonable prediction autogenous shrinkage of SCC.

The inconsistencies between predicted creep and shrinkage strain with experimental results clearly reveal the fact that none of the existing models is able to cover the broad range of SCC mix designs used today. Therefore, new approaches could be necessary to improve the situation. As an example, Long and Khayat [57] used a statistical model based on a factorial design approach considering SCC specific mix design parameters to predict mechanical properties, creep and shrinkage of SCC. The model proved to be valid for a wide range of SCC mix designs. Another approach for modelling drying shrinkage is presented in the following paragraph.

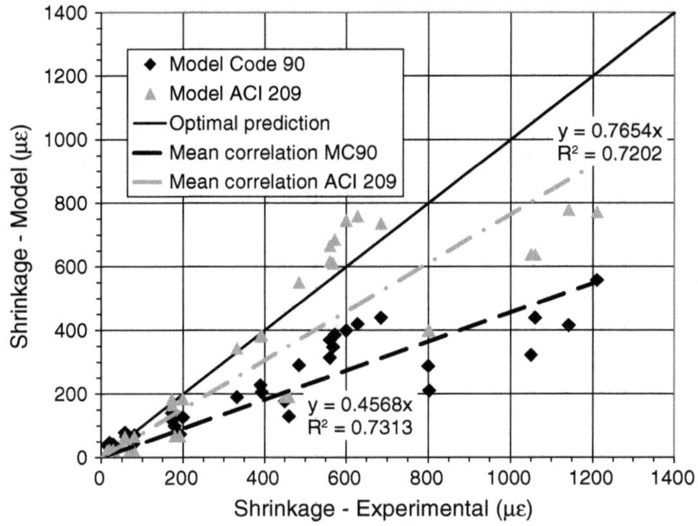

Fig. 3.15 Predicted shrinkage according to ACI 209 and MC 90 models compared to measured shrinkage of VC (W/CM = 0.42, cement content = 410 kg/m^3) and SCC with a mixture of limestone powder and fly ash as mineral additions (W/P by vol. = 0.31–0.39, powder content = 533–554 kg/m^3/[52])

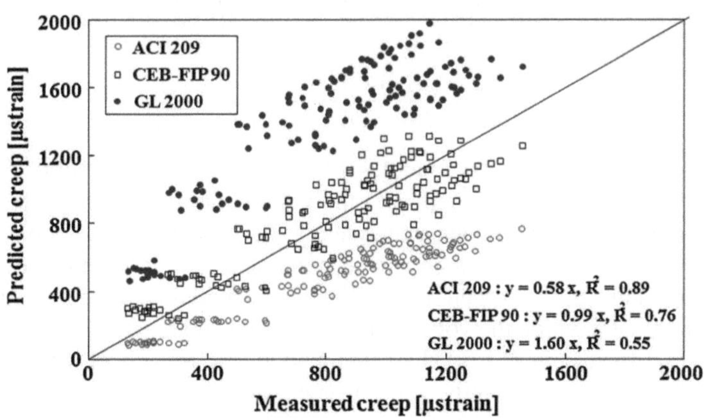

Fig. 3.16 Predicted creep according to ACI 209, CEB-FIP 90 and GL 2000 models compared to measured creep of SCC (W/P by vol. = 0.34–0.40, binder content = 440–500 kg/m^3/adapted from [53])

3.4.2 A Fundamental Approach to Modelling the Effect of Paste Content on Shrinkage

As observed in the previous sections, one reason for differences in creep and shrinkage of SCC and VC is the different paste and aggregate contents. The scope of this section is to show how composite models can be used to explain, and predict, these differences. The analysis is applied to results obtained in [7], but can be extended to other compositions, if the necessary shrinkage data and mechanical properties are available.

In [7] it was observed that:

- The modulus of elasticity of the SCC was slightly lower than the one of VC;
- Shrinkage of the SCC was significantly higher than the one of VC.

Even if W/CM of the cement paste in the two mixtures was not identical (W/CM 0.50 in the VC vs. W/CM 0.42 in the SCC), it is assumed here that these differences in elastic modulus, shrinkage, and possibly creep, are mainly due to the different cement paste volume of the two mixtures.

The elastic modulus of the cement paste of the SCC and the VC can be back calculated from the elastic modulus measured at 2, 7, and 28 days on the concrete mixtures.

The following equation developed by Hobbs [58, 59] allows calculating the elastic modulus of a concrete based on the elastic modulus of matrix and aggregates:

$$E_C = \frac{(E_A - E_M) \cdot \phi_A + E_A + E_M}{E_A + E_M + \phi_A \cdot (E_M - E_A)} \cdot E_M \tag{3.6}$$

where E_A (GPa) is the elastic modulus of the aggregate, E_M (GPa) the elastic modulus of the matrix, and ϕ_A (m^3/m^3) the volume fraction of the aggregate.

Equation 3.6 provides rather accurate elastic modulus prediction, see for example [60]. Equation 3.6 can be solved numerically to derive the value of E_M, the elastic modulus of the cement paste. The elastic modulus of the aggregates was assumed as 50 GPa, while the volume fractions of the aggregate were 0.746 for the VC and 0.662 for the SCC [7]. The calculated elastic moduli of the two cement pastes are similar. Consequently, the lower elastic modulus of the SCC mixture appears to be mainly a consequence of the higher cement paste content of this mixture.

In the following, the ratio between the shrinkage of SCC and VC is calculated based on a composite model and it is compared to the ratio between the experimental curves. More information about the calculation of shrinkage with composite models can be found in [58], while another application to VC and SCC is presented in [3].

The measured drying shrinkage curves at 70 % RH are shown in Fig. 3.17. The rate of shrinkage of SCC varies from about 1.2–1.6 times that of VC, depending on

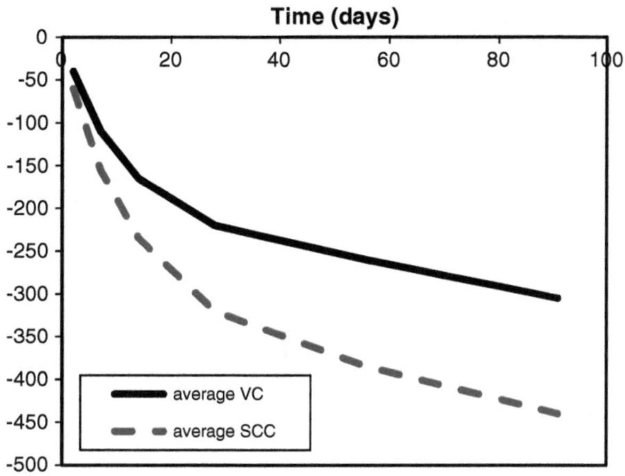

Fig. 3.17 Shrinkage at 70 % RH for VC and SCC mixtures (average of 2 samples). Data from [7]

age. This value is rather approximate since it was not calculated on the base of the actual shrinkage rate, but rather on the shrinkage difference between different ages.

According to another composite model developed by Hobbs [59], the shrinkage of the concrete, ε_C (m/m), can be calculated as:

$$\varepsilon_C = \frac{\varepsilon_P \cdot (1 - \phi_A) \cdot (G_P + G_A) + 2 \cdot \varepsilon_A \cdot \phi_A \cdot G_A}{G_P + G_A + \phi_A \cdot (G_A - G_P)} \tag{3.7}$$

where ε_P (m/m) is the shrinkage of the cement paste, ε_A (m/m) the shrinkage of the aggregates, G_A (GPa) the shear modulus of the aggregates, and G_P (GPa) the shear modulus of the paste.

If the object of the calculation is shrinkage development as a function of time, instead of final shrinkage, Eq. 3.7 can be used to calculate the rate of shrinkage. The actual shrinkage can be calculated by integrating this instantaneous rate of shrinkage [3, 60].

The shrinkage of the aggregates is considered negligible. This allows the calculation of the ratio between the shrinkage rates of the SCC and the VC concrete, by elimination of the shrinkage of the cement paste (which was not measured in [7]). This ratio depends only on G_A, G_P and ϕ_A of the two mixtures. The shear modulus of the cement paste as a function of age was calculated from the elastic modulus, assuming Poisson's ratio $\nu_P = 0.2$; the shear modulus of the aggregates was calculated from the elastic modulus (see previous section), assuming Poisson's ratio $\nu_A = 0.2$.

Figure 3.18 shows the ratio of shrinkage rate of SCC to VC according to both experiments and Hobbs's model. Considering the approximation involved in both calculations, the agreement between experiment and model is rather good. Consequently, the higher shrinkage of the SCC mixture can be attributed entirely to its higher cement paste content.

In conclusion, simple analytical composite models proved that the lower elastic modulus and higher shrinkage of the SCC tested in [7] were due to the higher paste

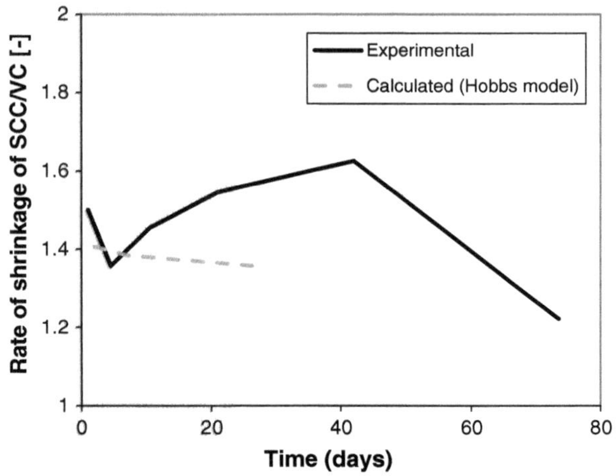

Fig. 3.18 Ratio of shrinkage rate of SCC to VC according to experiments by [7] and to Hobbs's model [58]

content of this mixture. The composite models can be successfully used to predict the behaviour of a series of concretes based on the same cement paste composition.

Similar composite models can be developed to study creep of concrete. It is anticipated that the higher creep exhibited by the SCC mixture [7] is a consequence mainly of the higher cement paste content. Finally, by combining results for elastic modulus, creep and shrinkage, it is expected that composite models can explain the different cracking risk of VC and SCC.

3.5 Conclusions

No unified conclusion can be drawn comparing the influence of paste volume and binder composition on creep. In general, SCC shows slightly higher creep coefficient than VC, and the influence of limestone powder on creep coefficient seems to be small. However, the use of GGBFS and fly ash can result in a decrease of the creep coefficient. The somewhat inconsistent results might be caused by differences in kinetics of both cement hydration and resulting strength development; the higher the strength increase after the load is applied, the lower the resulting creep coefficient. The current situation demands further research. In particular, data about creep in tension that can directly be compared with creep in compression are needed.

Autogenous shrinkage of SCC follows the same pattern as autogenous shrinkage of VC and HPC; it increases with increasing cement content and decreasing W/P by vol. Since the higher paste volume of SCC is usually achieved by using mineral additions, and not by increasing cement content, self-desiccation might be reduced depending on the type of mineral addition. Therefore, autogenous shrinkage of SCC with mineral additions is usually not higher than the one of VC of similar compressive strength produced with OPC. However, the higher paste volume results in a higher drying shrinkage of SCC compared to VC. This increase is slightly lessened by the use of mineral additions.

Due to reduced W/P by vol., finer cements, and higher volume of filler, SCC should be more susceptible to plastic shrinkage cracking than VC; however, the scarce results in the literature appear contradictory and more research is needed to clarify this issue. The differences between VC and SCC in cracking risk due to restrained shrinkage of hardened concrete depend on the degree of restraint and the drying velocity. When the degree of restraint and drying velocity are high, SCC exhibits a higher cracking risk than VC as a result of higher shrinkage. However, the importance of creep and E-modulus is increased in case of slow drying and/or low degree of restraint resulting in a relaxation of stress. Consequently, cracking risk of SCC is decreased and can be equal or even lower than the one of VC. The influence of mineral additions on the cracking risk is not clear. It seems to depend on the duration of curing.

The incorporation of SRA reduces or retards the total shrinkage and is able to delay time of cracking.

Several models enable the calculation of creep and shrinkage. Their accuracy in regard to creep and shrinkage differ. The main problem is that the models are designed for VC and do not take into account the influence of paste volume. Therefore, an adaptation of the existing models is necessary. However, the wide variety of mix designs and characteristics of mineral additions used in SCC tailored for various applications could restrict the accuracy of these models or require models that are rather adapted for particular types of SCC and mineral additions.

References

1. Bazant, Z.P., Wittmann, F.H.: Creep and Shrinkage in Concrete Structures. Wiley, New York (1982)
2. Heirman, G., Vandewalle, L., Van Gemert, D., Boel, V., Audenaert, K., De Schutter, G., Desmet, B., Vantomme, J.: Time-dependent deformations of limestone powder type self-compacting concrete. Eng. Struct. **30**, 2945–2956 (2008)
3. Leemann, A., Lura, P., Loser, R.: Shrinkage and creep of SCC - the influence of paste volume and binder composition. Const. Build. Mater. **25**, 2283–2289 (2011)
4. Turcry, P., Loukili, A., Haidar, K., Pijaudier-Cabot, G., Belarbi, A.: Cracking tendency of self-compacting concrete subjected to restrained shrinkage: experimental study and modeling. J. Mat. Civ. Eng. **18**, 46–54 (2006)
5. Loser, R., Leemann, A.: Self-compacting concrete: shrinkage and effects of shrinkage. cemsuisse report, Bern (in German) (2006)
6. Vieira, M., Bettencourt, A.: Deformability of hardened SCC. In: Wallevik, O., Nielsson, I. (eds.) Proceedings of the 3rd International RILEM Symposium on SCC, Reykjavik, pp. 637–644. RILEM Publications S.A.R.L, Bagneux (2003)
7. Loser, R., Leemann, A.: Shrinkage and restrained shrinkage cracking of self-compacting concrete compared to conventionally vibrated concrete. Mater. Struct. **42**, 71–82 (2009)
8. Poppe, A.M., De Schutter, G.: Creep and shrinkage of self-compacting concrete. In: Yu, Z., Shi, C., Khayat, K.H., Xie, Y. (eds.) Proceedings of the 1st International RILEM Symposium on SCC, pp. 329–336. RILEM Publications S.A.R.L, Bagneux (2005)

9. Lowke, D., Schießl, P.: Effect of powder content and viscosity agent on creep and shrinkage of self-compacting concrete. In: Sakata, T., Sato, M., Nakamura, M. (eds.) Creep, Shrinkage and Durability Mechanics of Concrete and Concrete Structures – Proceedings of 8th CONCREEP, pp. 655–661. Ise-Shima, Japan (2009)
10. Persson, B.: A comparison between mechanical properties of self-compacting concrete and the corresponding properties of normal concrete. Cem. Concr. Res. **31**, 193–198 (2001)
11. Wüstholz, T., Reinhardt, H.W.: Deformation behaviour of self-compacting concrete under tensile loading. Mater. Struct. **40**, 965–977 (2007)
12. Gutsch, A.W.: Properties of early age concrete - experiments and modelling. Mater. Struct. **35**, 76–79 (2002)
13. Atrushi, D.S.: Tensile and compressive creep of early age concrete: testing and modelling. Dissertation, Trondheim (2003)
14. Li, H., Wee, T.H., Wong. S.F.: Early-age shrinkage of blended cement concrete. ACI Mat. J. **99**, 3–10 (2002)
15. Bissonnette, B., Pigeon, M., Vaysburd, A.M.: Tensile creep of concrete: study of its sensitivity to basic parameters. ACI Mat. J. **104**, 360–368 (2007)
16. Hwang, S.D., Khayat, K.H.: Effect of mix design on restrained shrinkage of self-consolidating concrete. Mater. Struct. **43**, 367–380 (2010)
17. Rozière, E., Granger, S., Turcry, P., Loukili, A.: Influence of paste volume on shrinkage cracking and fracture properties of self-compacting concrete. Cem. Conc. Comp. **29**, 626–636 (2007)
18. Hammer, T.A.: Measurement methods for testing of early age autogenous strain. In: Kovler, K., Bentur, A. (eds.) RILEM Proceedings of Conference on Early Age Cracking in Cementitious Systems EAC'01, Haifa, pp. 217–228 (2002)
19. Craeye, B., De Schutter, G., Desmet, B., Vantomme, J., Heirman, G., Vandewalle, L., Cizer, Ö., Aggoun, S., Kadri, E.H.: Effect of mineral filler type on autogenous shrinkage of self-compacting concrete. Cem. Concr. Res. **40**, 908–913 (2010)
20. Khayat, K.H., Long, W.J.: Shrinkage of precast, prestressed self-consolidating concrete. ACI Mat. J. **107**, 231–238 (2010)
21. Pons, G., Proust, E., Assié, S.: Creep and shrinkage of self-compacting concrete: a different behaviour compared with vibrated concrete? In: Wallevik, Ó., Nielsson, I. (eds.) Proceedings of the 3rd International RILEM Symposium on SCC, Reykjavik, pp. 645–654. RILEM Publications S.A.R.L, Bagneux (2003)
22. Esping, O.: Effect of limestone filler BET(H2O)-area on the fresh and hardened properties of self-compacting concrete. Cem. Concr. Res. **38**, 938–944 (2008)
23. Termkhajornkit, P., Nawa, T., Nakai, M., Sito, T.: Effect of fly ash on autogenous shrinkage. Cem. Conc. Res. **35**, 473–482 (2005)
24. Piérard, J., Dieryck, V., Desmyter, J.: Autogenous shrinkage of self-compacting concrete. In: Shah, S. (ed.) 4th International RILEM Symposium on SCC, Chicago (2005)
25. Kovler, K., Jensen, O.M.: Internal curing of concrete. In: State-of-the-Art Report of RILEM Technical Committee 196-ICC. RILEM Report 41, RILEM Publications. S.A.R.L., Bagneux (2007)
26. Ghoddousi, P., Monir Abbasi, A.: Influence of aggregate grading and cement paste volume on drying shrinkage of self-consolidating concrete. In: 3rd North American Conference on the Design and Use of Self-Consolidating Concrete, Chicago (2008)
27. Leemann, A., Hoffmann, C.: Properties of self-compacting and conventional concrete - differences and similarities. Mag. Conc. Res. **57**, 315–319 (2005)
28. Khatib, J.M.: Performance of self-compacting concrete containing fly ash. Const. Build. Mater. **22**, 1963–1971 (2008)
29. Gesoğlu, M., Güneyisi, E., Özbay, E.: Properties of self-compacting concretes made with binary, ternary and quaternary cementitious blends of fly ash, blast furnace slag and silica fume. Const. Build. Mater. **23**, 1847–1854 (2009)
30. Van Bui, K., Montgomery, D.: Drying shrinkage of self-compacting concrete containing milled limestone. In: Skarendahl, Å., Petersson, Ö. (eds.) Proceedings of the 1st International

RILEM Symposium on SCC, Stockholm, pp. 227–238. RILEM Publications S.A.R.L, Bagneux (1998)
31. Hammer, T.A.: The influence of some mix design parameters on drying shrinkage of SCC. In: De Schutter, G., Boel, V. (eds.) Proceedings of the 5th International RILEM Symposium on SCC, Ghent, pp. 559–564. RILEM Publications S.A.R.L, Bagneux (2007)
32. Vikan, H., Hammer, T.A., Kjellsen, O.: Drying shrinkage of SCC - influence of the composition of ternary composite cements. In: Khayat, K.H., Feys, D.F. (eds.) Proceedings of the 6th International RILEM Symposium on SCC, Montreal, pp. 271–282. Springer, Berlin (2010)
33. See, H.T., Attiogbe, E.K.: Performance of self-consolidating concrete under restrained shrinkage. In: Gardner, N.J., Weiss, W.J. (eds.) Shrinkage and Creep of Concrete, ACI 227, pp. 303–315. Michigan, Farmington Hills (2005)
34. Hwang, S.D., Khayat, K.H.: Effect of mixture composition on restrained shrinkage cracking of self-consolidating concrete used in repair. ACI Mat. J. **105**, 499–509 (2008)
35. Saito, K., Kinoshita, M., Umehara, H., Yoshida, R.: Properties of low-shrinkage, high-strength SCC using shrinkage-reducing admixture, blast furnace slag and limestone aggregate. In: Khayat, K.H., Feys, D.F. (eds.) Design, Production and Placement of Self-Consolidating Concrete, Proceedings of SCC2010, Montreal, pp. 283–293. Springer, Berlin (2010)
36. Bentz, D.P., Peltz, M.A., Snyder, K.A., Davis, J.M.: Verdict: viscosity enhancers reducing diffusion in concrete. Concr. Int. **31**, 31–36 (2009)
37. Gram, H.E., Pentti, P.: Properties of SCC: especially early age and long term shrinkage and salt frost resistance. In: Skarendahl, Å., Petersson, Ö. (eds.) Proceedings of the 1st International RILEM Symposium on Self-compacting Concrete, Stockholm, pp. 211–225. RILEM Publications S.A.R.L, Bagneux (1999)
38. Turcry, P., Loukili, A.: Evaluation of plastic shrinkage cracking of self-consolidating concrete. ACI. Mat. J. **10**, 272–279 (2006)
39. Holt, E., Schodet, O.: Self-compacting concrete: early age shrinkage. Internal Report, VTT, Building and Transport, RTE40-IR-21, Finland (2002)
40. Persson, B.: Plastic shrinkage of self-compacting concrete. In: Jensen, O.M., Geiker, M., Stang, H. (eds.) Proceedings of the Knud Hojgaard Conference, Lyngby, Report R-155 DTU, pp. 43–57 (2005)
41. Lura, P., Leemann, A.: unpublished data
42. Hammer, T.A.: Cracking susceptibility due to volume changes of self-compacting concrete (SCC). In: Wallevik, O., Nielsson, I. (eds.) Proceedings of the 3rd International RILEM Symposium on SCC, Reykjavik, pp. 553–557. RILEM Publications S.A.R.L, Bagneux (2003)
43. Van Itterbeeck, P., Cauberg, N., Parmentier, B., Vandewalle, L., Lesage, K.: Evaluation of the cracking potential of young self-compacting concrete. In: Khayat, K.H., Feys, D.F. (eds.) Proceedings of the 6th International RILEM Symposium on SCC, Montréal, pp. 991–1001. Springer, Berlin (2010)
44. Tongaroonsri, S., Tangtermsirikul, S.: Effect of mineral admixtures and curing periods on shrinkage and cracking age under restrained condition. Const. Build. Mater. **23**, 1050–1056 (2009)
45. Fernández-Gomez, J., Agranati Landsberger, G.: Evaluation of shrinkage prediction models for self-consolidating concrete. ACI Mat. J. **104**, 464–473 (2007)
46. ACI Committee 209: Prediction of creep, shrinkage and temperature effects in concrete structures (209R-92). Farmington Hills, American Concrete Institute (1990)
47. CEB-FIP: Model code 1990 – design code. Thomas Telford, London (1993)
48. EHE: Instrucción de hormigón estructural', Comisón Permanente del Hormigón, 5th Edición, Ministerio de Fomento, Madrid (2002)
49. Bazant, Z.P., Baweja, S.: Creep and shrinkage prediction model for analysis and design of concrete structures - Model B3. Mater. Struct. **28**, 357–365 (1995)
50. Gardner, N.J., Lockman, M.: Design provision for drying shrinkage and creep of normal-strength concrete. ACI Mater. J. **98**, 159–167 (2001)

51. Eurocode 2: Design of concrete structures. Part 1: General rules and rules for buildings. CEN, Brussels (1992)
52. Mazzotti, C.: Long term behaviour of self-compacting concrete: comparison between experimental results and predicting models. In: Morinconi, G. (ed.) 2nd ACI Workshop: The New Boundaries of Structural Concrete, pp. 1–8. Ancona, Italy (2011)
53. Long, W.J., Khayat, K.H.: Creep of prestressed self-consolidating concrete. ACI Mat. J., **108**(5), 476–484 (2011)
54. Jonasson, J., Hedlund, H.: An engineering model for creep and shrinkage in HPC. In: Baroghel, V., Aïtcin, P. (eds.) Proceedings of the International RILEM Workshop on Shrinkage of Concrete, RILEM, France, pp. 507–529. RILEM Publications S.A.R.L, Bagneux (2000)
55. Long, W.J., Khayat, K.H., Xing, F.: Prediction on autogenous shrinkage of self-consolidating concrete. Adv. Mater. Res., ICAMMP Shenzhen, China, Adv. Compos. 150–151, 288–292 (2011)
56. Tazawa, E., Miyazawa, S.: Influence of constituent and composition on autogenous shrinkage of cementitious materials. Mag. Concr. Res. **49**, 15–22 (1997)
57. Long, W.J., Khayat, K.H.: Statistical models to predict mechanical and visco-elastic properties of SCC designated for precast/prestressed applications. In: Proceedings of the 2nd International Symposium on Design, Performance and Use of Self-Consolidating Concrete, Beijing (2009)
58. Hobbs, D.W.: Bulk modulus shrinkage and thermal expansion of a two phase material. Nature **222**, 849–851 (1969)
59. Hobbs, D.W.: Influence of aggregate restraint on the shrinkage of concrete. ACI J. **71**, 445–450 (1974)
60. Lura, P.: Autogenous deformation and internal curing of concrete. Dissertation, Delft University

Chapter 4
Bond Properties of Self-Compacting Concrete

Kamal H. Khayat and Pieter Desnerck

4.1 Introduction

Proper force transfer between the reinforcement and surrounding concrete is one of the most significant factors affecting the structural behaviour of reinforced concrete structures. Due to the importance of this issue, there is lots of research devoted to investigating the bond properties of conventional vibrated concrete mixtures. This chapter presents a comprehensive review of different aspects of bond properties of self-compacting concrete (SCC).

In the first part of this chapter, bond strength to reinforcing bars and pre-stressing strands is reviewed, including the top-bar effect for various types of SCC mixtures. The effect of using chemical admixtures, such as viscosity-modifying admixtures (VMAs), to reduce the top-bar effect is discussed. The effect of using supplementary cementitious materials (SCMs) and fillers on bond strength characteristics of SCC is reviewed. Bond strength between successive lifts of SCC in multi-layer casting is reviewed highlighting the effect of elapsed time between the castings of different layers and the thixotropy of the lower lift on bond in multi-layer casting. Finally, bond strength characteristics between SCC mixtures used in repair applications and existing hardened concrete are discussed.

The bulk of the literature agrees that bond properties of SCC to embedded reinforcement, pre-stressing strands, and hardened concrete is higher than that of vibrated concrete (VC). The rheological properties of the SCC, including the yield

K. H. Khayat (✉)
Civil, Architectural and Environmental Engineering, Missouri University of Science and Technology, Rolla, MO 65409-0710, USA
e-mail: khayatk@mst.edu

P. Desnerck
Ghent University, Ghent, Belgium

P. Desnerck
Center for Infrastructure Engineering Studies, Missouri University of Science and Technology, Rolla, MO 65409-0710, USA

K. H. Khayat and G. De Schutter (eds.), *Mechanical Properties of Self-Compacting Concrete*, RILEM State-of-the-Art Reports 14,
DOI: 10.1007/978-3-319-03245-0_4, © RILEM 2014

stress, plastic viscosity, and hence static stability play a key role in achieving the desired bond properties for SCC mixtures.

4.2 Bond Between Reinforcement and SCC

Compared to vertically embedded reinforcement, horizontal reinforcing bars have larger area under which bleed water could accumulate and weaken the interfacial bond properties. Surface settlement resulting from the lack of static stability of concrete after placement can also have greater influence on bond with horizontal rebars than vertical ones. Therefore, the top-bar effect is usually more pronounced in horizontal reinforcements than in the vertical bars, for VC, as reported in Fig. 4.1 [1].

Menezes et al. [3] analysed the bond behaviour of SCC in comparison to VC using pull-out and beam tests according to RILEM procedures. The testing program considered bar sizes of 10 and 16 mm and VC and SCC with compressive strengths of 30 and 60 MPa. The average bond strength was calculated by mean of the bond stress measured for slippage values of 0.01, 0.1, and 1.0 mm according to RILEM recommendations [2]. It was reported that for SCC and VC specimens with 30 MPa compressive strength, the pull-out and beam tests exhibited similar results based on which it was proposed that for normal strength concrete, the evaluation of bond strength can be performed using the simple pull-out test. Based on the obtained results, they reported that for 30 MPa specimens, most of the specimens (both the pull-out and beam specimens) had slip failure. From the results it was observed that the bond stress between the concrete and reinforcement is better in the case of SCC mixtures and small bar diameters, as indicated in Fig. 4.2.

In the case of specimens with a 60 MPa compressive strength, all the pull-out specimens failed due to splitting and failure in the beam specimens occurred due to the yield of the steel rebars. It was also observed that for these specimens, there was no significant difference between the bond properties of SCC and VC mixtures, which may be due in the high quality of the concrete (Fig. 4.3). For both compressive strength levels, it was observed that the total bond strength decreased as a result of the increase in bar diameter. The study concluded that the bond behaviour presented by the SCC was similar to that of the VC and was even better in some cases.

Forughi-Asl et al. [4] investigated the bond strength in SCC mixtures using the pull-out tests according to RILEM/CEB/FIP Standard and the Rehm and Eligehausen pull-out test. Based on comparing the critical bond strength corresponding to 0.25 mm slip, it was observed that in SCC, compressive and bond strength can develop slower than in the conventional concrete. This may be caused by the retarding effect of the carboxylate-based superplasticiser. However, after 28 days, the SCC mixtures had higher bond strength compared to VC specimens. The authors also reported that the relationship between bond strength and compressive strength of normal concrete is more consistent than that of SCC (Table 4.1). Based on the results at 28 and 56 days, it was concluded that the bond strength of SCC was slightly higher than that of the corresponding VC with the same W/C.

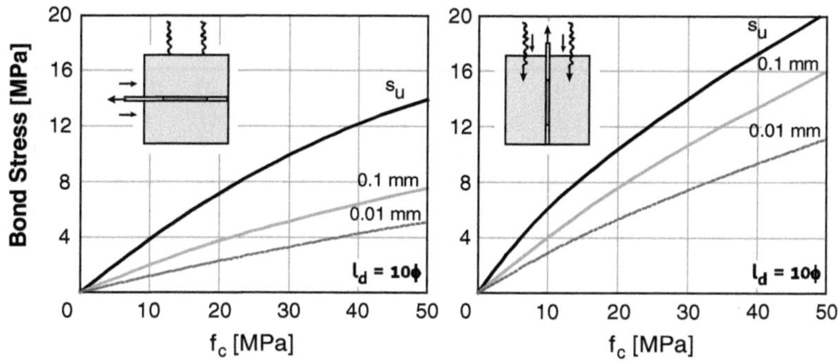

Fig. 4.1 Effect of rebar direction on bond strength (based on [1])

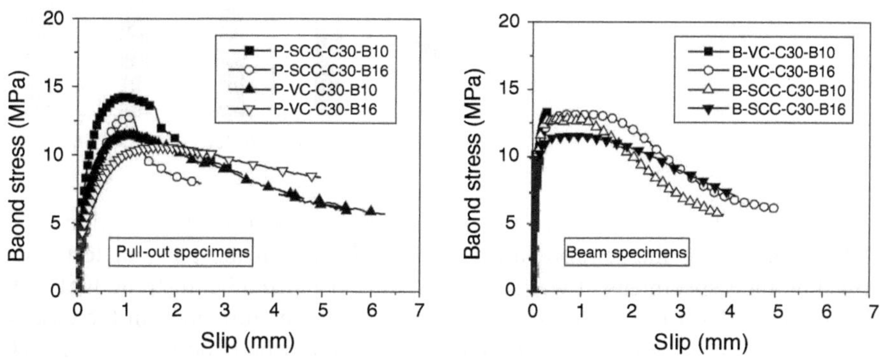

Fig. 4.2 Bond stress versus slip curves for the pull-out and beam specimens for 30 MPa mixtures [3]

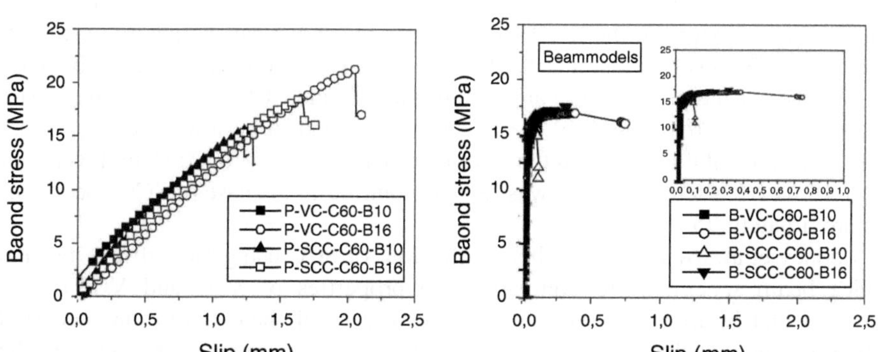

Fig. 4.3 Bond stress versus slip curves for the pull-out and beam specimens for 60 MPa mixtures [3]

Table 4.1 Comparison of normalised bond ratio $\tau/f'^{1/2}_c$ of SCC and VC [4]

	SCC					VC				
W/C (days)	0.30	0.40	0.45	0.50	0.60	0.30	0.40	0.45	0.50	0.60
3	1.22	1.18	1.19	1.10	1.02	1.50	1.41	1.37	1.25	1.16
7	1.39	1.36	1.35	1.28	1.26	1.59	1.46	1.48	1.39	1.35
28	1.77	1.75	1.70	1.64	1.65	1.69	1.72	1.65	1.54	1.53
56	1.83	1.70	1.66	1.54	1.51	1.78	1.67	1.59	1.50	1.47

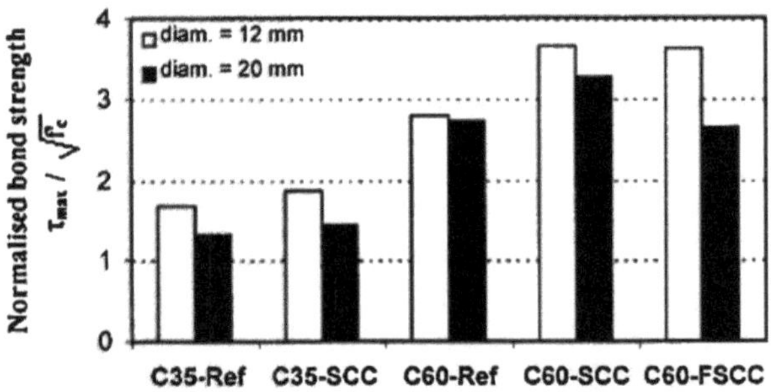

Fig. 4.4 Variation in normalised bond strengths of all mixtures [5]

Zhu et al. [5] have also studied bond strength and Interfacial Transition Zone (ITZ) properties of the reinforcement in SCC. Based on the pull-out tests performed on their specimens, it was stated that the normalized bond strength of SCC mixtures were 10–40 % higher than that of the reference mixtures. In the case of specimens with a compressive strength of 35 MPa, where both SCC and VC specimens had W/C of 0.68, the SCC specimens had slightly higher bond strengths. In the case of specimens with a higher compressive strength, where the W/C of the SCC and VC were 0.36 and 0.43, respectively, the SCC specimens exhibited higher bond strengths, especially for the 12 mm reinforcing steel bars. The difference between the bond strength was lower for bar diameters of 20 mm (Fig. 4.4). The results illustrate that the bond strength decreased with the increase in diameter of the reinforcing bar for both the VC and SCC mixtures. The elastic modulus and micro-strength of the ITZ were reported to be lower on the bottom side of a horizontal bar than on the top side (approximately 70–80 % in the case of VC). However, such difference was less pronounced in the case of SCC specimens (approximately 75–100 %).

Looney et al. [6] studied the bond behaviour of SCC using the pull-out test and splice beam specimens to compare bond properties of SCC and VC. It was reported that in all of the pull-out specimens, a bond shear failure has occurred which means that the reinforcing bar and associate concrete located between the transverse ribs of the bar pulls out of the specimen as a cylinder without splitting the remaining concrete. Based on their experimental results, it was reported that in normal strength concretes of 41 MPa target compressive strength, the normalized bond strength of the SCC specimens was approximately 15 % higher than that of

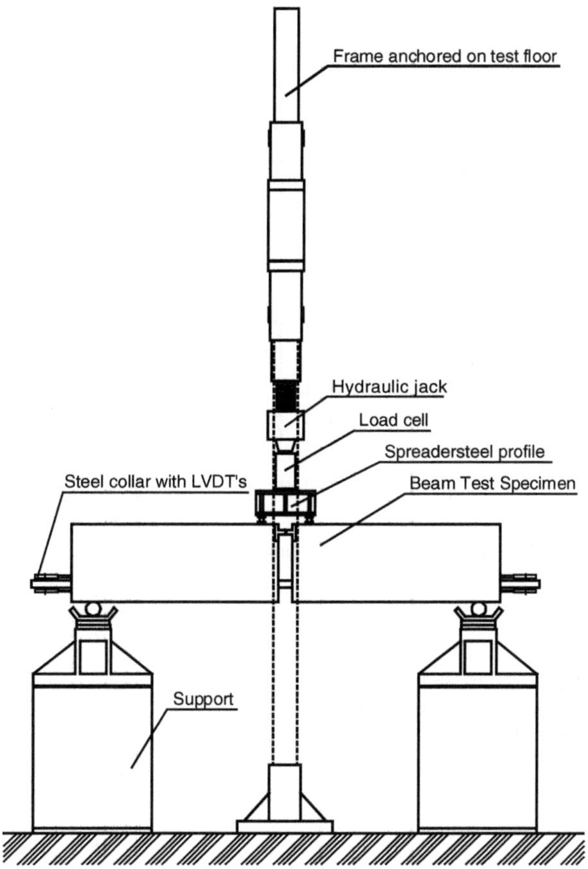

Fig. 4.5 Test set-up for beam test specimen [7]

VC. However, in the case of the high strength specimens with target compressive strength of 69 MPa, the normalized bond strength of SCC was approximately 7 % lower than that of VC. From the splice test specimens, it was also found that all of the beams experienced a splitting failure and similar load–deflection results were obtained for all of the specimens. The authors reported that SCC specimens of normal and high strength possess reinforcement bond strength comparable or slightly larger than that of VC.

Desnerck et al. [7] used beam tests to investigate bond properties of SCC and the effect of bar diameter on bond. The test setup and execution were according to RILEM Recommendations RC6 [8], details are shown in Fig. 4.5.

The authors reported that with a bond length of 10 times the bar diameter can undergo yielding or rupture. The investigated concrete had a compressive strength of about 60 MPa. Hence, a bond length of five times the bar diameter was recommended for the evaluation of bond strength. Two types of SCCs with compressive strength of 63 and 57 MPa were considered in this investigation

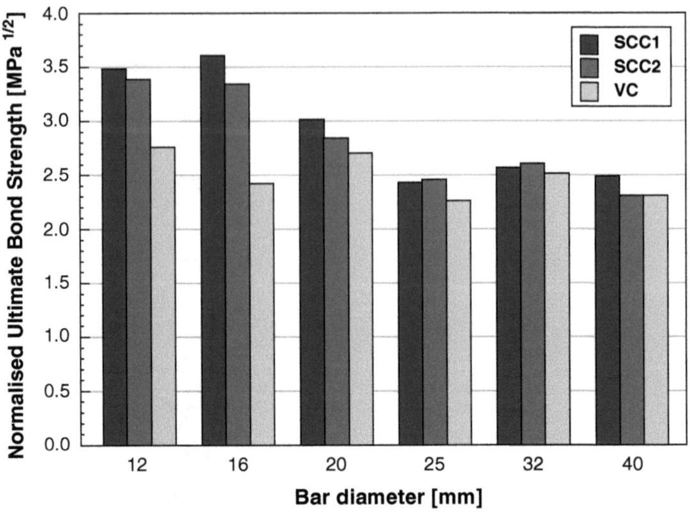

Fig. 4.6 Normalised ultimate bond strength for different diameters and concrete compositions [7]

Fig. 4.7 Pull out test set-up by [15] accordance with RILEM, RC6

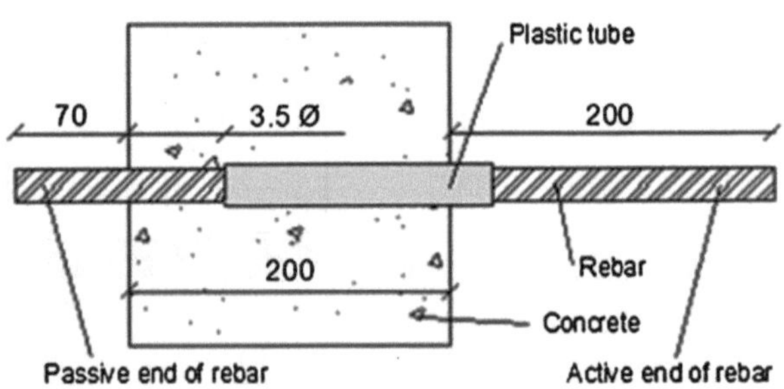

(SCC1 and SCC2). A conventional concrete with compressive strength comparable to that of SCC2 (52 MPa) was considered. Specimens with bar diameters of 12, 20, 25, 32 and 40 mm were tested (two specimens for each type, resulting in four individual measurements). Figure 4.6 shows the normalized ultimate bond strength of the various types of concrete and bar diameters used in this study.

Based on the experimental results, it was reported that the bond strength of SCC can be as high as that of VC when large bar diameters are studied. For smaller bar diameters, the bond strength of SCC was slightly higher, with the largest difference occurring for the smallest bar diameters.

Further research in Ghent University by Helincks et al. [15] used an experimental program to investigate the bond performance of powder-type SCC. Pull-out tests were carried out in accordance with RILEM, RC6, Part 2 recommendations (Fig. 4.7). In total, 72 pull-out specimens were tested, cast with different concrete mixtures and rebar diameters (8, 12, 16, and 20 mm). It was found that SCC showed normalised characteristic bond strength values as high as or higher than equivalent VC. When larger diameters up to 20 mm were used a decrease in average bond stress was observed.

Castel et al. [9] have also studied the bond behaviour of the SCC and reported no significant difference between SCC and VC in terms of transfer length irrespective of the compressive strength of the concrete. The authors reported that for both concrete types, the transfer length slightly decreased with the increase in compressive strength.

Ertzibengoa et al. [10] studied the bond behaviour of flat stainless steel rebars in SCC and VC mixtures. They reported that SCC allows for developing higher bond strength compared to VC. This influence was more pronounced for round rebars than for flat elements.

Pandurangan et al. [11] evaluated the effect of using SCC on bond strength and mode of bond failure of tension lap splices anchored in normal strength concrete. To meet this objective, full-scale NSC beam specimens were tested. Each beam was designed with bars spliced in a constant moment region at mid-span with various levels of stirrup confinement. Test results indicated that at low passive confinement by stirrups, the bond strength for SCC was found to be almost equal to that of VC. Moreover, for well confined concrete, bond strength in SCC was higher than VC.

Dehn et al. [12] studied the time-dependency of material properties and the bond behaviour between the reinforcing bars and SCC from pull-out tests. Their test results showed that the bond behaviour of SCC was better than the bond stresses of VC. By conducting pull-out tests, Brameshuber et al. [13] reported that the bond between SCC and reinforcement was comparable to that of VC.

Pull-out tests on steel reinforcing bars of 12 and 20 mm diameter were conducted by Sonebi et al. [14]. Results showed that the bond strength of SCC was about 18–38 % higher than that of regular concrete mixtures.

Chan et al. [16] also found that the SCC members had significantly higher bond strengths with reinforcing bars than did ordinary concrete members. They also reported that the reduction in bond strength due to bleeding and inhomogeneity in the ordinary concrete was prevented with the use of SCC.

In their study, Sonebi et al. [18] investigated the bond strength of rebars in SCC and VC. They concluded that the normalised bond strengths of the SCC mixes were about 27–65 % higher than that of VC, a finding they attribute to the lower water content and higher powder content of the SCC mixtures leading to reduced accumulation of bleeding water underneath the bars. The increase in the diameter of bars led to a reduction of the bond strength, confirming the results of Desnerck et al. [7]. Furthermore they found that an increase of the cover of the concrete from 20 to 50 mm resulted in an improvement of the bond strength for all mixtures but slightly better for SCC.

Ashtiani [19] investigated the bond properties between reinforcement and high strength self-compacting concrete (HSSCC) and vibrated high-strength concrete

(VHSC). Comparable concrete compressive strengths of ~ 90 MPa in both the HSSCC and VHSC were used. The other variables were the bar grade, diameter, bond length. Special attention was paid to the post-yield slip behaviour. It was found that the difference in ductility of bars with different grades of steel resulted in a different rate in reduction of axial tensile stress with respect to the bar diameter in in both the HSSCC and the VHSC. This consequently affects their bond performance especially in the post-yield range. The bond characteristics between reinforcement and self-compacting concrete were analysed experimentally by Zheng et al. [20]. The tests were conducted on 28 SCC specimens using pulled out by an electro hydraulic tension testing machine. Influence laws were found to be valid for the SCC-bond interaction.

Surong et al. [21] investigated the bond characteristics between reinforcement and self-compacting concrete under dynamic loads experimentally. The tests were conducted by pulling-out tests with strain gauges in the reinforcement. The concrete strength ranged from 47 to 71 MPa and testing was in accordance with the Chinese Concrete Structure Rest method standard (GB50152-92). The average bond stress was higher in the high strength concrete and bond stress laws were found applicable to the SCC samples.

The proceeding from the 10 yearly International symposium on Bond In Concrete [22] present several papers investigating the bond in SCC with different types of reinforcing bar as summarised in this section: Sonebi et al. investigated the bond strength between two grades of self-compacting concrete compared to VC using a beam-type pull-out specimens. The average bond strength mm was lower in the 20 mm bars compared with the 12 mm diameter for all samples. The average bond strength in HSCC was higher than NSCC and the top-bar factors varied approximately between 1.2 and 1.53. The Eurocode and fib 1990 predicted lower average bond strengths for top and bottom bars in the SCC samples. Kaffetzakis and Papanicolaou, investigate the steel-to-concrete bond in LWSCC using direct pull-out in accordance with EN 10080. The variables were bar diameter (12 and 16 mm), bond length (5Ø and 10Ø) and type of additive (LSP and SF). The results were compared with NWSCC. The majority of LWSCC specimens exhibited splitting failures at relatively low slip values. The addition of normal-weight sand led to an increase in the average bond stress but failed to enhance the energy absorption capacity. El-Sawy et al., evaluated the effect of the bar diameter, splice length and the confinement level provided by the transverse reinforcement on bond strength in SCC. All the descriptive equations in the ACI 408R-03 overestimated the predicted bond stress in SCC beams by 25–40 %. Nejadi and Aslani studied the behaviour of the SFRSCC matrix with deformed reinforcing steel. The average bond strength in the SFRSCC samples was similar to VC at large bar diameters while for smaller bar diameters it was slightly higher than in equivalent VC.

4.3 Bond Between Prestressing Strands and SCC

Bond between a strand and concrete is affected by the position of the embedded reinforcements and quality of the cast concrete. Bond to prestressed tendons can be influenced by the flow properties of the SCC, grading of the aggregate and content of fines in the matrix [17]. Khayat et al. [23] evaluated the uniformity of bond strength to prestressing strands and in situ mechanical properties of flowable concrete along experimental wall elements. Four SCC mixtures and two conventional flowable mixtures suitable for prestressed and precast applications were evaluated. The mixtures incorporate 20 % fly ash replacement and were used for casting experimental wall elements measuring 1.54 m in height, 1.1 m in length, and 0.2 m in width. Two types of VMAs and two high-range water reducers (HRWRs) were employed. Two of the walls were steam-cured, while the remaining elements were air-cured. Each wall had 16 prestressing strands, four per row positioned at four levels that were subjected to pull-out tests at 1 and 28 days of age. All mixtures developed a 1-day compressive strength greater than 40 MPa. Uniform distribution of in-place compressive strength and adequate bond to the prestressing strands were obtained with relatively small variations along experimental wall elements. The 1- and 28-day top-bar effect ratios varied between 0.9 and 1.9. The top-bar effect in the air-cured SCC was lower than in the steam cured highly flowable concrete. For steam curing, the top-bar effect was larger in the case of the SCC than in the case of the conventional flowable concrete at 1 day, but lower at 28 days. The coupled VMA type affected the level of static stability with direct implication on the top-bar effect. The distribution of compressive strength along the wall height for the stable SCC mixtures was quite uniform, with strengths along the wall height being within 6 % of that at the bottom. Attiogbe and Nmai [24] also showed that the highly stable nature of SCC can enhance the top-bar factor of SCC for reinforcing bars and prestressing strands.

Martí-Vargas et al. [25] analysed the transfer lengths and anchorage bond behaviour of 7-wire prestressing strands in SCC made with different cement contents, W/C and particle size distributions. The results were compared to the performance of VC with the same cement contents and W/C using bond strength testing [26, 27]. The results indicated that while SCC and VC showed a similar compressive strength, the tensile strength was higher in the SCC. Bond losses during release were consistently greater in the SCC, particularly when the inert addition content was high. This finding was attributed to more intense concrete shrinkage and its effect was calculated from stressed end slip. As a result, it was concluded that SCC design should make provision for greater prestressing loss, regardless of concrete strength. Also, the SCC made with lower doses of cement was more ductile in terms of free end slip during release. Nonetheless, both the transmission length and the final free end slip values were similar for SCC and the equivalent traditional concrete. Anchorage length was analysed both relative to reinforcement slip in the free end (anchorage length with slip, LA) and without slip (anchorage length without slip, LW). The anchorage length was greater in the SCC with low cement contents. At high cement doses, however, no differences were observed in anchorage length. Concretes with low W/C had a smaller anchorage length with slip than transmission length. This difference declined as the W/C rose. In addition, anchorage length without slip was consistently greater than anchorage length with slip.

4.4 Top-Bar Effect for Reinforcing Bars and Positioning Factor for Prestressing Strands

The reduction of bond strength to horizontally anchored or overlapped bars located in the upper sections of structural elements as opposed to those located near the bottom is known as the top-bar effect. A high top-bar factor necessitates an increase in the anchorage length and further contributes to the congestion of some structural sections.

In the tests conducted by Attiogbe et al. [28], SCC yielded similar top-bar factors to those of normal concrete with 102 to 152 mm of slump. This factor is defined as the bond strength of the bottom layer of reinforcing bars divided by the bond strength of the top layer. In a test using air-cured SCC and a VMA, the top-bar factor was actually lower than that of VC.

Valcuende and Parra [29] investigated bond strength to reinforcing bars of SCC and Vibrated Concrete (VC) made with different W/CM and compressive strength values. Three W/CM values of 0.45 for concrete with targeted f'_c of 42.5 MPa, 0.55 for concrete with targeted f'_c of 32.5–42.5 MPa, and 0.65 for concrete with targeted f'_c of 32.5 MPa were used. The W/CM was the same for the SCC and VC mixtures. Two types of specimens were tested: 200-mm cubes and 1,500-mm high columns, as shown in Fig. 4.8. In both tests, the bars were positioned perpendicular to the casting direction. The bar diameters were 16 and 12 mm for the cube and column specimens, respectively. The authors observed that the normalized mean bond strength is greater in SCC than in VC mixtures. However, the behaviour of both concretes tends to even out as the mechanical properties improved. They observed that in concretes with a 30 MPa compressive strength, the mean bond strength is about 30 % greater in SCC than in VC; whereas for the 60 MPa concretes, the difference decreased to less than 10 % (Fig. 4.9).

The same scatter in material characteristics between SCC and VC made with the same W/CM was also observed in the pull-out tests of Mendez et al. [3], where the normalised mean bond strength in 30 MPa specimens is higher in the case of SCC specimens comparing with VC ones. However, in the case of 60 MPa specimens, for 10 mm bars, the VC has stronger bond and for the 16 mm bars, higher bond strength was observed in SCC specimens.

As shown in Figs. 4.10 and 4.11, the top-bar measurements of 1,500 mm high columns show that the mean bond strength decreases as the distance from the bottom of the member increases for both the SCC and VC specimens [28]. This

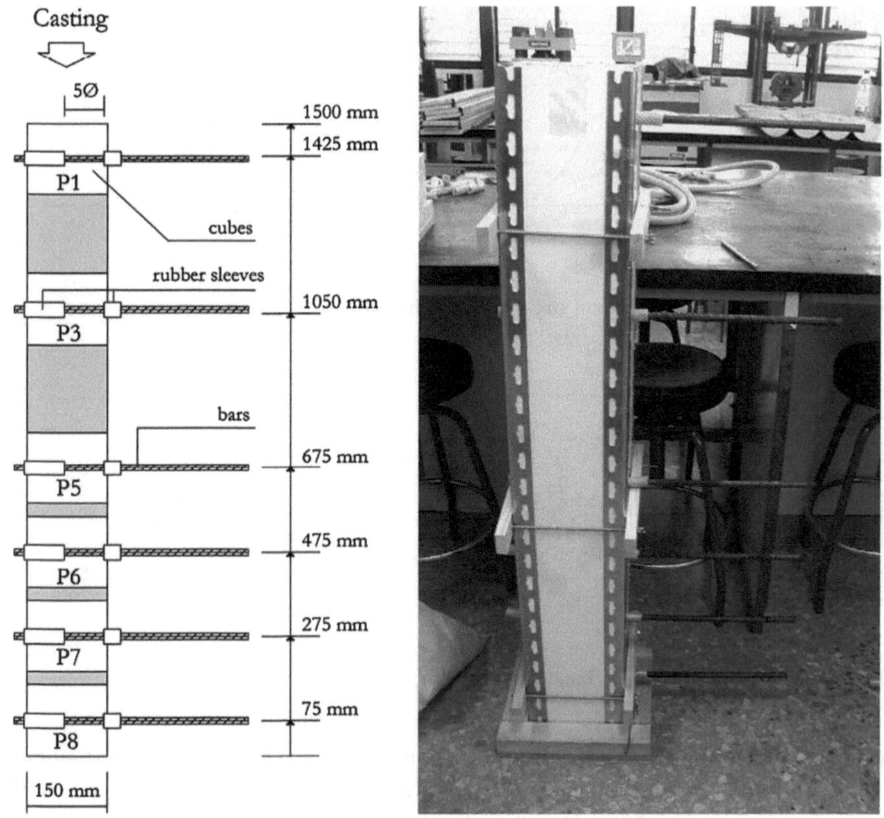

Fig. 4.8 Diagram and photograph of 1,500 mm high specimens [28]

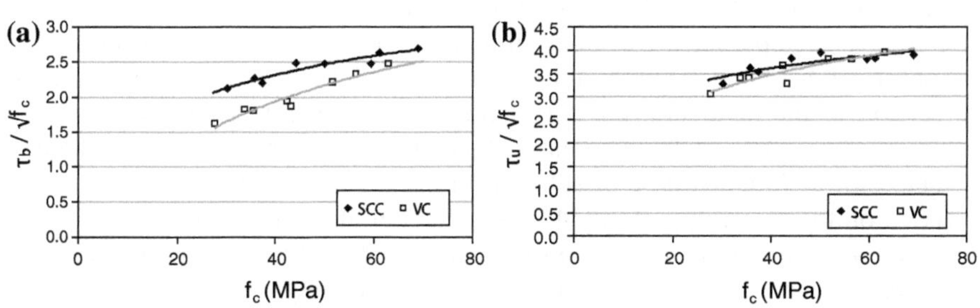

Fig. 4.9 Normalised mean bond stress (**a**) and normalised ultimate bond stress (**b**) [28]

decrease in strength in SCC specimens varies from 40 to 61 % and 79 to 86 % in the case of the VC. The better performance is traced the homogeneity of SCC.

Esfahani et al. [30] analysed the bond properties of reinforcing bars in SCC and VC by conducting pull-out tests. Their test series comprised of two groups of

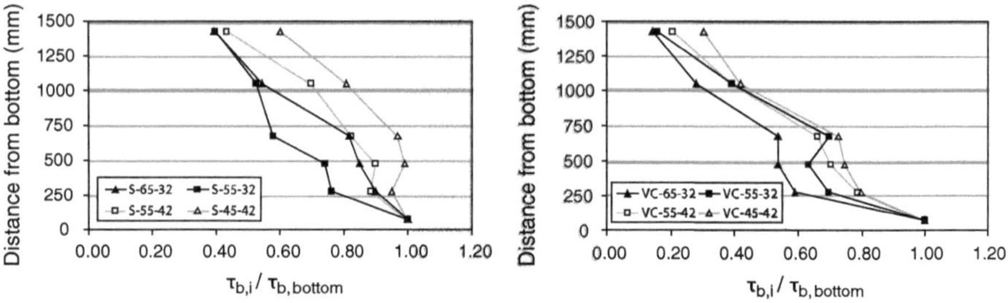

Fig. 4.10 Variation of mean bond stress with height [28]

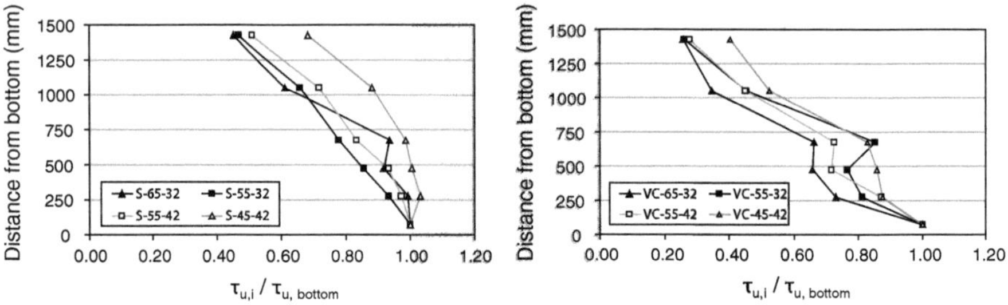

Fig. 4.11 Variation of ultimate bond stress with height [28]

columns with various cover depth to bar diameter ratios. Bars of 25 mm diameter were positioned perpendicular to the casting direction of the 900 mm high columns (Fig. 4.12). The bond length of the bars was equal to the thickness of the wall, being 100 mm. The compressive strengths of the SCC mixtures were 62 and 68 MPa and the compressive strengths of the VC columns were 58 and 61 MPa. All their specimens failed due to splitting of the concrete, and no pull-out was observed. By comparing the average ultimate bond strength of the specimens, it was observed that bond strength in the bottom was higher than that of rebars in the top section of the members (top-bar effect). Bond strength in mid-height of the specimens was reported to be greater than that for the bottom bars, which is attributed to the V-notch type of splitting in middle bars [30].

Based on the research performed by Esfahani et al. [30], Valcuende et al. [31] published a discussion and based on the performed analysis of variations, stated that lower top bar effect in VC compared to SCC is merely a random effect. It was also stated that as the bond behaviour in SCC is stiffer, in the absence of transverse reinforcement, and when the concrete cover of the reinforcing bars is smaller, early failure of bond can occur due to the splitting of the concrete cover, resulting in lower ultimate bond strength while the bond behaviour is stiffer. This means that if the specimen had more confinement caused by concrete cover or stirrups, it could show different fracture types and different results were probable.

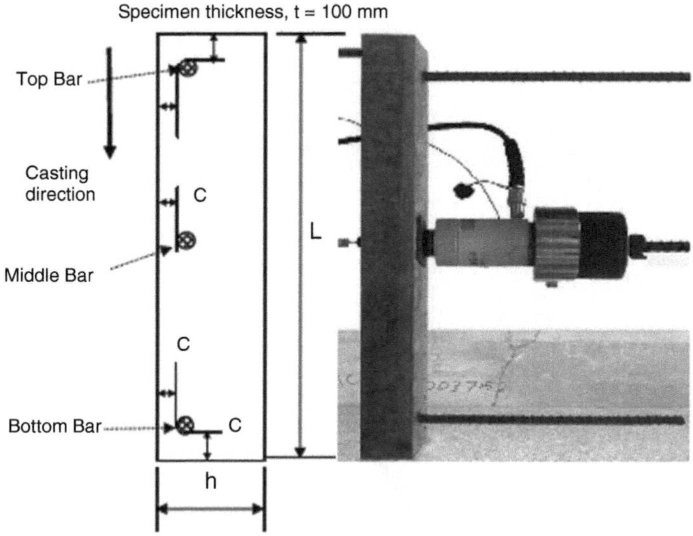

Fig. 4.12 Specimen and test set up [30]

Hossain et al. [32] performed direct pull-out tests on horizontally and vertically cast specimens in order to compare the bond properties of the SCC and VC including the top-bar effect (Fig. 4.13). Based on the ultimate bond strength results, it was observed that in horizontal specimens, the bond strength in the rebars placed in the middle of the specimens was higher than those located in both ends of the member. Although the depth of the horizontal specimens was no more than 200 mm, it was also reported that bars located at the bottom of the horizontally cast members exhibited higher bond comparing with those located in top parts of the specimens. In vertically cast specimens it was also observed that the bond strength decreases as the distance from the bottom increases.

Furthermore, for both the horizontal and vertical members, the top-bar effect was more pronounced in VC specimens. Comparing the overall results, it was also concluded that the normalized bond strength of SCC is higher than that of VC specimens.

Hassan et al. [33] compared the bond strength of SCC and VC based on results obtained from pull-out tests performed on bars embedded in full-scale heavily reinforced concrete sections. They tested the bond stress for bars located at three different heights of 150, 510, and 870 mm from the bottom of the member at different ages of 1, 3, 7, 14, and 28 days. Details of their tested section are shown in Fig. 4.14. The results of the pull-out tests for both the SCC and VC specimens showed pull-out failure of embedded bars without cracks or spalling of concrete cover which is because of the stirrups used for confinement.

Contrary to the results of Forughi et al. [4] in this research, it has been reported that for both the SCC and VC specimens, the bond strength developed very fast at

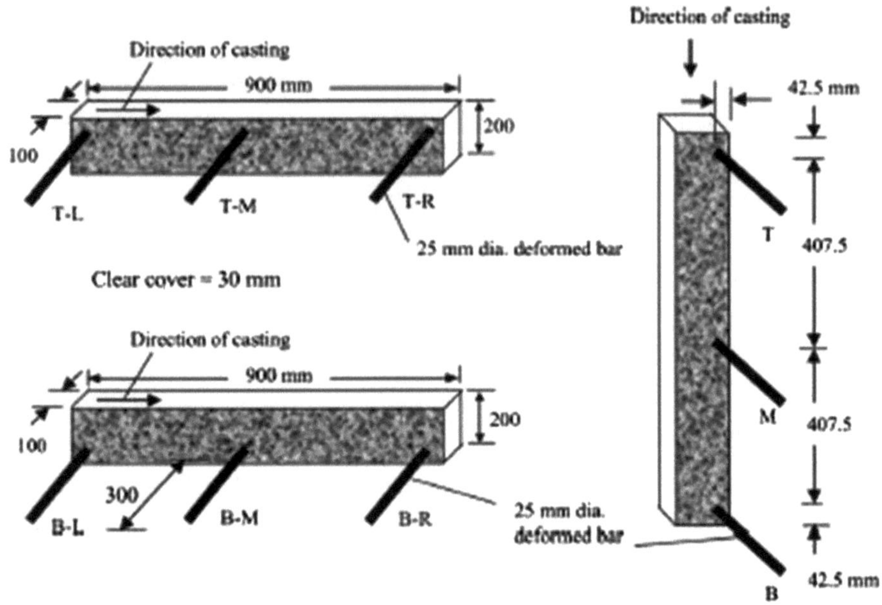

Fig. 4.13 Horizontal and vertical specimen details [32]

early age (up to 7 days) and then stagnated with very slow development up to 28 days, and no significant difference were noted between SCC and VC mixtures in terms of bond stress or compressive strength development with age. However, it should be noted that the normalised bond strength was slightly higher in SCC than that of VC at 3, 7, 14, and 28 days (Fig. 4.15).

From the results, it is observed that the ratio of normalized bond stress of SCC to that of VC was higher in top bars and late tested ages compared to bottom bars and early tested ages. They have also reported that the bond stress-slip relationship in this investigation did not show significant difference between SCC and VC mixtures. The two mixtures provided similar compressive strengths and were designed for high durability (high strength and low W/C), hence the fresh mixtures were stable and therefore the effect of bleeding, segregation and surface settlement that can influence the bond stress was reduced.

For both VC and SCC specimens, the bond stress was slightly higher for the bottom bars than that in the top and middle bars at all ages. For example, the bond strength was 20.5 MPa for the bottom bars compared to 19.8 MPa for the top bars in the SCC specimen at 28 days. Also, the bond strength was 20.6 MPa for the bottom bars compared to 18.9 MPa for the top bars in the VC specimen at 28 days. The ratio of the bond stress of bottom to top bars for 28 days is presented in Fig 4.16. The top-bar ratio was higher than 1.0 at all tested ages. Also, this ratio was slightly lower in the SCC specimen compared to the VC specimen at all ages which indicates that the top bar effect was somehow reduced in SCC compared to

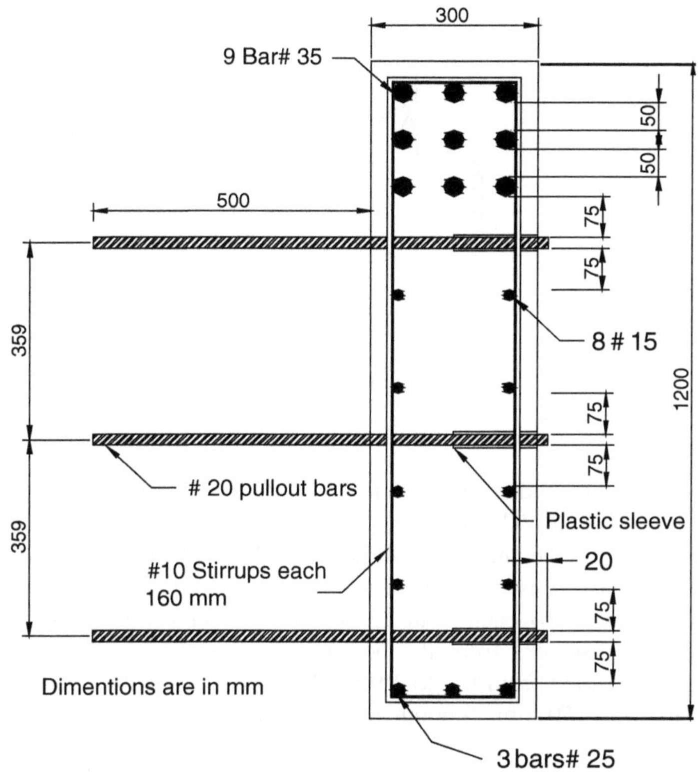

Fig. 4.14 Details of pull-out specimen [33]

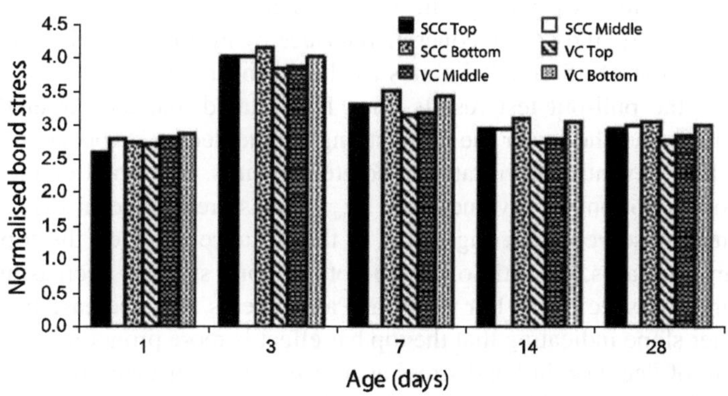

Fig. 4.15 Normalised bond stress with age in both SCC and VC pull-out specimens [33]

VC. This is attributed to the nature of SCC that ensures good filling ability with less bleeding, segregation and surface settlement compared to VC. However, those factors are largely minimized due to the use of high quality mixtures.

Fig. 4.16 Top bar factor at various free end slip for SCC and VC at 28 days [33]

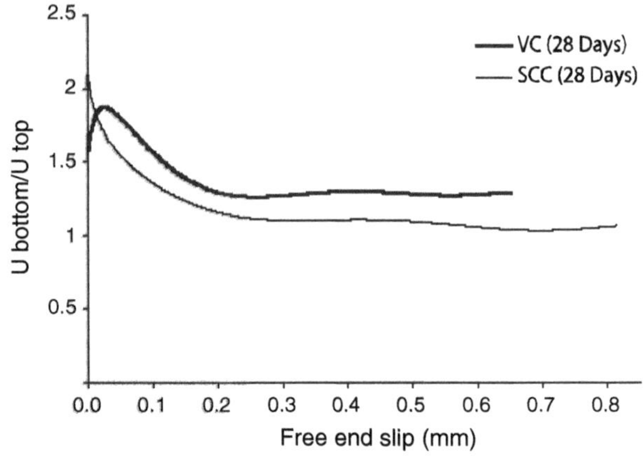

Söylev et al. [34] studied the effect of bar-placement conditions on the bond behaviour in SCC and VC specimens. They used smooth bars because of their higher sensitivity to the interface quality in order to better compare the bond qualities of the used concretes. Five deep panels (measuring 2000 × 200 × 150 mm) each with 13 horizontally embedded reinforcing bars were used. The distances from the bottom of the panels to the centre of the reinforcing bars varied from 0.1 to 1.9 m. The centre-to-centre spacing between adjacent bars was 150 mm. Smooth round steel bars ($\phi = 10$ mm) were used (Fig. 4.17).

Five different mix designs were adopted for the concrete: C20, C40, SCC40, C50, and SCC50. The mixes were named with respect to their compressive strength (20, 40 and 50 MPa). Samples were obtained from the concrete panels, measuring 2,000 mm, by sawing: the 150 mm part was used for pull-out test and the other part to the study steel–concrete interface by means of a video-microscope at a magnification of 25 and 175 times as shown in Fig. 4.18.

Based on the pull-out test results, they have stated that the larger the cover underneath the bar, the lower the bond strength. The decrease appears to be more important for conventional vibrated concrete mixtures. The bond efficiency ratio (ratio of top-to-bottom τ_{max} values) and τ_{max} values are plotted in Figs. 4.19 and 4.20. As it is observed from Fig. 4.19, as the distance between the top and the bottom bars increases, the ratio of top-to-bottom bond stress is decreasing in both the SCC and VC specimens. For the 40 MPa concretes, this decrease takes places with a higher slope indicating that the top bar effect is more pronounced in the case of VC. Rate of decrease in bond stress as a result of the distance from the bottom of the member is again higher in 50 MPa VC comparing with SCC. However, there is an increase in bond strength above the bottom-cast bar for the C50 specimen which is stated to be due to the free fall of the fresh concrete which is more likely to cause segregation because of the high amount of HRWR used without any incorporation of fine materials or VMA.

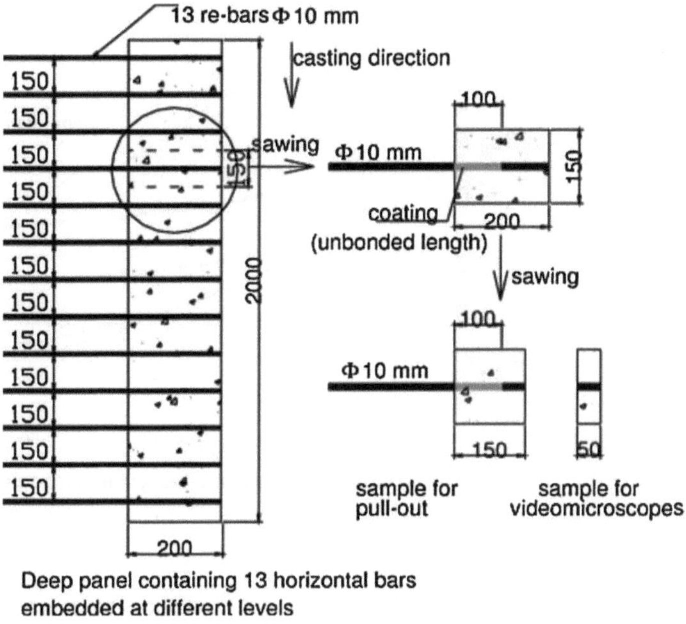

Fig. 4.17 Details of the panels and bar positions [34]

Fig. 4.18 Steel-concrete interface observed at video-microscope with an amplification of 25 times [34]

Another important finding is that for both SCC and VC specimens, the bond efficiency ratio increases as the compressive strength increases, indicating that the top-bar effect decreases as a result of the increase in compressive strength and concrete quality.

Based on the obtained results, the casting position factors (inverse of bond efficiency ratio) of these specimens are compared with the Eurocode [35] and

Fig. 4.19 Bond efficiency as a function of concrete cover underneath the bar [34]

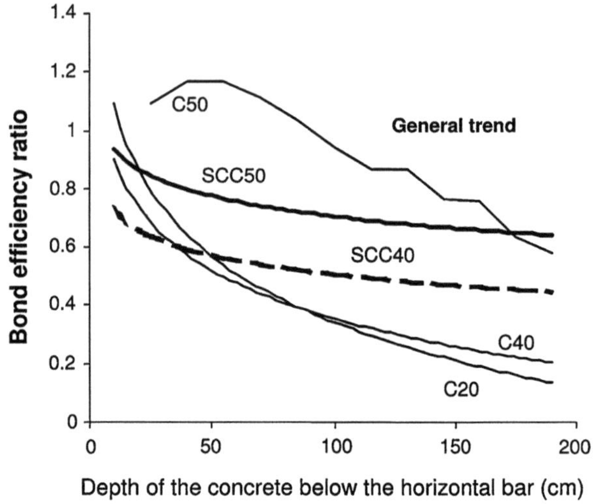

Fig. 4.20 Ultimate bond strength as a function of concrete cover underneath the bar [34]

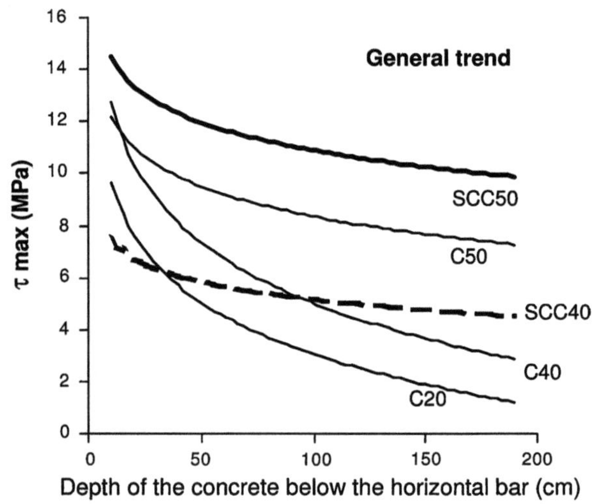

ACI 318 [36] recommendations. The linear function of the C40 has the highest slope and the C50 the lowest. The slope of the linear function of C20 is very close to that of C40 if the two top bars with a complete loss of bond are not considered. The ACI Code and Eurocode require the increase of the development lengths by 30 and 40 %, respectively, for reinforcing bars whenever the bar has at least 300 and 250 mm, respectively, of concrete underneath it. Code requirements are indicated in Fig. 4.21. Based on the video-microscopy analysis, they have also stated that the voids under the horizontal bars increases as the concrete cover (distance to bottom of member) increases [34].

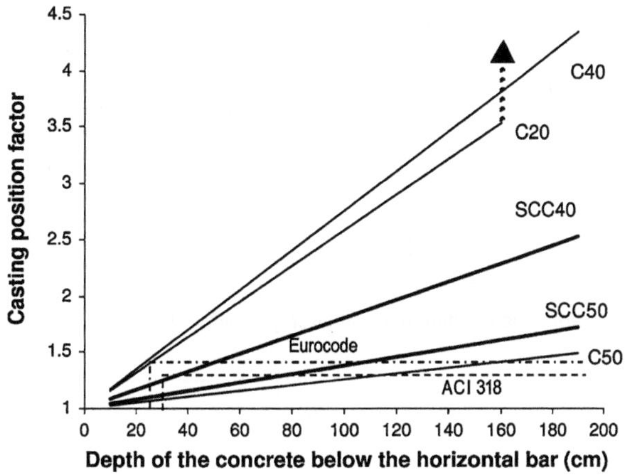

Fig. 4.21 Plots of the casting position factor as a function of concrete cover underneath the bar [34]

Desnerck [37] studied the top-bar effect through investigating the bond strength of bars at different heights in columns and in a wall segment. Specimens made with a VC and two types of SCCs were used. Top-bar testing showed a reduction in the bond strength with increasing height in the elements. The reduction was significantly higher for the VC compared to SCC. It was also reported that the casting position factors for VC were as high as 2.5 in contrasts with 1.0 and 1.5 for the SCC. The change of the casting position factor over the height of the elements cast with SCC was much more gradual in comparison with the VC elements (Fig. 4.22).

Khayat et al. [38] evaluated the uniformity of in situ mechanical properties of SCC used to cast experimental wall elements measuring 1.5 m in height. Eight SCC mixtures with slump flow values greater than 630 mm and a VC with a slump of 165 mm were investigated. The SCC mixtures incorporated various combinations of cementitious materials and chemical admixtures. The W/CM ranged between 0.37 and 0.42. Cores were drilled to evaluate the uniformity of compressive strength and modulus of elasticity along the height of each wall. The bond strength was determined for 12 horizontal reinforcing bars embedded at various heights of each wall. In general, variations in f'_c of cores tested from the top and bottom sections of the experimental walls were limited to 8 %. Slight reductions in in situ modulus of elasticity were observed between core samples drilled near the top and bottom portions of the walls. The maximum reduction in MOE between the top and bottom sections varied between 0 and 8 % for SCC mixtures and was 7 % for the control concrete, indicating uniform mechanical properties of the optimized SCC mixtures. The maximum top-bar factors near the upper sections of the 150-cm high walls for all but one SCC varied between 1.2 and 1.6, compared to approximately 2.0 for the control concrete [30]. Smaller variations in in situ f'_c and bond strength with height were obtained in SCC mixtures containing aggregates with maximum size 10-mm than those with maximum aggregate size 20-mm.

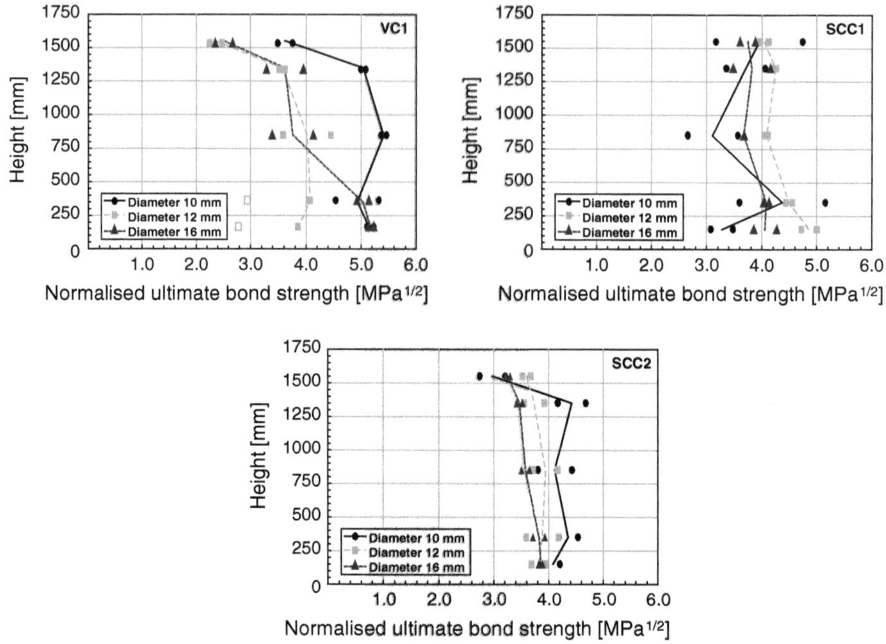

Fig. 4.22 Normalised ultimate bond as a function of height for VC1, SCC1 and SCC2 columns [37]

4.5 Effect of Viscosity Modifying Admixtures (VMA) on Bond Properties

The flowability and viscosity of the mixture influence the settlement of the plastic concrete and resistance to segregation and bleeding. VMAs are water-soluble polymers that increase the viscosity and cohesion of cement-based materials. The incorporation of a VMA can improve the stability of fresh concrete which can reduce the top-bar effect, as shown in Fig. 4.23 [39]. The anchorage length of the reinforcing bars embedded horizontally near the top and bottom of 500–1100 mm high column specimens was either 2.5 or 5 times the bar diameter. Regardless of the height of the cast specimen, the top-bar effect decreased considerably with the incorporation of VMA. As in the case of bleeding, settlement, and segregation, the top-bar factor was smaller in mixtures containing 0.07 % welan gum, by mass of cementitious materials, and no silica fume compared to those made with 0.035 % welan gum and 8 % silica fume.

Highly stable SCC mixtures incorporating proper concentrations of VMA (also referred to as VEA) were found to secure low top-bar factors [39–41]. The top-bar effect of SCC with slump flow values on the order of 650 mm was quite low, ranging between 1.22 and 1.35. These values were comparable to those obtained for rodded concrete with slump of 190 mm (1.25–1.40). As shown in Fig. 4.24, the

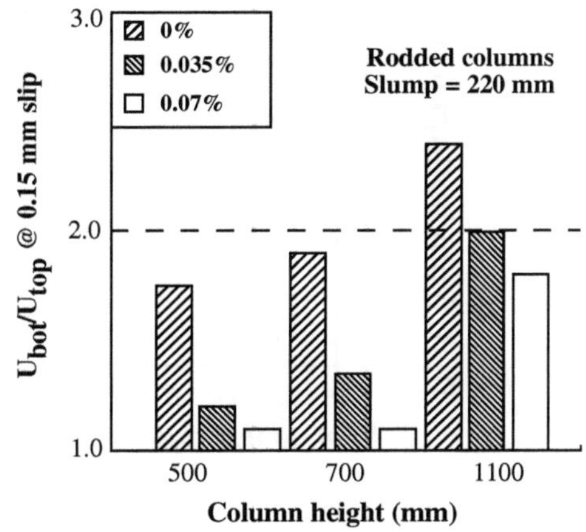

Fig. 4.23 Effect of welan gum dosage and column height on top-bar effect [39]

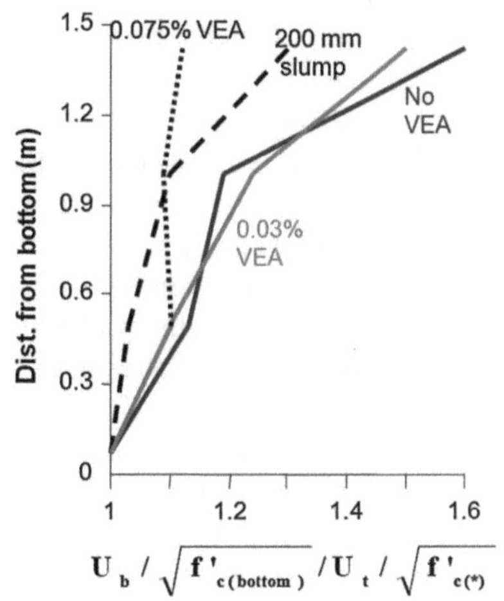

Fig. 4.24 Normalised bond strength in the height of the specimens [41]

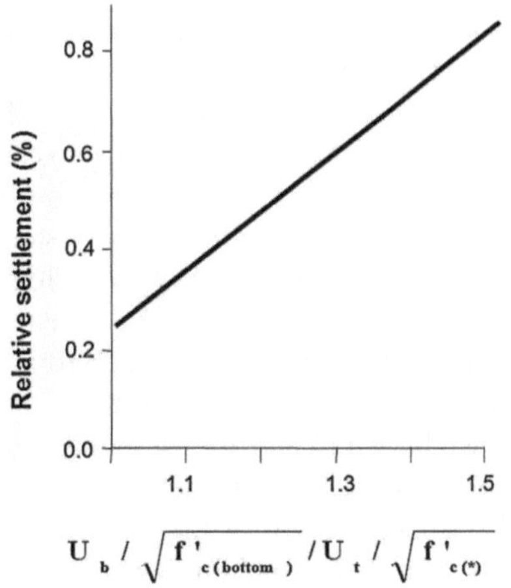

Fig. 4.25 Surface settlement as a function of *top-bar* ratio [41]

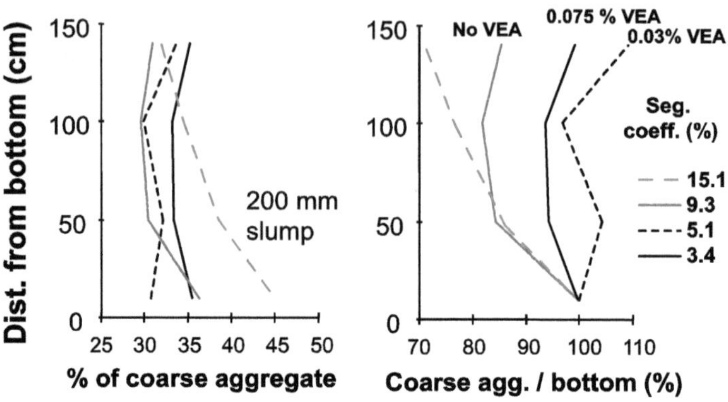

Fig. 4.26 Coarse aggregate distribution in the height of the specimens [41]

SCC mixture with 0.075 % of VMA had the highest uniformity in the bond strength. A relationship was established between surface settlement and the top-bar ratio normalised by in situ f_c values at different heights, as shown in Fig. 4.25 [41].

The segregation resistance of the SCC is shown in Fig. 4.26 in terms of changes in coarse aggregate content along the height of the cast wall elements. The increase in the welan gum VMA content of SCC with a given slump flow is shown to enhance the homogeneity of coarse aggregate distribution, which is consistent with the surface settlement results and the top-bar effect.

It is observed that the surface settlement of SCC of a constant slump flow is dependent on the plastic viscosity [42]. The viscosity was changed by changes in W/CM, as illustrated in Fig. 4.27. The surface settlement in this study was measured according to [43].

Khayat et al. [44] studied the bond strength of reinforcing bars (20 mm) and prestressing strands (12.5 m) along wall elements cast measuring 2,150 mm in length 1,540 mm in height and 200 mm in width. Flowable vibrated and SCC mixtures were tested with corresponding slump and slump flow of 210–230 mm and 650–700 mm, respectively. Table 4.2, summarizes the mixture properties of the specimens. The mixtures had a targeted 1-day compressive strength of 40 MPa and were proportioned with Type III cement with 20 % Class F fly ash substitution and a W/CM of 0.37. SCC mixtures were proportioned with a cellulose-based VMA (VMA-1) and a synthetic-based VMA (VMA-2). Polycarboxylate-based HRWR that can promote early strength gain for precast applications was used for the two superplasticiser types. Both air-curing and steam-curing were used to cast the six experimental wall elements. In total, 16 prestressing strands and 16 reinforcing bars grouped in four rows and positioned at four levels along the walls were tested for pull-out tests on the reinforcing bars at the ages of 1 and 28 days. Two bars for each age were pulled out in each wall. Cores samples were also taken at various heights to determine in situ compressive strengths.

Bond strength of the prestressing strands at the age of 1 day corresponding to 1.0 mm free end slip were calculated and used to prepare the top-bar effect

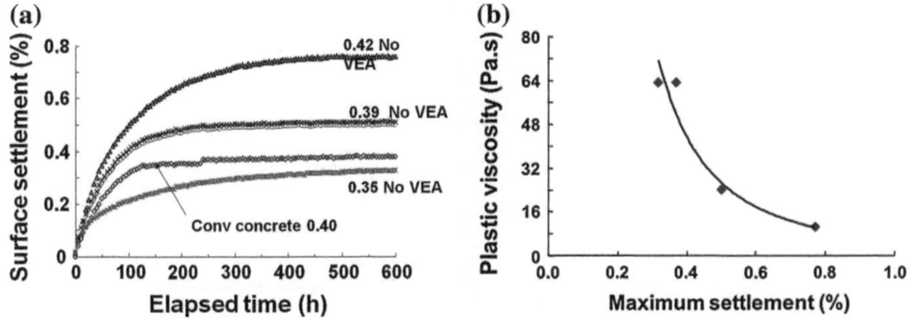

Fig. 4.27 Variation in surface settlement of the tested SCCs as function of VMA dosage (**a**) and relationship between maximum settlement and plastic viscosity (**b**): all SCC had constant slump flow values of approximately 650 mm [42]

Table 4.2 Mixture proportions of conventional flowable concrete and SCC mixtures [44]

Mixture	VC		SCC			
	1R-Air	2-Steam	3-Steam	4-Air	5-Air	6-Air
Type III cement (kg/m^3)	377	371	363	380	382	386
Class F Fly Ash (kg/m^3)	94	93	91	94	94.5	96
5-14 mm agg. (kg/m^3)	972	957	743	776	780	789
Sand (kg/m^3)	754	742	909	949	966	966
HRWR-1 (l/m^3)	1.581	2.167	3.078	–	–	–
HRWR-2 (l/m^3)	–	–	–	2.825	3.371	2.695
VMA-1 (l/m^3)	–	–	2.066	2.157	–	–
VMA-2 (l/m^3)	–	–	–	–	1.865	–
W/CM	0.37	0.37	0.37	0.37	0.38	0.37

diagram shown in Fig. 4.28. The variations of normalized top-bar ratios as a function of height at the age of 28 days are included in Fig. 4.29.

The maximum top-bar effect values for reinforcing bars in vibrated flowable concrete and SCC mixtures at 1 day were 1.28 and 1.45, respectively. The maximum top-bar effect ratios at 28 days for reinforcing bars in Mixtures 2 and 3 were 1.40 and 1.59, respectively (Fig. 4.29). The slightly better top-bar value for Mixture 4 reflects the improved consolidation and better stability obtained with VMA-1. In general, mixtures produced with the cellulose-based VMA (VMA-1) had considerably lower top-bar effect than those produced with the synthetic-based VMA (VMA-2); both mixtures were proportioned with the same HRWR.

The top-bar ratios obtained for the reinforcing bars at 28 days are compared in Fig. 4.30 to design recommendations offered by Jirsa and Breen [45]. Such recommendations were given for the variation of the top-bar factor with depth of concrete below spliced or anchored reinforcing bars for mixtures of various slump consistency levels. The results show that the two tested highly flowable concrete mixtures lie well within the design recommendations offered for concrete with

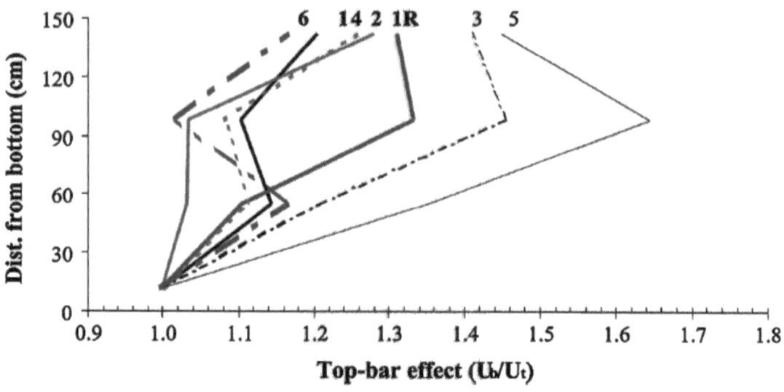

Fig. 4.28 Variations in top-bar ratio for reinforcing bars with height at 1 day [44]

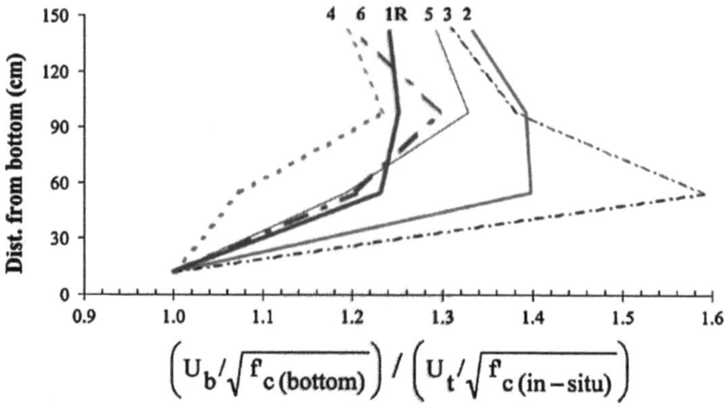

Fig. 4.29 Normalised top-bar ratio for reinforcing bars with height at 28 days [44]

slump consistency of 50 and 100 mm. For the majority of cases, the four SCC concrete mixtures can develop top-bar factors similar to those of concrete with slump of 100–150 mm cast with internal vibration.

Similar results were reported for the top-bar effect of concrete with prestressing strands, as shown in Fig. 4.31. The 1- and 28-day top-bar effect of prestressing strands was shown to vary between 0.9 and 1.9. The top-bar effect was again shown to be sensitive to the type of VMA. The top-bar effect was lower for the air-cured mixtures compared to the steam-cured concrete.

Long et al. [48] studied the effect of plastic viscosity and static stability on the bond behaviour of prestressing strands in various SCC mixtures. High Performance Concrete (HPC) of moderate plastic viscosity and five SCC mixtures of various viscosity levels ranging from approximately 10 to 150 Pa.s were used. This was done to evaluate the influence of viscosity on stability and bond strength. As shown in Table 4.3, the surface settlement values of the SCC mixtures ranged from 0.30 to 0.62 %. The maximum surface settlement of the HPC was 0.29 %.

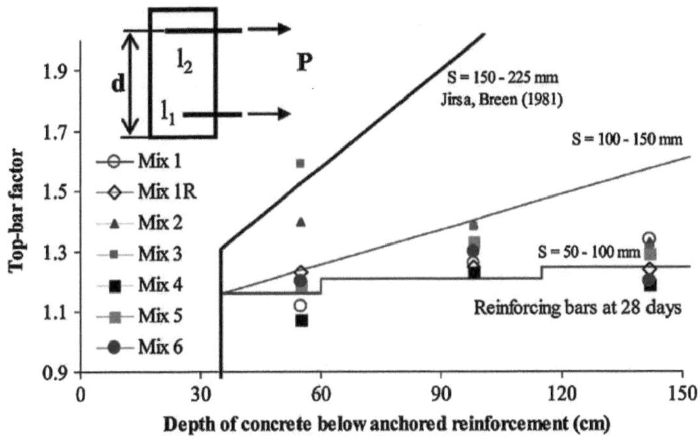

Fig. 4.30 Comparison of top-bar ratio of reinforcing bars at 28 days with the recommendations [45]

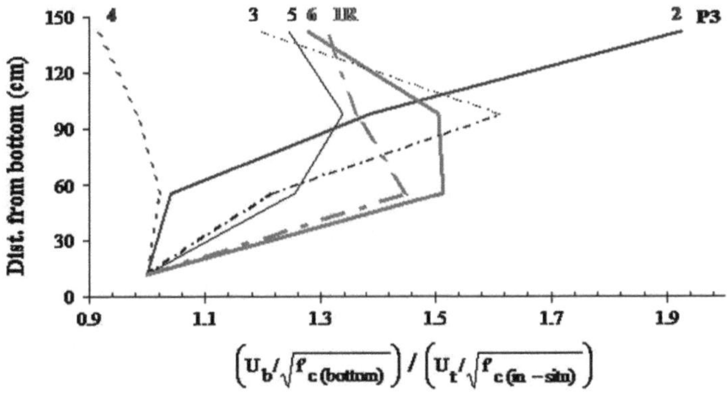

Fig. 4.31 Normalised *top-bar* ratio of prestressing strands versus height at 28 days [46]

Testing consisted of determining the maximum pull-out load versus end slip of strands horizontally embedded in the experimental wall elements. In total, 16 prestressing strands of 15.2 mm diameter were embedded at four heights in the wall elements. The load at free end slip of 1.0 mm was taken to calculate bond strength in the post elastic cracked region. Three cores of 95 mm diameter were also taken at heights corresponding to these of the embedded strands to calculate the variation of compressive strength at the height of specimens. As shown in Fig. 4.32, walls No. 4, 5, and 6, cast with relatively low surface settlement concrete, exhibited more homogeneous in place compressive strength compared to walls No. 1, 2, and 3 at 56 days.

Bond strengths of prestressing strands determined at 56 days are presented in Fig. 4.33. Walls No. 4 and 5 cast with stable SCC mixtures exhibited more

Table 4.3 Mixture proportioning and fresh properties of tested concrete [47]

Wall #	1	2	3	4	5	6
Mixture	SCC1	SCC2	SCC3	SCC4	SCC5	HPC
W/CM	0.34	0.40	0.40	0.34	0.34	0.34
Yield stress (Pa)	5	30	35	25	10	575
Plastic viscosity (Pa.s)	148	11	34	41	76	92
Slump flow (mm)	665	695	670	660	660	145
t_{50} (s) (ASTM C1611)	6.8	1.5	1.8	2.9	5.5	–
VSI (ASTM C1611)	0.5	1	1	0.5	0.5	0
Air content (%)	2.4	2.3	1.9	2.0	1.1	2.7
Filling capacity (%)	92	93	89	91	82	–
Max. surf settlement (%)	0.44	0.59	0.62	0.43	0.30	0.29

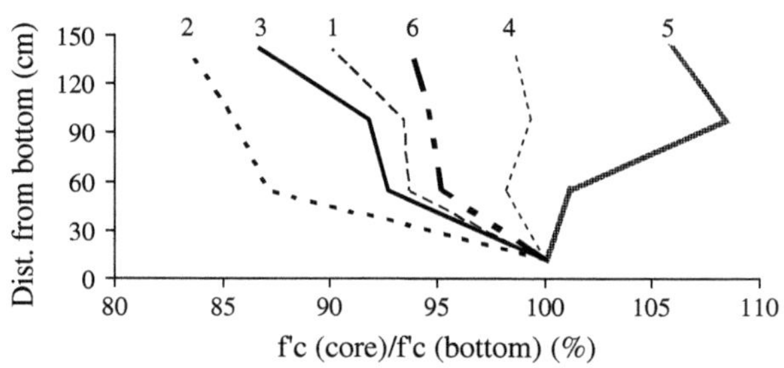

Fig. 4.32 Variations of in-place compressive strength with height [47]

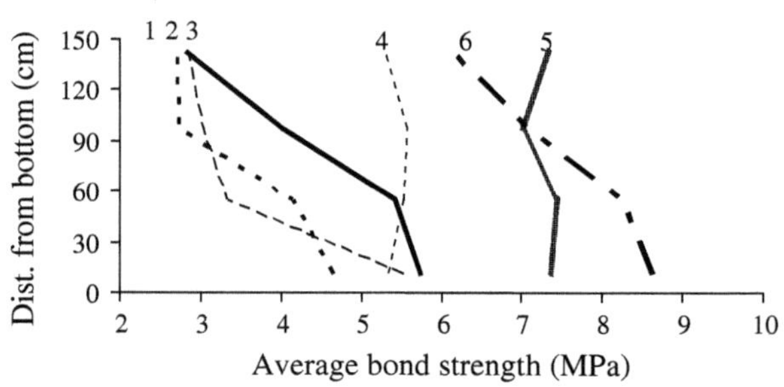

Fig. 4.33 Variation in bond strength of prestressing strands along wall height [47]

homogenous pull-out bond strengths compared to walls cast with unstable SCC. Walls No. 4 and 5 exhibited better homogeneity in terms of bond than observed for Wall No. 6 cast with HPC. It should be noted that Wall No. 1 exhibited relatively large variation in pull-out bond strength along the height. This can be attributed to the high plastic viscosity of the concrete (approximately 150 Pa.s), which seems to hinder self-consolidation. Among the six tested wall elements, Wall No. 4 had the highest degree of homogeneity of in situ bond strength. This SCC mixture developed relatively moderate yield stress (25 Pa) and plastic viscosity (40 Pa.s) and a surface settlement of 0.43 %.

Variations of the normalised modification factor of bond strength between the concrete and prestressing strands are illustrated in Fig. 4.34. In general, Walls No.

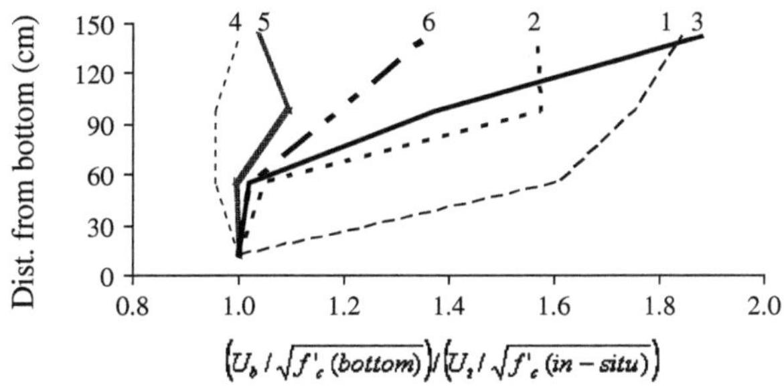

Fig. 4.34 Variation of modification factor of prestressing strands along wall height [47]

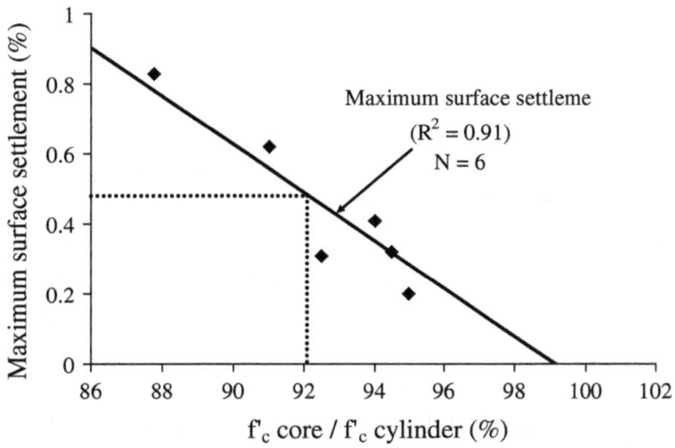

Fig. 4.35 Relationship between mean relative in-place compressive strength and maximum surface settlement (PVC settlement test) [47]

4, 5, and 6 casted with stable SCC and HPC exhibited lower modification factors of 1.00, 1.00, and 1.36, respectively, compared to 1.57 and 1.88 for Walls No. 2 and 3 casted with unstable mixtures, respectively. Wall No. 1 cast with SCC No. 4 exhibited relatively large spread in modification factor values along the height which is attributed to the lack of consolidation leading to undesirable bond between concrete and prestressing strand. It should be noted that Walls No. 4 and 5 exhibited lower modification factors than Wall No. 6 made with HPC mixture. Wall No. 6 had higher modification factor of 1.36 compared to 1.0–1.1 for Walls No. 4 and 5 cast with stable SCC.

Again, the surface settlement, affected by the viscosity of the SCC, is shown to have considerable influence on in-place compressive strength relative to the reference cylinders. As presented in Fig. 4.35, the mean relative in-place compressive strength increased with the decrease in maximum settlement ($R^2 = 0.91$). It is concluded that concrete with maximum settlement of 0.5 % can develop a relative in-place compressive strength (core/cylinder) higher than 0.92. The study indicates that although the SCC mixtures had similar workability levels in terms of slump flow consistencies, caisson filling capacity, and visual stability index, different levels of uniformity of in situ compressive strength and pull-out bond strength were obtained. Highly stable SCC was shown to secure more homogeneous in situ properties than HPC of normal consistency subjected to mechanical vibration [48].

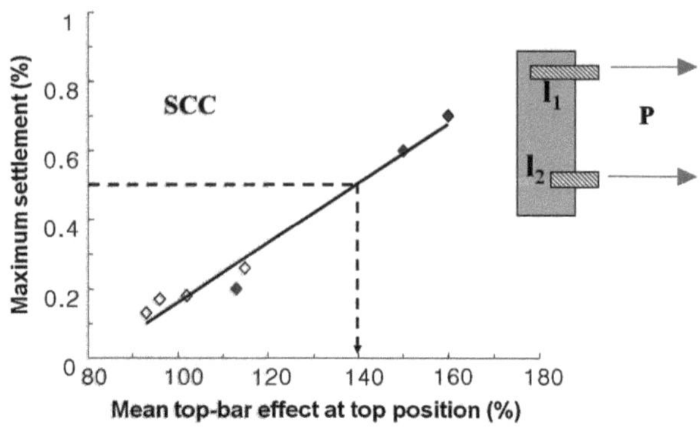

Fig. 4.36 Relation between surface settlement and *top-bar* effect [47]

Based on the results of Fig. 4.35, a maximum surface settlement of 0.5 % is recommended for SCC to ensure homogenous in situ properties. This value corresponds to a mean core-to-cylinder compressive strength higher than 90 % and a top-bar factor lower than or equal to 1.4 (Fig. 4.36). Such relationship demonstrates that regardless of the fluidity and composition of concrete and the height of concrete cast under the upper reinforcing bar, the top-bar factor is significantly affected by surface settlement. Such settlement depends on the extent of bleeding and segregation and can be effectively reduced by the incorporation of a VMA, even in the case of SCC.

Khayat and Mitchell recommended in the NCHRP 628 report [49] avoiding the use of highly viscous SCC with a plastic viscosity greater than 80 Pa.s, a t_{50} greater than 6 s should be avoided. This is recommended to prevent entrapment of air voids which can have an adverse effect on bond strength and in situ mechanical properties [49]. The authors recommend that SCC targeted for precast, prestressed applications should have maximum surface settlement, column segregation index, and percentage of static segregation of 0.5, 5, and 15 %, respectively, particularity for deep elements. Such concrete is expected to develop at least 90 % in situ relative compressive strength (core results) and has a modification factor (top-bar effect with prestressing strands) of 1.4. The rate of surface settlement at early-age of testing can be related to the maximum settlement. A rate of settlement of 0.16 % per hour determined after 30 min of testing is shown to correspond to a maximum surface settlement of 0.5 % [50].

4.6 Effect of SCM and Limestone Filler on Bond of SCC

Karatas et al. [51] compared the bond strength of tension lap-spliced bars embedded in VC and SCC. Four types of SCC with silica fume (SF) replacements levels of 5, 10, 15, and 20 % were used. In total, 15 full-scale beam specimens measuring 2,000 mm in length, 300 mm in height and 200 mm in width were tested. The beams had 20 mm reinforcing bars with a 300-mm splice length as tension reinforcement at mid-span. The authors found that the bond strength of the reinforcement embedded in SCC beams was higher than that of VC beams. The bond strength increased with the increase in SF replacement. Beam specimens produced from SCC containing 5 % SF had the highest normalized bond strength of 1.07 followed by SCC beams with 10 % SF, 15 % SF, VC beams, and 20 % SF.

Turk et al. [52] investigated the effect of using different types and contents of SCMs on bond strength of lap-spliced bars in SCC. Nine different types of concretes were adopted: VC with low slump of 70 mm and eight types of SCC with 25, 30, 35 and 40 % of Class F fly ash (FA), and 5, 10, 15 and 20 % of SF. The replacements of cement by an equal mass of FA or SF in SCC generally had a positive effect on the bond strength of reinforcing bar regardless of the content of SCM. Beam specimens made with SCC containing 5 % SF and 30 % FA had the highest normalized bond strength of 1.07. This was attributed to the superior filling capability of the SCC compared to the VC enabling more effectively coverage around the reinforcements and particle packing reducing the wall effect. Moreover, the beam specimens produced from SCC with SF had the greatest stiffness compared to the other beams. The bond tests were carried out using pull-out specimens, De Almeida et al. [53] obtained similar bond strengths with SCC and VC. Daoud et al. [54] obtained 5 % higher strengths with SCC.

Siad et al. (2009) [55] studied 12 types of concrete mixtures with compressive strength classes of 30, 50, and 70 MPa and three types of mineral admixtures (natural pozzolan, fly ash, and limestone filler). The concrete mixtures had a water-to-powder ratio (W/P) of 0.40, 0.52, and 0.70 respectively. The first set of mixtures contained 450 kg/m^3 of Portland cement and 70 kg/m^3 of natural pozzolan, fly ash, or limestone filler. The second and third sets of SCC mixtures contained 350 and 260 kg/m^3 of Portland cement, respectively, and 170 and 260 kg/m^3 of natural pozzolan, fly ash, or limestone filler, respectively. The three reference mixtures did not contain any mineral admixture. The authors investigated bond behaviour by conducting pull-out tests on 20 mm reinforcing bars at 28 and 90 days and 1 year. As expected, the ultimate bond strength increased with the compressive strength class and the time elapsed before the test, as illustrated in Fig. 4.37. The bond strength of 30 MPa specimens containing limestone filler at 28 days was higher than that obtained with the other specimens. The bond strength reported for the reference concrete was the largest followed by the specimen containing limestone filler at the age of 90 days for the 50 MPa concrete class.

4.7 Code Provisions for and Modelling of Bond to Reinforcement

In the literature, a lot of models to predict the ultimate bond strength, corresponding slip and equations to describe the bond stress-slip behaviour of reinforcing bars in VC can be found. In these models several parameters, such as bar

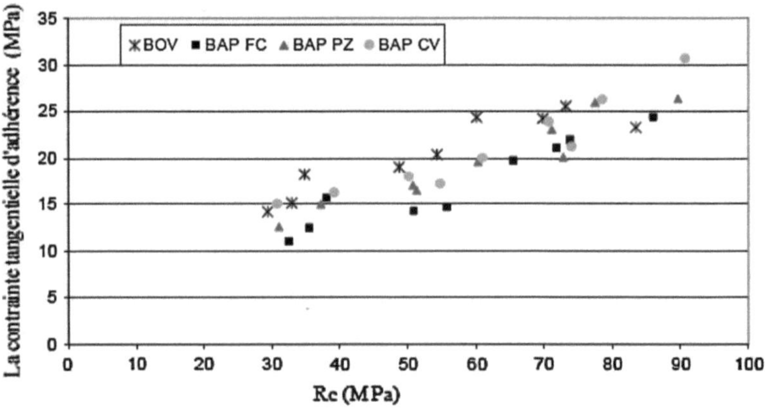

Fig. 4.37 Increase in the bond strength as a result of increase in the compressive strength [55]

diameter, concrete cover, and concrete compressive strength are incorporated. Codes mostly provide only formulas to determine design bond strengths or anchorage lengths and do not allow for the determination bond stress–slip relationships. Recently, studies were performed to verify the applicability of existing code provisions and models to SCC.

Desnerck et al. [37] used the results from their study to verify the applicability of the bond stress–slip relationship provided by CEB fib [54] (which is based on the Eligehausen model). A comparison between the model and the obtained experimental results revealed a rather poor agreement. Therefore, a new proposal was made based on the model of Eligehausen but with modified formulas to predict the ultimate bond strength and the corresponding slip as the concrete cover (c) to the bar diameter (d_b) ratio, and the clear rib spacing (c_O) of the bars showed to be of major importance. The equation to calculate the bond strength and ultimate slip for SCC were reported to be:

$$\tau_R = \left[1.762 + 0.514\frac{c}{\phi}\right].\sqrt{f_{cm}} \tag{4.1}$$

$$s_1 = 0.0032.c_O^2 + 0.041 \tag{4.2}$$

The validity of the model (determined for SCC containing limestone powder) was checked with additional tests results (not used in the determination of model). The agreement turned out to be good.

Based on pull-out tests on flat stainless steel bars, Ertzibengoa et al. [56] concluded that SCC allows for developing higher bond strength values compared to VC. However, the influence was more pronounced for round bars than for flat elements. When test results were compared to the bond model described in the CEB-FIP Model Code 1990, an adaptation of the existing model (modification of the proposed slip at ultimate bond strength value) was proposed for the analysed flat ribbed samples which allowed for an acceptably accurate prediction of their bond behaviour.

A study by Aslani and Nejadi [57] presented a bond strength model based on the experimental results from eight recent investigations on SCC and VC concrete specimens. The comparisons were based upon models of structural sections using pull-out tests to measure bond between the steel reinforcing bar and concrete. Further tests, to assess the top-bar effect on single bars in small prismatic specimens, were conducted using beam tests. The main variables were the steel bar diameter, concrete compressive strength, concrete type, curing age of the concrete, and height of the embedded bar along the formwork. It was found that the SCC had slightly higher bond strengths, but the existing code provisions were valid for SCC and VC. The proposed model showed similarities with the model proposed by Desnerck et al. but included the influence of the bond length of the rebar (l_d). A good agreement was obtained as shown in Fig. 4.38.

$$\tau_R = \left[0.672\left(\frac{c}{d_b}\right)^{0.6} + 4.8\frac{d_b}{l_d}\right] \cdot \left(f_c^{'}\right)^{0.55} \tag{4.3}$$

The ACI Code 318 and Eurocode (EC2) require the increase of the development length by 30 and 40 %, respectively, for reinforcing bars whenever the bar has at least 300 and 250 mm, respectively, of concrete underneath it. These values turned out to be adequate for the SCC specimen tested by Desnerck et al. in their study [37]. However, for the tested VC columns (height 1.70 m, width 0.45 m, depth 0.20 m, and with bars embedded at five different heights), the reduction in bond strength of the top bars with 1550 mm of concrete underneath compared to the bottom bars turned out to be much larger, as can be seen in Fig. 4.39 (in which CVC1 is VC). It is concluded that SCC that is properly designed to exhibit high stability can exhibit top-bar factors within recommended code provisions.

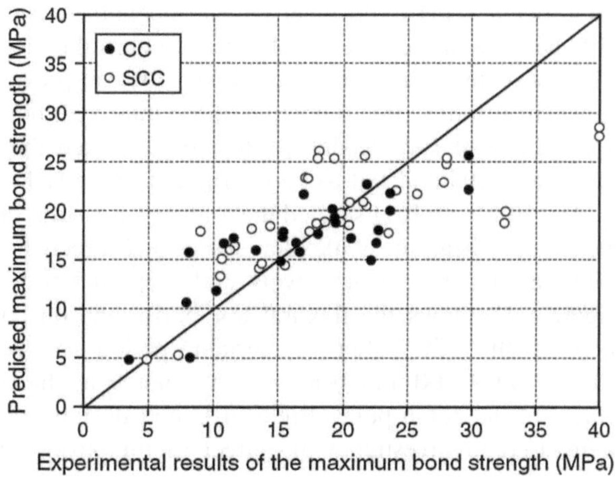

Fig. 4.38 Comparison of experimental results for SCC and VC with the predicted value based on the Aslani et al. model [57]

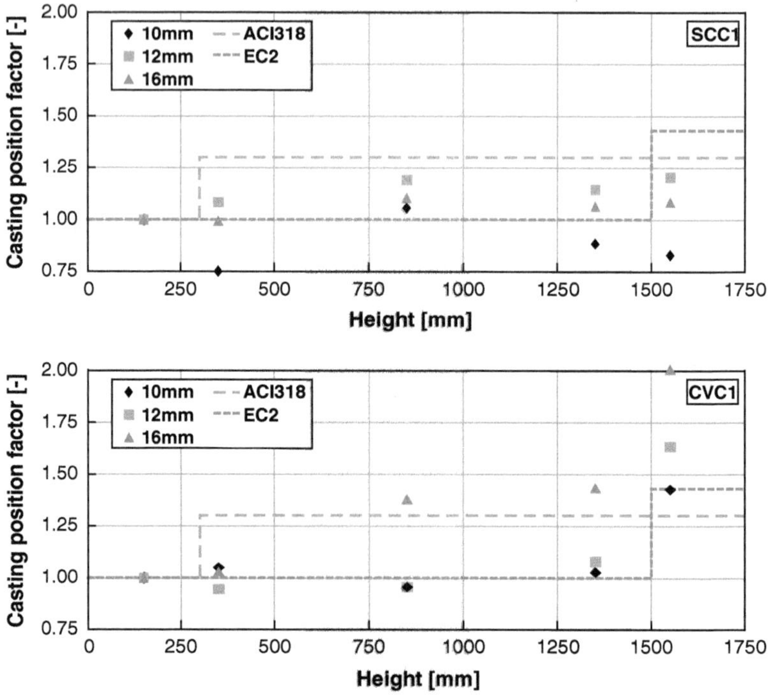

Fig. 4.39 Comparison of results of casting position factors with code provisions from ACI 318 and EC2 [37]

4.8 Multi-Layer Casting of SCC

During concrete placement, if a layer undergoes a high structural build-up at rest due to the thixotropic behaviour prior to the casting of a successive layer, a lift line could occur. This multi-layer casting phenomenon is problematic in the case of SCC because of the absence of vibration consolidation during casting. As a result, a high structural build-up of the lower lift can prevent proper intermixing of the upper lift resulting in a reduction of the interlayer bond strength (Fig. 4.40). Fold-lines represent surface defects that can weaken the bond strength between the successive lifts that fail to be intermixed properly [57–61].

Abd El Megdi [61] studied the multilayer casting of the SCC mixtures based on the slanted shear stress, flexural stress, and direct shear tests with specimens cast with various thixotropic levels and elapsed periods between successive lifts. Figure 4.41 shows the flexural test performed on composite beams cast in two steps with SCC after a certain period of rest. The specimens were 100 × 100 × 400 mm beams with a notch at the mid-span which ensures the failure at the notch point. Eight SCC mix designs with various thixotropic levels were used to investigate the effect of thixotropy on multilayer casting of the

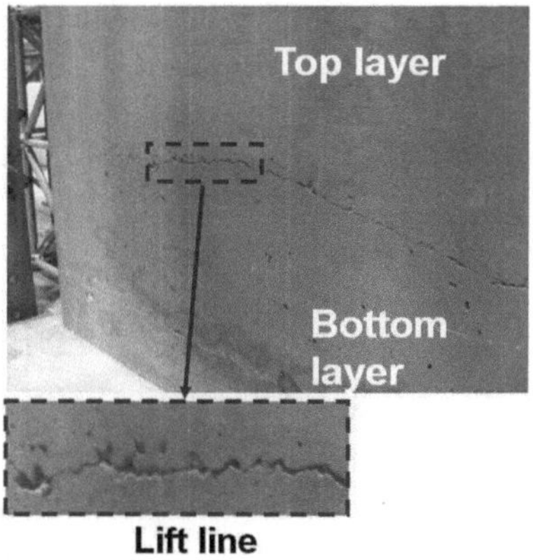

Fig. 4.40 Presence of lift lines caused by multi-layer casting of SCC resulting from delay of casting of the *top layer* of SCC [courtesy of K. H. Khayat]

concrete elements. For each type of SCC mixture, the mould was filled up to the notch point at the centre of the span and after certain delay times the other half of the mould was filled with the same SCC mixture. Delay times of 15, 30, 45, and 60 min were selected in order to investigate the effect of delay time between casting the two parts on the bond between the two layers. For each type of the SCC mixture, three reference specimens were cast in one lift of concreting (no delay time) as well as three specimens made for each delay time.

The residual bond strength between the two layers of concrete for each delay time was calculated by dividing the flexural strength of a specimen of the same delay time to the flexural strength of the reference specimen with no delay time in casting. Figure 4.42 shows that the thixotropy of the SCC can negatively affect bond between the different lifts of SCC. It was also observed that for a certain SCC mixture of known thixotropy, the residual bond strength decreases with increase in delay time in casting two distinct layers. Thixotropy was evaluated using the structural build-up at rest of the yield stress using the portable vane test [62] or inclined plane test [63]. The results indicated that as the static yield stress increases (as a function of increase in rest time for a given SCC), the residual bond strength between the two layers decreases. As shown in Fig. 4.42, the residual bond strength decreases sharply as a result of the increase in static yield stress of the bottom layer at the time of casting of the top layer.

The author [61] proposed the following model for predicting the residual bond strength in flexure as a function of thixotropy and delay time between casting the two layers:

$$RB_f(\%) = -0.0004\,\mathrm{DT}\,A_{thix1} - 0.2816\,\mathrm{DT} + 100 \qquad (4.4)$$

where:

RB_f (%) = Residual bond strength under flexure stress; DT = Actual delay time between layers in minute; A_{thix1} = Thixotropy index, in Pa determined using the portable vane method after 15 min of rest.

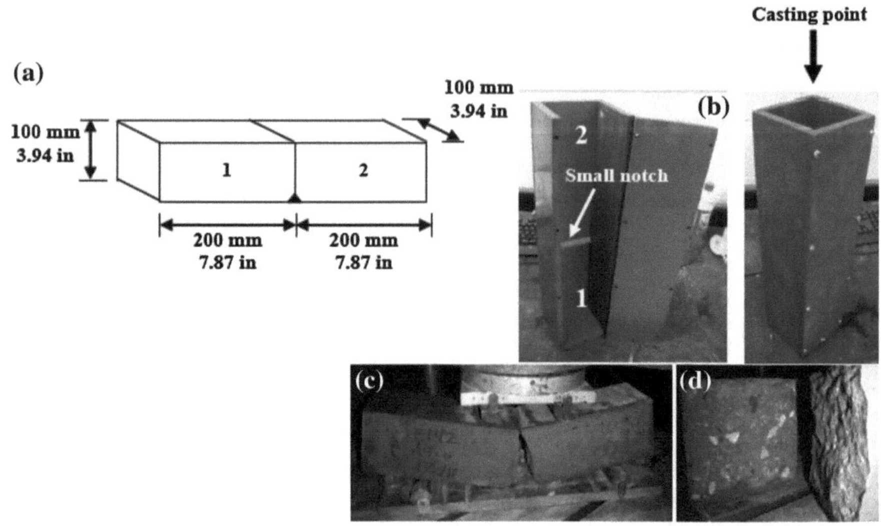

Fig. 4.41 Casting and testing flexural beams with distinct casting *fold line* [61]

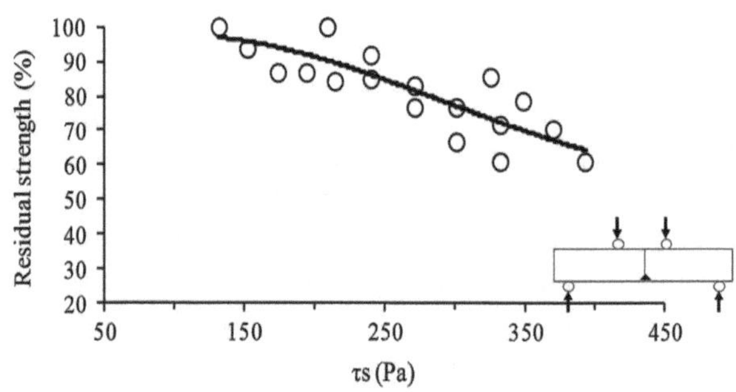

Fig. 4.42 Decreaseour in residual bond strength as a result of structural build up [61]

Figure 4.43 presents contour diagrams of residual flexure strength of the composite beams as a function of structural build-up of the concrete and waiting time between successive layers. Based on such statistical model, Abd El Megdi [61] proposed the following model for estimating the critical time delay between successive castings that can result in 10 % reduction in residual bond strength in flexure:

$$t_c = (0.38\, RB_f(\%\,) - 38.46)\ln A_{thix2} - 5.6\, RB_f(\%\,) + 560.32 \tag{4.5}$$

where:

t_c = critical delay time between layers in minute; RB_f (%) = Residual bond strength under flexure stress; A_{thix2} = thixotropy index in Pa.Pa/min, which is the static yield stress after 15 min of rest multiplied by the rate of gain in static yield stress in the first 60 min.

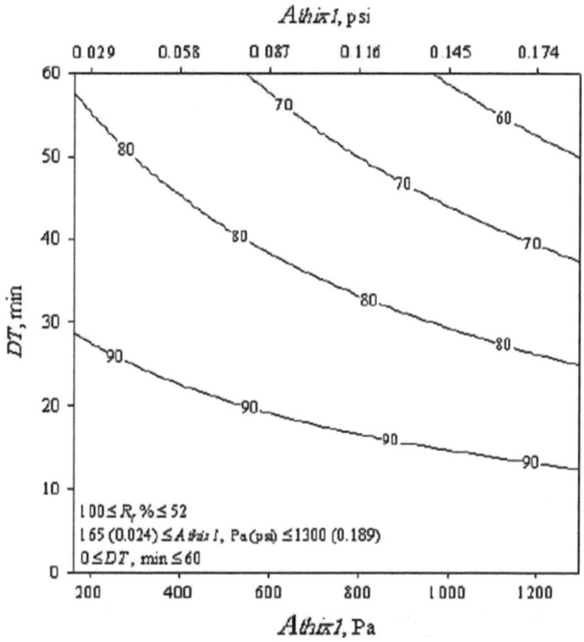

Fig. 4.43 Model of residual flexure resistance as a function of thixotropy of concrete and waiting time to cast successive layers [61]

Based on the evaluation of three mechanical strength characteristics and water permeability testing, Abd El Megdi [61] found out that the direct shear stress tests (Fig. 4.44) showed the sharpest decrease in critical rest time required to maintain 90 % of residual strength compared to monolithically cast specimens. The sharpest drop in properties due to the presence of the lift line testing was obtained in the case of the water permeability test. Thixotropy was determined using the portable vane test with the thixotropic index taken as the product of the static yield stress at rest and the rate of increase in yield stress at early age (usually within 60 min).

Roussel and Cussigh [56] also investigated the effect of structural build-up at rest of SCC on bond between successive lifts. The structural build-up at rest is in part reversible; i.e., due to thixotropy, and may be in part non-reversible, i.e., due to cement hydration. The authors prepared four SCC mixtures of different thixotropic levels. As shown in Fig. 4.45, small slabs with dimensions of 200 × 400 × 450 mm were cast in two layers with various delay times to determine the effect of delay time on bond strength between the two SCC layers. The thickness of each layer was 100 mm and delay times varied between 30 and 180 min. After 28 days of curing, cylindrical core samples were drilled from the slabs and prepared for test. Shear strength at the interface was determined, and the average shear strength of the three extracted samples was reported for each delay time. It was reported that for all the mixtures, the interface between the two layers could be visually identified when the delay between the two layers was more than 60 min.

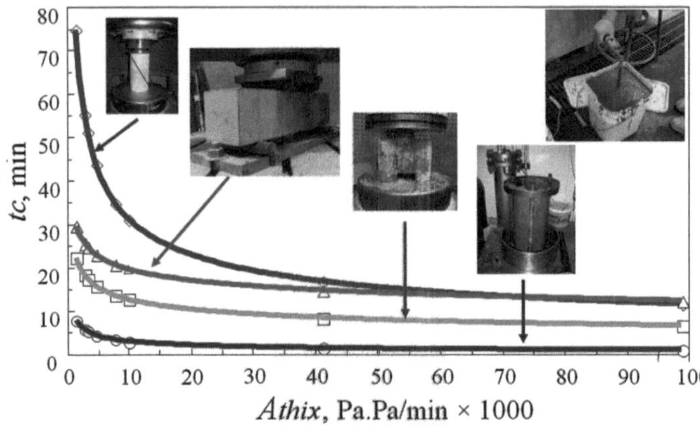

Fig. 4.44 Relation between the thixotropy index and critical time required to maintain 90 % of the residual property across *lift lines* subjected, from *left* to *right*, to slant shear stress, flexural stress, direct shear stress, and water permeability testing [61]

Based on the results presented in Fig. 4.46, the overall mechanical strength was reported to decrease with the increase in the delay between castings of the two layers. It was also observed that for the specimens with the highest thixotropy value (SCC3 and SCC4), there was almost no decrease in mechanical strength as a result of increase in the structuration rate of the bottom material. Decrease in mechanical strength was observed to be more in the case of specimens with lower thixotropy levels (SCC1 and SCC2).

The surface roughness of the fresh concrete at the bottom layer resulting from the floatation of coarse aggregate was taken into consideration in assessing the interlayer bond strength [56]. The authors found that even if the two layers do not remix at the time of casting of the top layer, the surface roughness of the first layer can be sufficient to ensure that the final mechanical strength may not be affected by distinct-layer casting. Thixotropy of the constitutive cement paste increases the stability of the mixture, thus increasing the presence of the coarse aggregate particles at the surface of the concrete. The authors [49] suggested that the distinct-layer casting problem may only occur in the case of a smooth interface between the two layers (i.e. no coarse particles at the surface of the first layer) and thus in the case of slightly unstable mixtures. The following equation was also proposed for predicting the critical delay time t^{c}_{rest} between casting the two layers (Fig. 4.47):

$$t^{C}_{rest} = \frac{\sqrt{(\rho g h)^2/12 + (2\mu_p V/h)^2}}{A_{thix}} \qquad (4.6)$$

where:

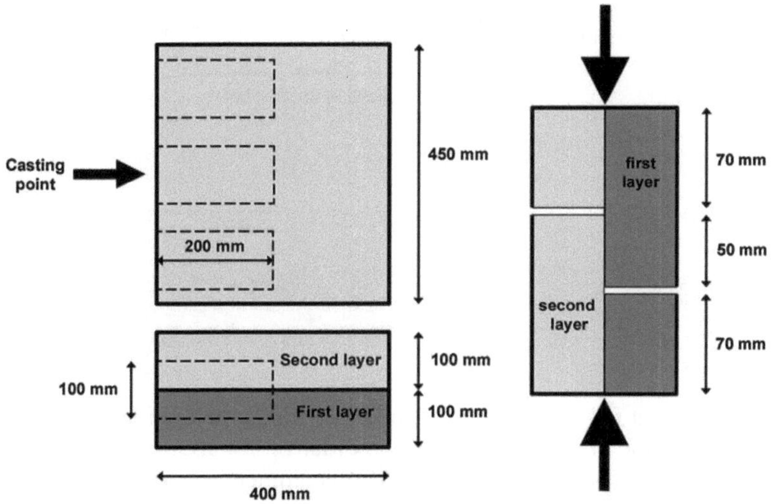

Fig. 4.45 *Left* Casting protocol and sample sawing and extraction (*dashed line*); *Right* Preparation of the sample for the mechanical shear strength measurements [56]

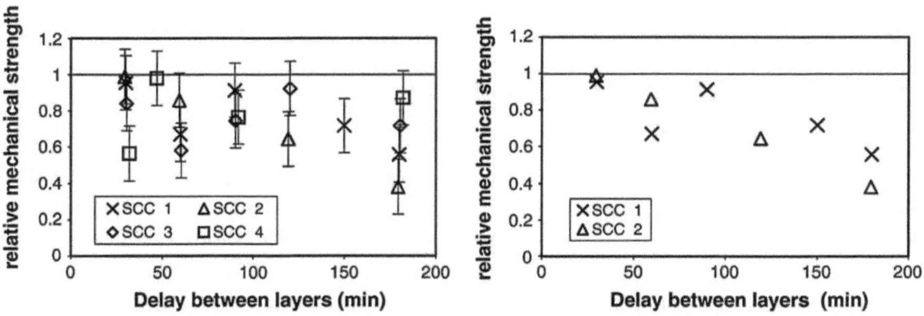

Fig. 4.46 Relative mechanical strength as a function of delay time [56]

ρ is the density of the second layer, h is the thickness of the second layer, μ_p is the plastic viscosity of the SCC (Pa.s), and A_{thix} is the structuration rate of the material (Pa/s).

From a practical point of view, it can be shown that, even for the most viscous SCC, the shear stress at the interface can be neglected compared to the effect of weight of the second layer when the thickness of the second layer becomes greater than 100 mm [56]. In this case, the expression of the critical delay between the two layers simplifies to:

$$T_c = \frac{\rho g h}{3.5 A_{thix}} \tag{4.7}$$

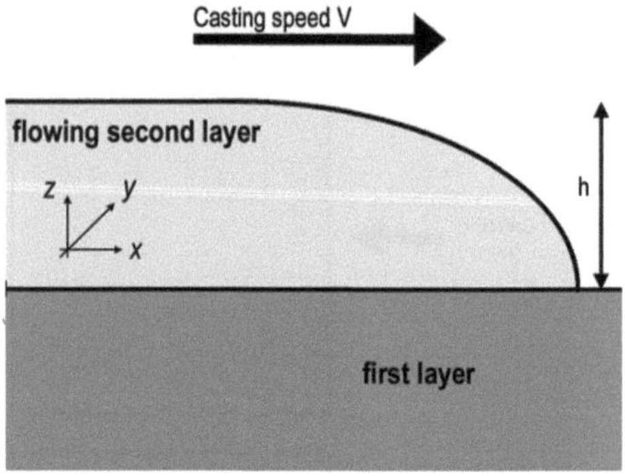

Fig. 4.47 Distinct-layers casting process and notations [56]

This means that for traditional SCC mixtures with structuration rate of the order of 0.3–0.5 Pa/s, the critical delay is of the order of 20–30 min.

Based on these results, the authors [56] recommended that for wall castings, SCC should be highly thixotropic to decrease formwork pressure and to increase stability of the mixture. Inversely, for slab casting, the main difficulties arise from the risk of distinct-layer casting, and the SCC should be as non-thixotropic as possible (low structuration rate: lower than 0.1 Pa/s for example).

4.9 Bond to Existing Hardened Concrete

Bouksani et al. [64] investigated the interfacial properties of layered beams composed of an existing concrete substrate and an overlay repair material made of SCC. The interface substrate surface had two rough and smooth surface textures and had three moisture conditions at the time of casting. Three types of overlay materials measuring 50 mm in thickness were investigated: ordinary vibrated concrete (OC), SCC, and SCC with silica fume, SCCSF. Using a grooving tool on the fresh concrete, grooves measuring 15–20 mm in depth were made to roughen the substrate surface. Before placing the overlay material, three different saturation levels were considered: dry substrate (SD) after 24 h at 105 °C, saturated with wet surface (SSW) after immersion in water for 2 days, and saturated with dry surface (SSD). The mixture composition and mechanical characteristics of the different overlay materials are given in Table 4.4.

A four point bending test of the layered repair beams was carried out. According to the flexural strength results presented in Fig. 4.48, it can be concluded that in order to ensure a macro-roughness necessary to create mechanical interlocking, a rough substrate surface is necessary. The SCC overlay was shown to be very sensitive to the smooth substrate interface regardless of the moisture condition of the substrate. Repair materials made with SCC with silica fume resulted in a slight increase in flexural strength, especially in the case of SSD

Table 4.4 Composition and main properties of the overlay materials [64]

Composition [kg/m^3]	VC	SCC	SCC-SF
Sand 0/3	630	770	800
Aggregate 3/8	300	380	400
Aggregate 8/15	750	400	450
Cement CEM II	400	500	550
Water	200	200	240
Superplasticiser	–	7	7.7
Silica fume	–	–	55
W/CM	0.5	0.4	0.4
Slump flow (mm)	–	710	690
Compressive strength (MPa)	36	42	45
Tensile strength (MPa)	3.4	4	4.8
Young's Modulus (MPa)	24850	35000	38000
Poisson's ratio (-)	0.2	0.2	0.2

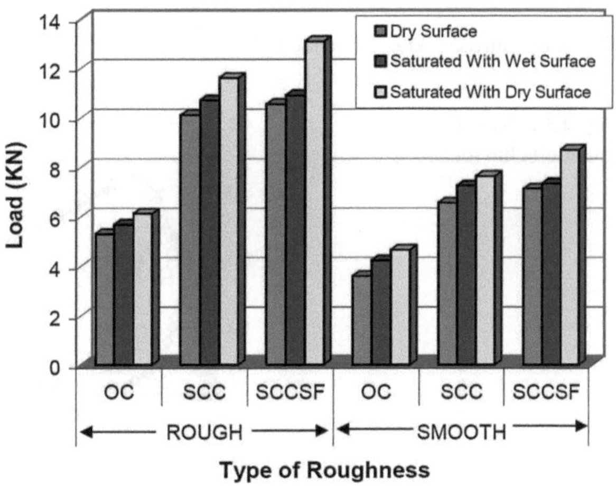

Fig. 4.48 Effect of the roughness and moisture surface of the substrate on bond strength between the existing concrete substrate and repair materials made with different concrete mixtures [64]

conditions. The SSD condition led to the highest bond strength, and flexural strength of the composite beams.

Kharchi et al. [65] carried out a similar investigation to evaluate the influence of the repair material composition on the flexural response of retrofitted beam elements and draw similar conclusions to those reported in [64]. Concrete substrate beam had dimensions of 50 × 100 × 400 mm and were cured for 5 months before repair. The repair material was cast on the top of the substrate. Two types of roughness were considered, smooth and rough surface. The rough surface of the old concrete substrate was obtained in fresh state by using a chisel to remove slurry cement from the coarse aggregate surface. Three states of moisture were also adopted as in [64]. The substrates were cast with vibrated concrete. Four repair materials were investigated. They included a VC and three SCC mixtures;

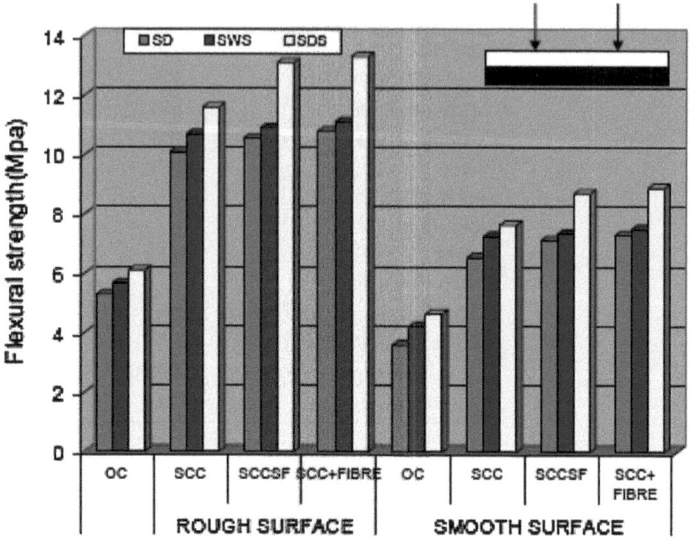

Fig. 4.49 Surface state influence [65]

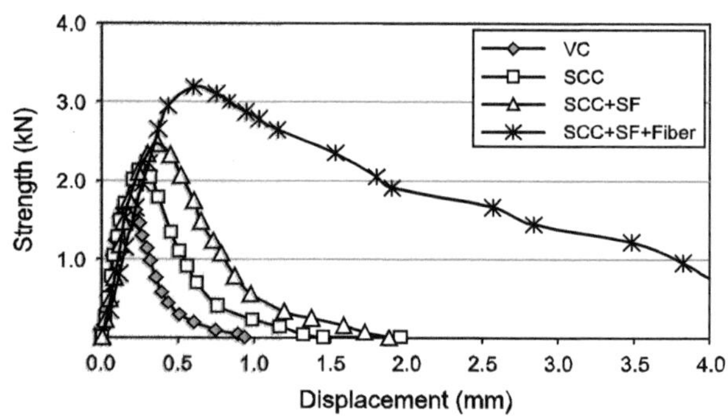

Fig. 4.50 Notched beam fracture behaviour [65]

one made without any silica fume, one with 10 % silica fume, and the third one with 10 % silica fume and synthetic fibers. The slump flow values were 690 mm for the first two SCC mixtures and 650 mm for the third mixture. The t_{50} values were 4.5, 4.5, and 7.7 s, respectively.

The results of the bending tests are presented in Fig. 4.49. The results of the notched beam fracture behaviour are presented in Fig. 4.50. The performance of the SCC-material was better compared to repaired beams using VC. The high stability and uniformity of the deformability of the SCC are attributed to the good adhesion. The surface preparation had a direct impact on the behaviour of the double layer elements. Roughness and moisture conditions are again shown to have significant influence on bond between the two materials. SSD conditions were again shown to secure high bond compared to dry surface and saturated with wet surface conditions. Surface preparation with roughened surfaces resulted in behaviour as monolithic beams. The addition of silica fume improved the flexural strength slightly, and the incorporation of synthetic fibers led also to slight improvement in strength. The fibers increased ductility and resistance to crack propagation, which should enhance considerable the structural integrity of the overlay repair system.

4.10 Conclusions

The review presented in this chapter shows that bond between SCC and reinforcement is not less than that bond of VC and in some cases higher values are reported. This may be attributed to the superior stability and filling ability of SCC that can also result in better encapsulation of reinforcement and existing surfaces.

Despite the high fluidity of SCC, high static stability after placement and until the onset of setting is necessary to secure more homogenous in situ properties and a denser matrix at the interface between the cement paste and the reinforcement. Such bond can be significantly affected by excessive segregation found in poorly designed SCC.

Static stability of the SCC is critical in reducing the top-bar effect to embedded reinforcing steel and prestressing strands. Static stability can be expressed in terms of the maximum surface settlement percentage and static segregation determined from the column segregation test. Such values should be limited to 0.5 and 15 %, respectively, particularly in deep elements, in order to ensure a relatively low top-bar effect of the SCC. Highly flowable SCC can develop at least 90 % in situ relative compressive strength and modification factor (or top-bar effect to prestressing strands) of 1.4 for horizontally embedded prestressing strands. The increase in plastic viscosity of SCC at a given yield stress (or slump flow value) can lead to a greater resistance to surface settlement and segregation. However, high plastic viscosity can hinder self-consolidation and can lead to entrapment of air-voids during casting with negative implication on the bond strength. Upper limits to plastic viscosity and t_{50} to ensure adequate self-consolidation should be 80 Pa.s and 6 s, respectively.

The incorporation of SCMs and fillers can also enhance the bond strength of SCC. Incorporating SCMs, such as SF and FA and limestone powder are shown to enhance the bond properties of SCC.

Thixotropy and surface roughness seem to be the most important concrete properties affecting the multi-layer casting of SCC. An optimized combination of low thixotropic behaviour of the SCC, in the case of horizontal casting, and the limitation of the delay time between casting different layers are essential in enhancing the bond strength in a multi-layer casting. The residual strength in shear is especially affected by the presence of such defect, compared to flexural and compressive modes of stress application. Water permeability across lift lines is also highly affected by the thixotropy of the lower concrete lift at the time of casting and time lag between successive layers.

Due to its enhanced filling ability and self-consolidating properties, properly proportioned SCC used as a repair material is shown to develop greater strength to existing surfaces than repair overlay made with VC. Such bond is also affected by the roughness of the substrate and the moisture condition of the existing concrete at the time of casting, as is the case for conventional repair materials.

References

1. Models TGB (2000) Bond of reinforcement in concrete, CEB-fib, fib Bulletin No. 10, Lausanne, p 427
2. RILEM: Technical recommendations for the testing and use of construction materials: RC6, bond test for reinforcing steel: 2. pull-out test. Mater. Struct. **3**(15), 175–178 (1970)
3. Menezes de Almeida Filho, F., El Debs, M.K., Lucia, A., El Debs, H.C.: Bond-slip behavior of self-compacting concrete and vibrated concrete using pull-out and beam tests. Mater. Struct. **41**, 1073–1089 (2008)
4. Foroughi-Asl, A., Dilmaghani, S., Famili, H.: Bond strength of reinforcement steel in self-compacting concrete. Int. J. Civ. Eng. **6**(1), 24–33 (2008)
5. Zhu, W., Sonebi, M., Bartos, P.J.M.: Bond and interfacial properties of reinforce-ment in self-compacting concrete. Mater. Struct. **37**, 442–448 (2004)
6. Looney, T.J., Arezoumandi, M., Volz, J.S., Myers, J.J.: An experimental study on bond strength of reinforcing steel in self-consolidating concrete. Int. J. Concr. Struct. Mater. **6**(3), 187–197 (2012)
7. Desnerck, P., De Schutter, G., Taerwe, L.: Bond behaviour of reinforcing bars in self-compacting concrete: experimental determination by using beam tests. RILEM Mater. Struct. **43**, 53–62 (2010)
8. RILEM: Technical recommendations for the testing and use of construction materials: RC6, bond test for reinforcing steel: 1. beam test. Mater. Struct. **3**(15), 169–174 (1970)
9. Castel, A., Vidal, T., Viriyametanont, K., Raoul, F.: Effect of reinforcing bar orientation and location on bond with self-consolidating concrete. ACI Struct. J. **103**(4), 559–567 (2006)
10. Ertzibengoa, D., Matthys, S., Taerwe, L.: Bond behaviour of flat stainless steel re-bars in concrete. Mater. Struct. **45**, 1639–1653 (2012)
11. Pandurangan, K., Kothandaraman, S., Sreedaran, D.: A study on the bond strength of tension lap splices in self-compacting concrete. Mater. Struct. **43**, 1113–1121 (2010)
12. Dehn, F., Hoalschemacher, K., Weibe, D.: Self-compacting concrete (SCC) time development of the material properties and the bond behavior, Selbstverdichtendem Beton (2000)
13. Brameshuber, W., Stephan, V., Christian, T.: Self compacting concrete in the pre-cast element plant. BFT **1**, 80–88 (2001)
14. Sonebi, M., Bartos, P.J.M., Zhu, W., Gibbs, J., Tamimi, A.: Properties of hardened concrete. BriteEuRam Contract No. BRPR-CT96-0366, Task 4 May, p. 51. (2000)
15. Helincks, P., Boel, V., De Corte, W., De Schutter, G., Desnerck, P.: Structural behaviour of powder-type self-compacting convcrete: bond performance and shear capacity. Eng. Struct. **48**, 121–132 (2013)
16. Chan, Y., Chen, Y., Liu, Y.: Development of bond strength of reinforcement steel in self-consolidating concrete. ACI Struct. J. **100**(4), 490–498 (2003)
17. Holschemacher, K., Klug, Y.: A database for the evaluation of hardened properties of SCC. Leipzig annual civil engineering report No. 7, pp. 123–134. Universität Leipzig (2002)
18. Sonebi, M., Cleland, D., Anuar, I.: Bond strength between reinforcement and self-compacting concrete. In: Cairns, J.W., Metelli, G., Plizzari, GA. (eds.) Proceedings of 4th Symposium on Bond in Concrete: Bond, An-Chorage, Detailing, vol. 1, pp. 267–272. Brescia, 17–20 June 2012

19. Ashtiani, M.S., Dhakal, R.P., Scott, A.: Post-yield bond behaviour of deformed bars in high-strength self-compacting concrete. Constr Build Mater **44**, 236–248 (2013)
20. Zheng, JL, Zhuang, JP: Experiment research on bond properties between self-compacting concrete and reinforcement. Gongcheng Lixue/Eng. Mech. **30**(2), 112–117 (2013)
21. Surong L, Jinping Z, Yi W: Experimental study of bond stress distribution between reinforcement and different strength self-compacting concrete bond under dynamic load. Adv. Mater. Res. **446–449**, 330–333 (2012) (ISSN:1022-6680)
22. Cairns J.W., Metelli M., Plizzari G.A. (eds.) Proceeding of the Fourth International Symposium on Bond in Concrete, Brescia, June 2012
23. Khayat, K.H., Petrov, N., Attiogbe, E.K., See, H.T.: Uniformity of bond strength of prestressing strands in conventional flowable and self-consolidating concrete mixtures. In: Wallevik O. (ed.) Proceedings, 3rd International Symposium on Self-Compacting Concrete, pp. 703–712. Reykjavik, Aug 2003
24. Attiogbe, E., Nmai, C.: Engineering properties of self-consolidating concrete for high performance concrete. Proceedings, High Performance Concrete Symposium and Bridge Conference, 49th PCI Annual Convention and Exhibition, Orlando (2003)
25. Martí-Vargas, J.R.,Serna-Ros, P., Arbeláez, C.A., Rigueira-Víctor, J.W.: Transfer and anchorage bond behavior in self-compacting concrete. Materiales de Construcción **56**(284), 27–42 (2006)
26. Martí-Vargas, J.R.: Experimental study on bond of prestressing strand in high-strength concrete. Ph.D. Thesis, (in Spanish), UMI Dissertation Services, ISBN 0-493-55092-5 (2002)
27. Martí-Vargas, J.R., Serna-Ros, P., Fernández-Prada, M.A., Miguel-Sosa, P.F., Arbeláez, C.A.: Test method for determination of the transmission and anchorage lengths in prestressed reinforcement. Mag. Concr. Res. **58**(1), 21–29 (2006)
28. Attiogbe, E., See, H., Daczko, J.: Engineering properties of self-consolidating con-crete. In: Proceedings, 1st North American Conference on the design and use of Self-Consolidating Concrete, Center for advanced Cement-Based Materials, Nov 2002
29. Valcuende, M., Parra, C.: Bond behaviour of reinforcement in self-compacting concretes. Constr. Build. Mater. **23**, 162–170 (2009)
30. Esfahani, M.R., Esfahani, M.R.: Lachemi, and M., Kianoush, M.R., "Top-bar effect of steel bars in self-consolidating concrete (SCC)". Cement Concr. Compos. **30**, 52–60 (2008)
31. Valcuende, M., Parra, C., Balasch, B.S.:Cement and Concrete Composites. In: Esfahani M.R., Lachemi M., Kianoush M.R (eds.) A Discussion of the Paper ''Top-Bar Effect of Steel Bars in Self-Consolidating Concrete (SCC)", vol. 30, p. 1020 (2008)
32. Hossain, K.M.A., Lachemi, M.: Bond behavior of self-consolidating concrete with mineral and chemical admixtures. ASCE J. Mater. Civ. Eng. **20**, 608–616 (2008)
33. Hassan, A.A.A., Hossain, K.M.A., Lachemi, M.: Bond strength of deformed bars in large reinforced concrete members cast with industrial self-consolidating concrete mixture. Constr. Build. Mater. **24**, 520–530 (2010)
34. Soylev, T.A., Raoul, F.: Effects of bar-placement conditions on steel-concrete bond. Mater. Struct. **39**, 211–220 (2006)
35. Eurocode 2: Design of concrete structures—Part 1: general rules and rules for buildings, p. 222 (2001)
36. ACI 318: Building code requirements for structural concrete (ACI 318-02) and commentary (ACI 318R-02), ACI Committee 318 Structural Building Code, American Concrete Institute, p. 445 (2002)
37. Desnerck, P.: Compressive, bond and shear behaviour of powder-type self-compacting concrete. Ph.D. Thesis, Universiteit Gent, p. 343 (2011)
38. Khayat, K.H., Manai, K., Trudel, A.: In-situ mechanical properties of wall elements cast using self-consolidating concrete. ACI Mater. J. **94**(6), 491–500 (1997)
39. Khayat, K.H.: Use of viscosity-modifying admixture to reduce top-bar effect of anchored bars cast with fluid concrete. ACI Mater. J. **95**(2), 158–167 (1998)
40. Khayat, K.H.: Viscosity-enhancing admixtures for cement-based materials—an over-view. Cem. Concr. Compos. **20**(2–3), 171–188 (1998)

41. Petrov, N., Khayat, K.H., Tagnit-Hamou, A.: Influence of additions of viscosity-enhancing admixture on homogeneity of self-consolidating concrete (in French). In: Progress in Concrete, ACI Québec and Eastern Ontario Chapter Annual Meeting, Montréal, p. 8, Nov 1999
42. Khayat, K.H., Pavate, T., Jolicoeur, C., Vanhove, Y.: Conductivity approach to evaluate stability of fresh concrete, SCC 2005. In: Shah S.P. (ed.) Proceedings 2nd North American Conference on the Design and Use of Self-Consolidating Concrete and the 4th International RILEM Symposium on Self-Compacting Concrete, Evanston, pp. 831–838 (2005)
43. Assaad, J., Khayat, K.H., Daczko, J.: Evaluation of static stability of self-consolidating concrete. ACI Mater. J. **101**(3), 207–215 (2004)
44. Khayat, K.H., Attiogbe, E., See. H.: Effect of admixture combination on top-bar effect in highly flowable and self-consolidating concrete. ACI-SP247 (Publication No 247-4), p. 15 (2007)
45. Jirsa, J.O., Breen, J.E.: Influence of casting position and shear on development and splice length—design recommendations. Research report 242-3F, Center for transportation research, University of Texas, p. 42, Nov 1981
46. Khayat, K.H., Petrov, N., Attiogbe, E., See, H.: Homogeneity of bond strength of along wall elements cast with flowable and self-consolidating concrete. In: Wallevik O. (ed.) RILEM Proceedings 33, 3rd International Symposium on Self-Compacting Concrete, Reykjavik, pp. 703–712, Aug 2003
47. Long, W.J., Lemieux, G., Khayat, K.H., Xing, F.: Uniformity of bond strength of prestressing strands in precast prestressed self-compacting concrete mixtures. In: Proceedings 11th International Symposium on Structural Engineering, Guangzhou, pp. 569–574 (2010)
48. Khayat, K.H., Mitchell, D.: NCHRP report 628; Self-consolidating concrete for pre-cast, prestressed concrete bridge elements (2009)
49. Hwang, S.-D., Khayat, H.K., Bonneau, O.: Performance-based specifications of self-consolidating concrete used in structural applications. ACI Mater. J. **92**(6), 625–633 (2006)
50. Karatas, M., Turk, K., Ulucan, Z.C.: Investigation of bond between lap-spliced steel bar and self-compacting concrete: the role of silica fume. Can. J. Civ. Eng. **37**, 420–428 (2010)
51. Turk, K., Karatas, M., Ulucan, Z.C.: Effect of the use of different types and dosages of mineral additions on the bond strength of lap-spliced bars in self-compacting concrete. Mater. Struct. **43**, 557–570 (2010)
52. De Almeida, F., De Nardin, F.M., El Debs, S.: Evaluation of the bond strength of self-compacting concrete in pull-out tests. In: Proceedings 2nd North American Conference on the Design and Use of Self-Consolidating Concrete and 4th International RILEM Symposium on Self-Compacting Concrete, Chicago (2005)
53. CEB/FIP: CEB/fib Model Code 1990—design code; CEB/FIP, p. 437 (1993)
54. Daoud A., Lorrain M., Laborderie C.: Anchorage and cracking behaviour of self-compacting concrete. In: Proceedings 3rd RILEM International Symposium on Self-Compacting Concrete, Reykjavik (2003)
55. Said, H., Mouli, M., Mesbah, A.M., Khelafi, H.: Influence du type d'addition dans les bétons autoplaçants sur l'adhérence béton-ancrage. In: Proceedings 1st International Conference on sustainable Built Environment Infrastructures in Developing Countries, ENSET Oran, Algeria (2009)
56. Erzibengoa Gaztelumendi, D., Matthys, S., Taerwe, L.: Bond interaction between flat stainless steel rebars and traditional and self-compacting concrete. In: Cairns J.W., Metelli G., Plizzari (eds.) Proceedings of Bond in Concrete, pp. 885–892 (2012)
57. Aslani, F., Nejadi, S: Bond behavior of reinforcement in conventional and self-compacting concrete. Adv. Struct. Eng.**15**(12), 2033–2051 (2012)
58. Roussel, N., Cussigh, F.: Distinct-layer casting of SCC: the mechanical consequences of thixotropy. Cem. Concr. Res. **38**, 624–632 (2008)
59. Roussel, N.: A thixotropic model for fresh fluid concretes: theory and applications. In: Proceedings 5th International RILEM Symposium on Self-Compacting Concrete, Ghent, pp. 267–272 (2007)

60. Khayat, K.H., Omran, A.F., Al Magdi, W.: Evaluation of thixotropy of self-consolidating concrete and influence on concrete performance. In: Proceedings 3rd Iberian Congress on Self Compacting Concrete, Madrid, pp. 3–16 (2012)
61. Abd El Megid, W.: Effect of rheology on surface quality and performance of SCC. Ph.D. Dissertation, University of Sherbrooke (2012)
62. Omran, A.F., Naji, S., Khayat, K.H.: Portable vane test to assess structural build-up at rest of self-consolidating concrete. ACI Mater. J. **108**(6), 628–637 (2011)
63. Khayat, K.H., Omran, A.F., Pavate, T.V.: Inclined plane test method to determine structural build-up at rest of self-consolidating concrete. ACI Mater. J. **107**(5), 515–522 (2010)
64. Bouksani, O., Kharchi, F., Benhadji, M., Belhamel, F.: Influence of the roughness and moisture of the substrate surface on the bond between old and new concrete. Contemp. Eng. Sci. **3**(3), 139–147 (2010)
65. Kharchi, F., Benhadji, M., Bouksani. O.: Repair of concrete structures with SCC. World Acad. Sci. Eng. Technol. **58**, p. 122 (2011)

Chapter 5
Structural Behaviour of SCC

Mohamed Lachemi, Assem Hassan, Claudio Mazzotti and Mohamed Sonebi

5.1 Introduction

The high workability and non-segregating nature of self-compacting concrete (SCC) offer a valuable solution for accelerating the placement rate and reducing the labour demand needed for vibration and surface finishing. While some of the fresh properties of SCC may differ significantly from those of conventional vibrated concrete (VC), the engineering properties of SCC are fairly similar to that of VC. However, some researchers have observed a distinct difference between the two types of concrete in engineering properties such as shear and bond strength [1–4]. This difference is especially evident when the proportions of coarse aggregate (size and shape) or paste content of SCC mixtures are different from those of VC. However, in applications where those engineering properties are of special consideration for designers and/or engineers, the same test methods and procedures applied for VC should be used to verify that the desired SCC performance is achieved.

This chapter introduces the structural performance of SCC and discusses various engineering properties such as uniaxial compression strength, flexural strength, shear strength, ductility and seismic consideration, cracking behaviour and compressive membrane action for structural members cast with SCC mixture.

M. Lachemi (✉)
Ryerson University, Toronto, Ontario M5B 2K3, Canada
e-mail: mlachemi@ryerson.ca

A. Hassan
Memorial University of Newfoundland, Newfoundland, Canada

C. Mazzotti
University of Bologna, Bologna, Italy

M. Sonebi
Queen's University Belfast, Belfast, UK

K. H. Khayat and G. De Schutter (eds.), *Mechanical Properties of Self-Compacting Concrete*, RILEM State-of-the-Art Reports 14,
DOI: 10.1007/978-3-319-03245-0_5, © RILEM 2014

5.2 Axial Loading

In practice, most columns are subjected to eccentric loads (a) because the centroid of the column and the centre of gravity of the load from the carried member do not coincide, or (b) due to rigid frame action. Design codes account for this by factoring a minimum eccentricity into the computation of the capacity of the column. However, testing the behaviour of columns under uniaxial loads is key to determining the maximum column capacity. Significant research has been carried out on the structural behaviour of SCC columns under uniaxial compression. In general, SCC columns showed slightly different behaviour compared to VC columns in terms of maximum axial strength, strain at maximum load, stiffness, and ductility.

5.2.1 Axial Strength

No significant difference was reported between SCC and VC columns in terms of maximum axial strength P_{max}. Maximum axial strength for both SCC and VC columns was dependent on the concrete compressive strength, and was always higher than the corresponding theoretical column capacity P_0 [5–8]. P_{max}/P_0 decreased as the concrete compressive strength increased in both SCC and VC columns. However, with the increase in concrete strength, some researchers reported a reduction of P_{max}/P_0 ratio in SCC columns compared to VC columns [6, 7].

5.2.2 Strain at Maximum Load

SCC mixture usually contains a lower W/C and/or extra supplementary cementing materials (SCM) than VC mixture. Since the low W/C and the use of SCM in the concrete mixture warrants stronger and denser matrix, the strain associated with the maximum load for SCC columns tested under uniaxial compression is expected to be lower than that of VC counterparts. At the same time, higher content of the coarse aggregate in the general mixture also plays a significant role in increasing strength, stiffness and reducing concrete strain. For this reason, an SCC mixture that contains a relatively lower amount of coarse aggregate (compared to VC mixture) may be characterized by a higher strain at peak load.

Lin et al. [5] reported a slight reduction of the strain at peak load in SCC compared to VC (an average of 0.00335 for VC and 0.00308 for SCC) when testing columns under uniaxial compression. The SCC in their investigation had comparable compressive strengths and approximately the same amount of coarse aggregate as VC, but it contained more supplementary cementitious materials and a lower W/C than VC (0.6–0.48 for VC and 0.51–0.40 for SCC mixtures). The slight reduction of the strain at peak load in SCC was attributed to the stronger and denser matrix of SCC (compared to VC) due to the lower W/C and the increased amount of SCM.

Khayat et al. [8] tested axially loaded SCC and VC columns with comparable compressive strengths. The SCC mixture in their investigation contained a lower W/C (0.5 for VC and 0.42 for SCC) and extra SCM. The amount of coarse aggregate in their SCC mixture was lower than that in the VC mixture. Their results showed that the strain at maximum load was higher in SCC columns than that in VC (an average of 0.0029 for VC and 0.004 for SCC). The lower amount of coarse aggregate in the SCC mixture than VC could be the main reason behind the higher strain at peak load in SCC columns.

Another study conducted by Lachemi et al. [9] confirmed the results of the Khayat et al. study [8], which showed higher strain at peak load in SCC columns. Lachemi et al. tested steel tube columns filled with SCC and VC mixtures. Both mixtures contained no SCM and had comparable compressive strengths and W/C ratios. The main difference between them was the reduction of coarse aggregate content in SCC compared to VC. The result of their investigation showed higher axial and transverse strain in SCC columns than in their VC counterparts.

5.2.3 Stiffness

The stiffness of concrete columns under uniaxial compression depends on several factors including W/C ratio, the use of SCM, and the amount of coarse aggregate. When a comparable amount of coarse aggregate is maintained in both VC and SCC mixtures, SCC columns are expected to exhibit higher stiffness than VC columns before peak load when tested under uniaxial compression. Meanwhile, using a lower amount of coarse aggregate in the SCC mixture may significantly reduce column stiffness.

Lin et al. [5] tested the stiffness of SCC columns under uniaxial compression and compared the results with those of VC columns. They define the stiffness in their investigation E_{test} as the secant modulus of elasticity of concrete corresponding to 45 % of the compressive strength of the confined concrete column. $E_{test}/\sqrt{f'_c}$ values for VC and SCC are listed in Table 5.1. The results of their study showed that the average stiffness of SCC was 1.20 times that of VC. As both SCC and VC mixtures in their investigation contained comparable coarse aggregate content, the higher stiffness of the SCC columns was attributed to the use of SCM and the low W/C ratio in the SCC mixture. Lin et al. also found that the influence of the amount of coarse aggregate on stiffness appears to be more significant than the addition of SCM.

Other researchers [6, 7] have reported lower stiffness of SCC compared to VC when testing concrete columns under uniaxial compression. Although the SCC in

Table 5.1 Modulus of elasticity and ductility of VC and SCC [5]

VC specimen no.	$E_{test}/\sqrt{f_c'}$	Z_{50VC}	SCC specimen no.	$E_{test}/\sqrt{f_c'}$	Z_{50SCC}
N1	3127.96	–	S1	3807.16	–
N2	3248.43	–	S2	3842.75	–
N3	3490.96	–	S3	4165.59	–
N4	4105.21	37.913	S4	4357.64	31.965
N5	2528.99	50.186	S5	4430.57	33.661
N6	3115.55	36.93	S6	3747.53	23.111
N7	3722.13	49.761	S7	4237.15	35.293
N8	3240.05	48.648	S8	4340.94	41.799
N9	3483.77	33.447	S9	3740.52	24.341
N10	2217.99	48.814	S10	4024.46	38.992
N11	3599.74	47.165	S11	3941.92	55.748
N12	4378.17	20.813	S12	3317.94	17.655
N13	3275.4	11.145	S13	2101.71	10.239
N14	3054.06	22.08	S14	4695.49	19.923
N15	3476.01	50.505	S15	4199.72	35.135
N16	3673.63	49.13	S16	3419.89	33.775

their investigation had a lower W/C ratio and contained higher SCM, the lower amount of coarse aggregate in SCC was the main reason behind the reduction of column stiffness.

5.2.4 Ductility and Seismic Consideration

Ductility is the ability of concrete to sustain significant inelastic deformation prior to failure. It is a desirable structural property as it allows absorption of energy for impact load in case of earthquake, dynamic impact or explosion, and provides warning before structural failure.

SCC is expected to have better ductility than VC mixture due to better particle gradation, fewer voids, and a denser matrix structure [5]. In seismic design of reinforced concrete members, it is necessary to allow for relatively large ductility. Therefore, members constructed with SCC are expected to have similar or even better seismic behaviour than those constructed with VC.

Restrepo et al. [10] assessed the seismic performance of columns built using high-performance reinforcing steel. SCC with a specified compressive strength of 55 MPa was used throughout their tests. The result of their investigation indicated that the use of SCC in the construction of the test units did not appear to have any detrimental effect in terms of early concrete cover spalling, bar anchorage or any other way.

Huang [11] studied the earthquake resistance of SCC and VC frames under low-cyclic loading. His investigation focused on how the material characteristics and

Table 5.2 Characteristic parameter of SCC frame and VC frame when subjected to low-cyclic loading [11]

Frame type	Cracking load P_{cr} (kN)	Cracking displ. Δ_{cr} (mm)	Yield load P_y (kN)	Yield displ. Δ_y (mm)	Peak load P_{max} (kN)	Peak displ. Δ_{max} (mm)	Failure load P_f (kN)	Failure displ. Δf (mm)
SCC	63.11	2.2	162.75	12.062	208.47	25.6	177.2	56.6
VC	48.61	1.3	177.27	8.931	210.7	30.4	193.7	53.6

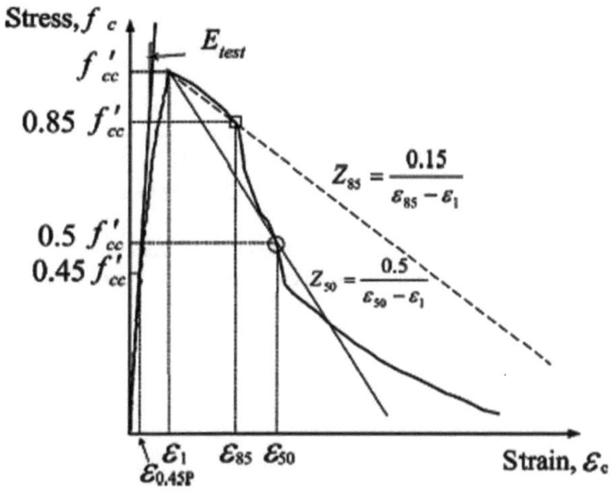

Fig. 5.1 Z_{50} index in the stress-strain curve [5]

constructability of SCC affect earthquake resistance behaviour of the frame. The 28 days compressive strength for SCC and VC was 45.1 MPa and 54.7 MPa, respectively. The SCC mixture had extra SCM (FA) and lower coarse aggregate content than the VC mixture. Huang found that the behaviour and the failure mode (column shear failure) of SCC and VC frames are similar. The observed small difference in failure mode was due to the fact that the SCC mixture was more compact and had better crack resistance, irrespective of the smaller compression strength. This was indicated in the cracking and yield displacements which were larger in the SCC frame than in the VC frame (Table 5.2).

Lin et al. [5] compared the ductility of SCC columns with that of VC columns. The SCC and VC mixtures in their study had approximately the same amount of coarse aggregate, but the SCC contained more SCM and less water content than the VC. Their results indicated that the descending range of the stress-strain curve of SCC columns exhibited better ductility than VC columns. Also, the load dropped more gradually in SCC columns than in VC columns after peak load.

Lin et al. used two methods to compare and present the ductility. First, they compared an index Z_{50} to reflect the slope of the descending branch of the stress-strain curve (Fig. 5.1).

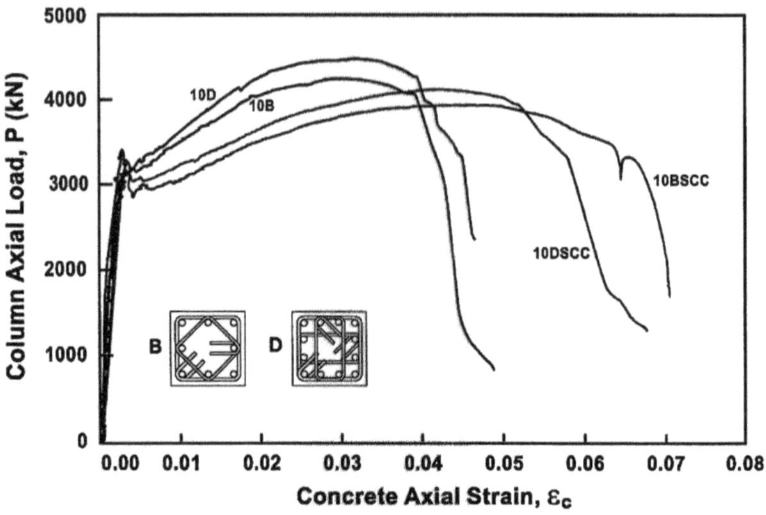

Fig. 5.2 Comparison of concrete axial load versus axial strain for VC and SCC columns (the number 10 corresponds to concrete with a compressive strength of 40 MPa) [8]

It is defined as:

$$Z_{50} = \frac{0.50}{\varepsilon_{50} - \varepsilon_1}. \tag{5.1}$$

While the smaller Z_{50} value indicates better concrete ductility, most SCC specimens had a smaller Z_{50} value than their companion VC specimens, which indicates that the ductility of SCC is better than that of VC. The average ratio of Z_{50}, SCC to Z_{50}, VC was 0.782 (Table 5.1).

The second method used by Lin et al. to compare the ductility was to calculate a ductility index μ which is defined as A_u/A_p, where A_u is the area under the stress-strain curve before the stress drops to 50 % of the maximum stress, and A_p is the area under the stress-strain curve up to peak stress. As the larger value of μ indicates better ductility, each SCC column had a larger μ value than its companion VC column and the average of μ_{SCC}/μ_{VC} was 1.324.

Khayat et al. [8] tested the ductility of SCC and VC columns by calculating the ratio of the axial strain when the stress drops to 0.5 f'_c (ε_{C50}) to the strain corresponding to maximum axial load carried by confined concrete ε_{C2}. Both the SCC and VC mixtures in their investigation had comparable compressive strengths. The SCC contained lower coarse aggregate content and W/C ratio and had more SCM than the VC. The result of the $\varepsilon_{C50}/\varepsilon_{C2}$ ratio in their investigation indicated that the SCC columns had greater ductility than the VC columns (62 and 33 % greater ductility for the 10 B SCC versus 10 B, and 10 D SCC versus 10 D, respectively, Fig. 5.2) when same stirrups arrangement was adopted. Khayat et al. attributed the higher ductility of SCC columns to the lower elastic modulus of the SCC

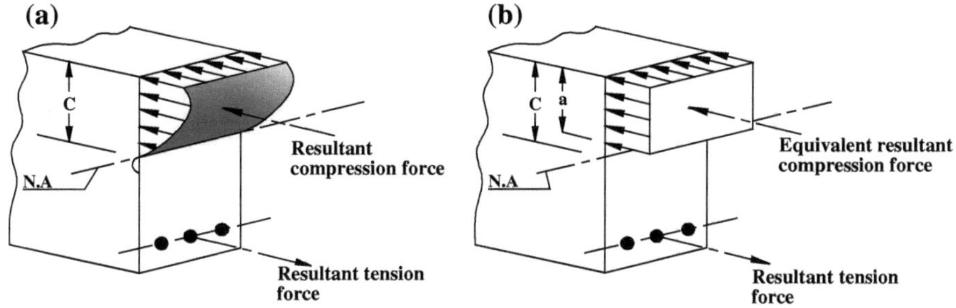

Fig. 5.3 Internal forces in the concrete cross section **a** actual compressive stresses, and **b** equivalent rectangular stresses block

(30.3 GPa compared to 37.0 GPa for the VC), which was mainly due to the lower coarse aggregate content in the SCC mixture.

Galano and Vignoli [12] tested 60 eccentrically-loaded slender columns made with SCC and VC mixtures using different target compressive strengths. The tested columns were 100 x 100 mm in cross section and 2000 mm in length. SCC and VC mixtures had comparable 28 day compressive strength and comparable W/C ratio in both normal and high strength columns. SCC mixture in medium strength columns had silica fume and less coarse aggregate content compared to VC mixture in similar strength columns. On the other hand, SCC mixture in high strength columns had insignificant differences in silica fume and coarse aggregate contents compared to VC mixture. The results showed that the ductility of slender columns was higher in SCC compared to VC mixtures for columns made with normal-strength concrete. For high strength concrete, no notable differences were observed between SCC and VC columns.

5.3 Flexural Strength

The ultimate flexural resistance of a concrete beam is calculated using the laws of statics by summing the moment about an axis through the point of application of compressive force (C) or tension force (T) (Fig. 5.3a). The compressive stress distribution in Fig. 5.3a is directly related to the stress-strain curves, or is assumed to be any shape that results in the prediction of strength that is in substantial agreement with the results of comprehensive tests. Rather than using a representative stress-strain curve (such as Fig. 5.3a), other diagrams expressed mathematically in terms of certain constants (Fig. 5.3b) can also be used to calculate the ultimate flexural strength, provided they adequately predict test results.

Since no significant differences in the stress-strain curve distribution were reported between SCC and VC [5,9], the computation of the ultimate flexural resistance of SCC beams should not be different than that of VC beams, and can be calculated using the same techniques provided in the related design codes.

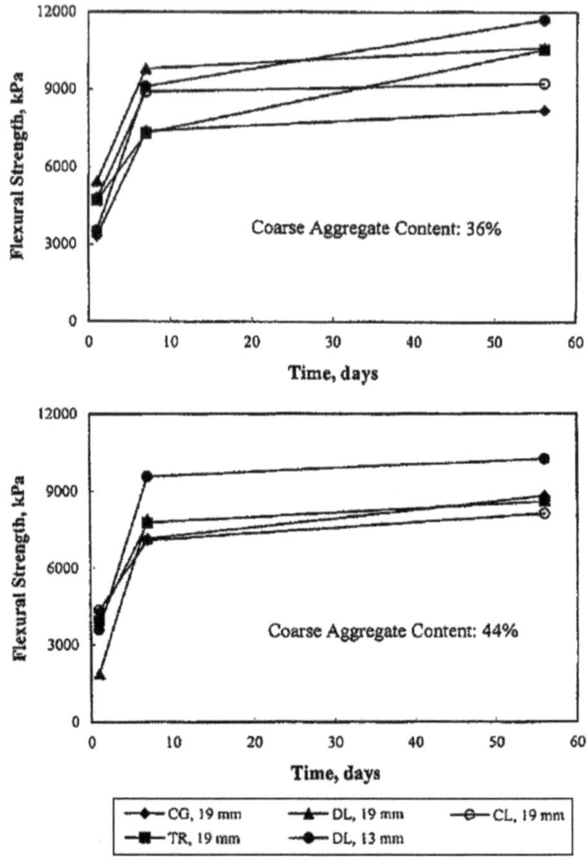

Fig. 5.4 Flexural strength development of concretes with 36 and 44 % aggregate content [13]

The flexural resistance of SCC beams, like that of VC beams, mainly depends on the W/C ratio, SCM, coarse aggregate size and volume, and the quality of the interface between the aggregate and cement paste. Using SCM with low W/C ratio improves the concrete microstructure, increases the density and strength of the matrix, and increases flexural strength. The coarse aggregate also plays a significant role in flexural strength. Increasing coarse aggregate content appears to result in a reduction in flexural strength for a given aggregate size. This may be explained by the interface transition zone around the aggregate, which is potentially weaker in tension than mortar or aggregate. With more aggregate added into concrete mixtures, more interfaces are formed in the hardened concrete and more reduction of flexural strength occurs [13, 14].

Figure 5.4 compares the effect of coarse aggregate content, type, and volume on the concrete flexural strength [13]. The figure indicates that a reduction up to 22 % in flexural strength was observed by increasing the aggregate content from 36 to 44 %. Fig. 5.4 also indicates that the reduction in the coarse aggregate size (13 mm compared to 19 mm) resulted in an increase of flexural strength.

SCC is usually developed by increasing the fines content at the expense of coarse aggregate content. It is characterized by dense, less permeable structure due to the low W/C ratio and the use of SCM. In addition, the maximum size of coarse aggregate in SCC is usually small to facilitate the concrete flowability. Therefore, the flexural strength of SCC may be higher than that of VC with similar mixture proportions.

Zanuy et al. [15] studied the transverse behaviour of the top slab of box-girder bridges, made of SCC, under fatigue flexural loads. No significant differences between the measured fatigue flexural strength and the expressions adopted for normal concrete were observed; in fact, available S–N curves were able to predict the fatigue strength safely. Though the number of tests is limited, SCC seems to be able to increase the negative frictional strength, thus modifying the cyclic tension-stiffening behaviour.

De Pauw et al. [16] studied the applicability of SCC for the production of precast/prestressed beams. 12 m length SCC and VC beams were tested up to failure in a four-point bending test. The results indicated that high strength prestressing strands and SCC are suitable for the manufacturing of prestressed concrete beams. The results also showed that, beams made with SCC exhibited higher deformation and their cracks appeared at lower loads and were more and wider compared to those of VC beams. However, De Pauw et al. noted in their results that the low temperature may have influenced the development of the mechanical properties of SCC beams, since all beams were cast between 0 and 6 °C.

Hassan et al. [17] tested full-scale SCC and VC beams under three point bending loading. Part of their study was a comparison of the first flexural cracking moment of both SCC and VC and a validation of the performance of several Code-based equations [ACI, Canadian (CSA), Australian (AS) and European (EC2) Codes] in predicting the first flexural cracking moment. The tested beams varied from 150 to 750 mm in depth and 1050–4500 mm in length. Both the SCC and VC mixtures were prepared from the same materials and had similar compressive strengths (around 46 MPa). The main difference between the two mixtures was the coarse aggregate content; the VC mixture contained 25 % more than the SCC. The beam designation in their investigation included a combination of letters and numeric: SCC or VC to indicate the concrete type, 1 or 2 to indicate the longitudinal reinforcement ratio, and 150, 250, 363, 500 or 750 to designate the total height. For example, a SCC beam having 1 % ρ_w with a total height of 750 mm is designated as 1SCC750.

The results showed that there is no difference between SCC and VC beams in terms of the first flexural cracking moment. For both concretes, the first flexural cracks were observed at about 20 and 35 % of the failure load for small and large size beams, respectively. The results also showed that for large size beams, both ACI and CSA equations underestimated the first flexural cracking load, while the AS and EC2 equations overestimated this value. However, for shallow beams, all four Code-based equations predicted values closed to those obtained from

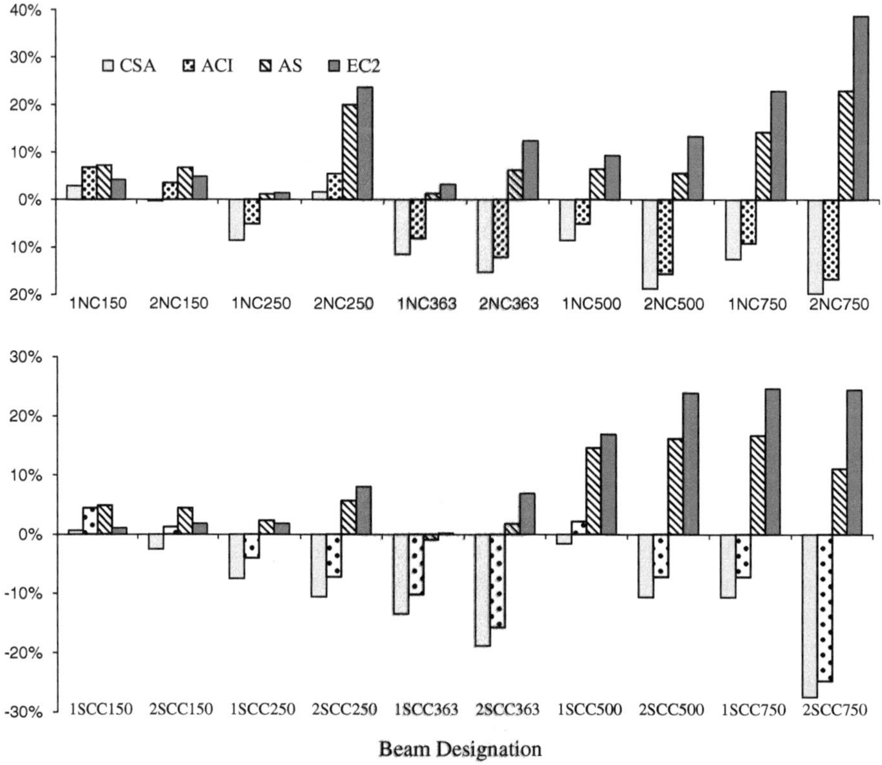

Fig. 5.5 Error percentage of prediction for first cracking load by Codes with respect to experiments Hassan et al. [17]

experiments (Fig. 5.5). Maghsoudi [18] performed similar tests on beam-column connections made of medium-strength SCC. He also found that ACI and CSA equations slightly underestimate the flexural cracking moment.

5.4 Cracked Behaviour

Sonebi et al. [19] investigated the flexural behaviour of concrete beams (200 mm width × 300 mm depth × 3800 mm length) cast with SCC, VC, and reinforced SCC (SCC60, RC60, and FSCC60 respectively) with standard compressive strengths of 60 MPa. The SCC mixture contained extra SCM (slag), lower W/C ratio and less coarse aggregate than the VC mixture. The results of their investigation showed that at service load, more and slightly wider cracks with greater penetration occurred in VC beams than in SCC beams (Fig. 5.6). The mode of failure and load deflection response of beams cast with SCC and VC were similar. It was also observed that the ultimate moment capacity of the SCC beam was

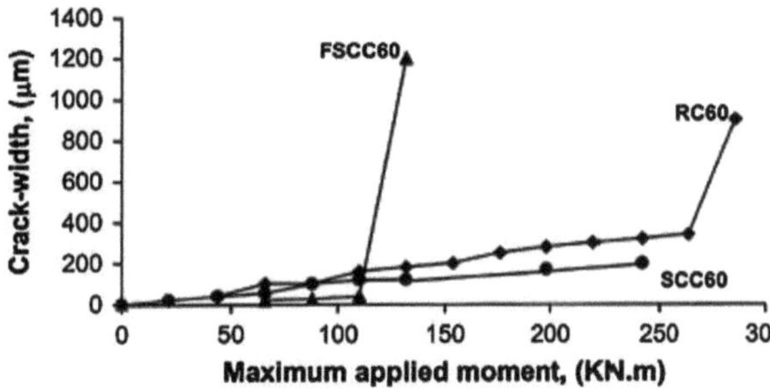

Fig. 5.6 Variation of crack width versus applied moment [19]

comparable with that of the VC beam and the maximum deflection of the SCC beam was slightly higher.

Luo and Zheng [20] investigated the cracking loads, flexural capacities and failure modes of reinforced SCC beams and compared the results with VC beams. The beams were 150 mm wide, 300 mm deep and 3000 mm long. All the tested SCC and VC mixtures had comparable compressive strengths, with an average of 58 MPa. The SCC mixtures contained extra SCM (fly ash) and lower coarse aggregate content than the VC mixtures. The W/B ratios for the SCC mixtures were 0.28 and 0.32 compared to 0.35 and 0.42 for the VC mixtures. The results indicated that in the process of flexural failure, there are no obvious differences between the behaviour of SCC and VC beams. The results also showed that the ultimate moment of the SCC beam was a little larger than that of the VC and the crack moment of the SCC beams was slightly lower.

De Corte and Boel [21] studied the flexural resistance and cracking of reinforced SCC/VC beams with varying reinforcement ratios. The SCC and VC mixtures had comparable compressive strength and the nominal maximum size of the coarse aggregates for both mixtures was 16 mm. Six reinforced concrete beams measuring 2400 mm in length, 150 mm in width, and 200 mm in depth were tested under static and dynamic tests. The beams were simply supported with span length of 2000 mm and a two-point load device was used to transfer the applied load. The maximum failure loads for VC and SCC showed excellent agreement with less than 5 % difference. The crack width evolution in SCC beams was somewhat slower compared to VC beams. This remark can also be found in Kumar et al. [22] which performed similar tests. In addition, it was found [21] that the crack width estimation formulation of Eurocode 2 gives slightly underestimated average crack width values, especially for lower reinforcement ratios.

Cuenca et al. [23] investigated the cracking behaviour of reinforced SCC/VC beams subject to shear load. The SCC and VC shared the same raw materials and had similar compressive strength (50 MPa); the SCC mixture was characterized by smaller amount of gravel, compensated by cement and sand; furthermore, also FRSCC was considered (mix C). Three reinforced I concrete beams, measuring 6000 mm in length, 500 mm in width, and 700 mm in depth were cast and tested; during flexural tests, shear cracks opening was measured by means of an optical technique and results show that both SCC/VC mixes provided for very similar results (Fig. 5.7), irrespective of the higher content of fines of SCC.

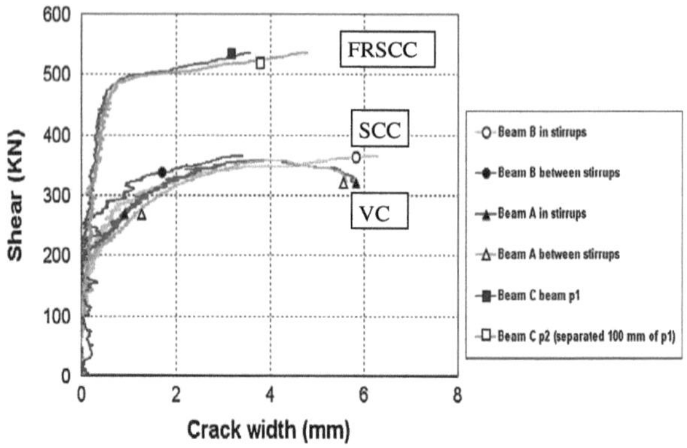

Fig. 5.7 Crack opening vs shear load for **a** VC, **b** SCC and **c** FR-SCC beams [23]

5.5 Shear Strength

Shear strength of a concrete beam is derived from the contributions of compression shear zone (ranging from 20 to 40 %), aggregate interlock mechanism (35 to 50 %), and the dowel action of longitudinal reinforcement (15 to 25 %). The dowel action force reaches its capacity first, transferring the shear to the aggregate interlock. The aggregate interlock is next to fail, transmitting all shear to the concrete in the compression zone, which then fails explosively.

The aggregate interlock that is considered the major component of the shear transfer is greatly influenced by beam size. As the depth of the beam increases, crack widths at points above the main longitudinal reinforcement tend to increase as well. This leads to a reduction in aggregate interlock across the crack, resulting in reduced shear stress [24]. The aggregate interlock is also influenced by the longitudinal reinforcement ratio. A higher percentage of longitudinal reinforcing steel keeps the cracks from opening wider, increasing the aggregate interlock and allowing the concrete to resist more shear [25].

Aggregate type also influences the capacity of the aggregate interlock. For this reason, the shear strength of lightweight concrete (which normally contains light and relatively weak aggregate) will be less than that of normal weight concrete, although their compressive strengths may be the same. In lightweight concrete, the cracks penetrate the coarse aggregate (which is weaker than the mortar), forming a smoother surface along the diagonal failure crack. In that case, the shear transfers along the diagonal crack by means of friction, which is less than in the case of aggregate interlock shear transfer. The shear transfer in high strength concrete is similar to that in lightweight concrete. In high strength concrete the cracks also penetrate the coarse aggregate, as it is weaker than the relatively stronger mortar.The presence of large size aggregate is also believed to affect the aggregate interlock [2]. It extends the path of the shear crack around the aggregate and subsequently improves post-cracking shear resistance (Fig. 5.8).

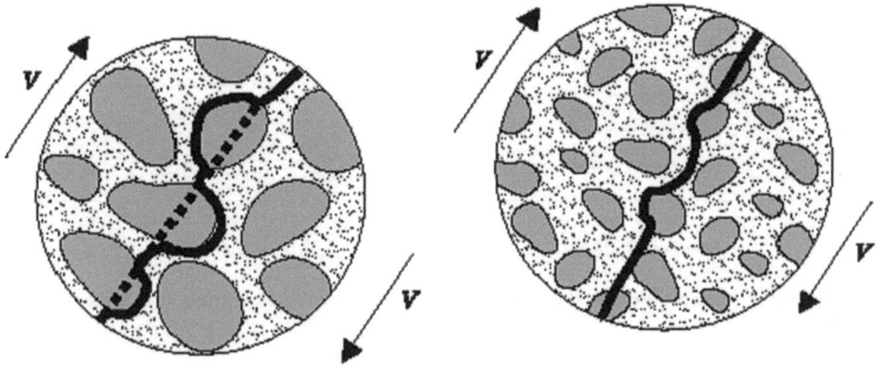

Fig. 5.8 Influence of coarse aggregate on crack shear plane [2]

Since SCC is normally produced by reducing the coarse aggregate content in the mixture, its shear strength is expected to be less than that of VC. In addition, the substantial difference in the rheology of the cement paste matrix in SCC (compared to VC) affects average aggregate diameter, aggregate spacing and shear resistance [26].

Lachemi et al. [2] investigated the shear resistance of 18 flexurally reinforced SCC beams without shear reinforcements and compared the results to VC beams. Their investigation was based on the assumption that SCC mixtures have a comparatively smaller amount of coarse aggregate content, which may reduce shear resistance by reducing the aggregate interlock between fracture surfaces. The tested beams varied in depth (from 150 to 300 mm) and had an effective span of 800 mm. The test parameters included concrete type, maximum size of coarse aggregate and coarse aggregate content.

The results of their investigation showed that SCC (with the same maximum size of coarse aggregate but a lower coarse aggregate content) had similar shear resistance in the pre-cracking stage as VC. The results also revealed the development of lower post-cracking shear resistance in SCC due to lesser aggregate interlock as a consequence of a lower quantity of coarse aggregate.

Kim et al. [27] measured the shear stress of 48 push-off samples to evaluate the aggregate interlock of the SCC and VC. Their study involved an investigation of the influence of aggregate and paste volumes on shear capacity. The aggregate interlock was evaluated based on the crack slip, crack width, and shear stress. The results showed that the SCC samples exhibited less aggregate interlock than the VC samples. As the crack width increased, the decreasing value of the normalised shear stress in SCC samples indicated a decrease in aggregate interlock.

Cattaneo et al. [28] investigated the shear strength of SCC beams with and without shear reinforcement under four-point bending tests considering four shear arm ratios (a/d = 1.5, 2.5, 3.5, 4.5). The outcomes of their research were compared to both VC beams experimental results and to standard design equations. The results indicated that VC and SCC had similar stiffness up to cracking, while

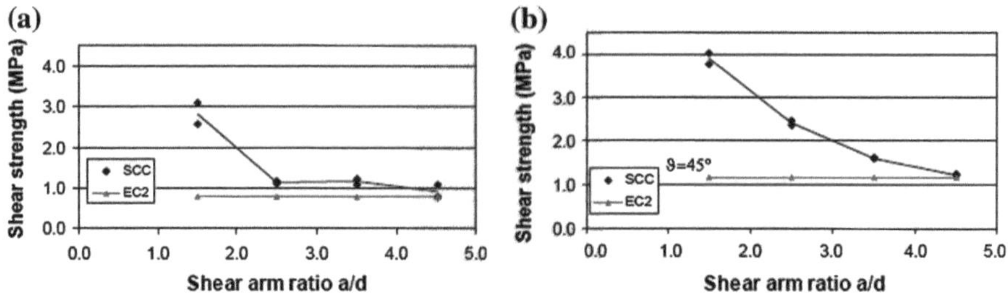

Fig. 5.9 Shear strength Versus shear arm ratio: **a** beams without stirrups and **b** beams with stirrups

in the cracked state SCC exhibits a more stiff behaviour associated with smaller deflection, denoting a more brittle behaviour. The comparison between EC2 predictions and the experimental shear strength for SCC beams (with and without shear reinforcement) showed that EC2 furnishes a lower bound with values closer to those experimentally observed by increasing the shear arm ratio (Fig. 5.9).

Hassan et al. [29] studied the behaviour of full-scale SCC and VC beams in shear. Twenty flexurally reinforced concrete beams, with no shear reinforcement, were tested under mid-span concentrated load until shear failure occurred. The tested beams had 150, 250, 363, 500, and 750-mm depths and 1050, 1750, 2340, 3200 and 4500 mm lengths respectively. The test parameters included coarse aggregate content (which was lower in the SCC), beam depth and longitudinal reinforcing steel ratio (1 and 2 %). Performance was evaluated based on crack patterns, crack widths, loads at the first diagonal cracking, ultimate shear resistance, and failure modes. The results indicated that the ultimate shear strength of SCC beams was slightly lower than that of VC beams, and that the difference was more pronounced with the reduction of longitudinal steel reinforcement and the increase in beam depth (Fig. 5.10). The results also indicated that there are no significant differences between SCC and VC beams in terms of crack widths, crack heights, crack angles or overall failure mode.

Choulli et al. [30] investigated the shear behaviour of full-scale prestressed I-beams made with high strength SCC (around 90 MPa) and compared the results with those of VC. The results obtained from their study showed that there is a reduction in shear capacity for beams made with SCC of about 10 %, in comparison with beams made with VC having the same compressive strength. They attributed this reduction to the fact that the crack surfaces in SCC were smoother than those of VC, due to the high amount of paste in SCC, which reduces the contribution of the aggregate interlock to shear strength. The results also showed that the SCC provides more ductility for structural members than VC. With regard to the Code-based equations, Eurocode 2, ACI 318-02, and the Spanish EHE Code were clearly conservative in their estimations of the ultimate shear load for all beams specimens.

Hassan et al. [31] investigated the shear strength, mid-span deflection, and cracking characteristics of corroded reinforced concrete beams made with SCC/VC beam. Their investigation was carried out at three different corrosion

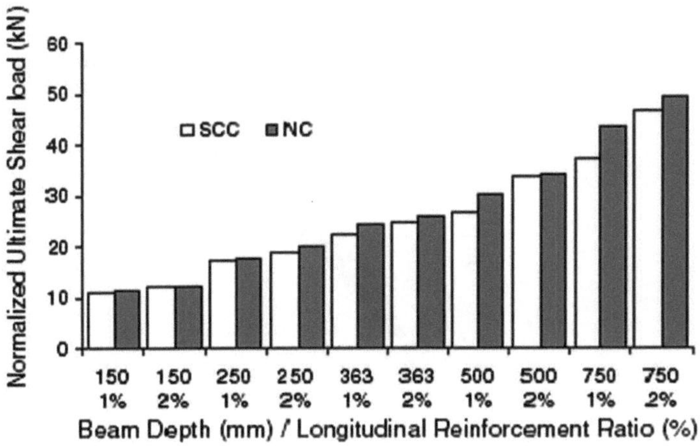

Fig. 5.10 Normalised ultimate shear load for SCC and VC beams [29]

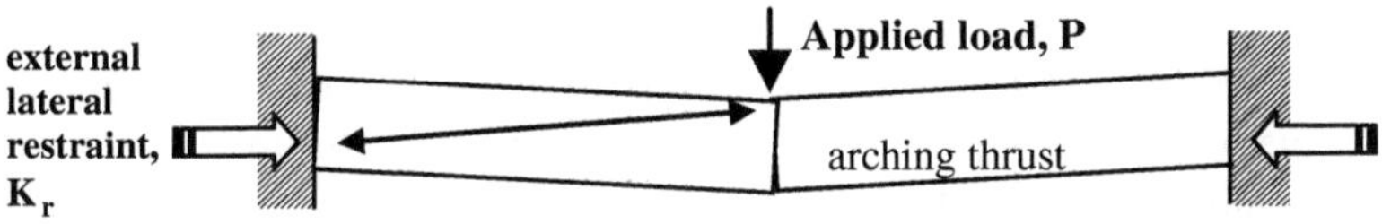

Fig. 5.11 Compressive membrane action in laterally restrained reinforced concrete

levels; zero corrosion (0 % mass loss), moderate corrosion (10 % mass loss), and severe corrosion level (30 % mass loss). The results indicated that at zero and moderate corrosion levels, SCC and VC beams failed in shear (as they were originally designed to). However, SCC beams exhibited lower shear capacity than VC beams. The post-diagonal cracking shear resistance and ductility of SCC beams were also lower compared to VC beams in both zero and moderate corrosion levels. This was due to the development of less aggregate interlock as a consequence of the presence of less coarse aggregate in SCC mixture compared to VC mixture. Hassan et al. also concluded that both ACI and EC2 can be used for the calculation of deflection in shear-dominated beam failure, even at moderate corrosion levels.

5.6 Compressive Membrane Action of SCC

The arching thrust develops when a slab is restrained against longitudinal expansion (Fig. 5.11). With the development of tension cracks at mid-span and at the supports the beam tries to expand longitudinally but as it is restrained, arching forces are induced which develop as the deformation increases.

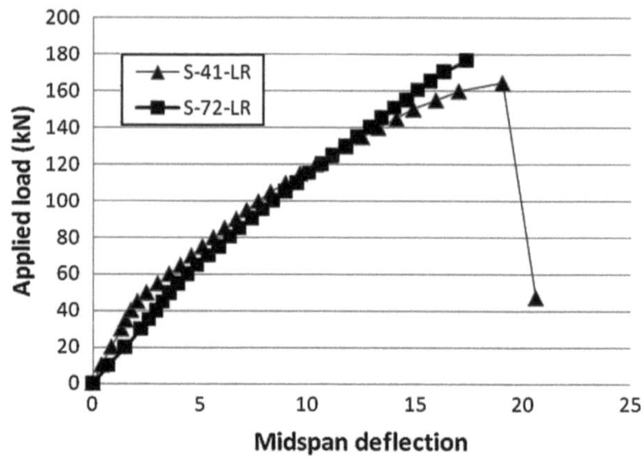

Fig. 5.12 Test results for midspan deflection versus applied load for the failure tests

This phenomenon is generally referred to as Compressive Membrane Action (CMA). The extent of the enhancement due to CMA, over and above the flexural strength, depends on the degree of restraint provided by the surrounding structure and the concrete compressive strength [32–34]. The bending capacity of an unrestrained concrete slab strips is affected primarily by the amount and strength of the reinforcement and only marginally by the concrete strength. In contrast, laterally restrained slab strips generally fail by crushing of the concrete at the hinges and so the capacity is significantly influenced by the concrete compressive strength.

For SCC, Taylor et al. [35] investigated the arching action of slabs with two grades of SCC (SCC40 and SCC60) containing limestone powder and ground granulated blast furnace slag (GGBS) in laterally restrained slabs made with steel reinforcing bars. The compressive strengths at 28 d of both grades of SCC were 41 and 72 MPa, respectively. The steel reinforcing bars were deformed high yield 12 mm diameter bars with yield strength of 503 N/mm^2. A line load was applied across the midspan of each test slab and the loading arrangement is shown in Fig 5.11. End restraint was provided by a self-straining stiff steel frame.

In this study, it was reported that the failure then occurred by crushing in the compression zone. The slab restraint enabled the development of a negative support moment and corresponding cracks. The results of applied load vs. vertical deflection at midspan are shown in Fig. 5.12 for both of the test slabs.

Taylor et al. [35] compared the experimental results of the ultimate capacity of the tested slabs with those predicted by the arching theory and the standard flexural theory and found that both predictions were conservative. They also concluded that the prediction using the arching strength was more accurate than the standard flexural theory prediction.

Taylor and Sonebi [36] also studied the arching action of two SCC slabs contained 0.6 % Glass fiber reinforced polymer (GFRP) reinforcement in a single bottom layer. The compressive strengths at 28 d were 49 and 69 MPa, respectively. Again, predictions obtained by using the arching theory gives a far better estimate for the ultimate capacity compared to standard flexural theory.

Fig. 5.13 Thompson Bridge cast with SCC and basalt fiber reinforced (bridge deck slab)

Taylor and Sonebi [36] reported that the ultimate strength of the laterally restrained slabs was more dependent on concrete compressive strength than the strength of the GFRP reinforcement. The GFRP reinforced slabs showed similar behaviour to previously tested steel reinforced SCC slabs [35] and this was due to compressive membrane action.

Thompson's Bridge was the first bridge casted with SCC in Northern Ireland (UK) using compressive membrane action. Since the slabs are restrained against lateral expansion by the supporting beams, the application of vertical loading, such as a wheel load, results in compressive membrane action/arching action (CMA). It was a mid-span section constructed with basalt-fiber-reinforced polymer (BFRP) bars of 12 mm diameter and the remaining slab had 12 mm steel reinforcement using SCC with grade C50 (Fig. 5.13) [37]. Experimental tests carried out on the bridge confirmed the excellent mechanical performances of the system.

5.7 Concluding Remarks

The brief description contained in the previous sections of the most relevant aspects of the structural behaviour of SCC, allowed to draw some final remarks. As a general statement, SCC is characterized by a structural behaviour similar to that of VC; nevertheless, some appreciable differences can be found.

In terms of deformation, stiffness and flexural strength some differences can be found with respect to VC, due to the usually smaller amount of coarse aggregate and denser cement matrix; reducing the coarse aggregate quantity, in fact, leads to an increase of the strain at peak, to a reduction of the axial stiffness of structural elements and to an increase of the flexural strength. At the same time, introduction of fines or fillers produces a denser cement matrix which can positively affect the stiffness and the flexural strength (improving the ITZ).

Ductility of SCC is in general higher than that of VC, even though the difference is not so relevant. If the concrete is confined by means of stirrups, SCC shows larger ductility due to its smaller elastic modulus.

When considering reinforced structural elements under flexure made of SCC, the structural behaviour is very similar to that of corresponding elements made of VC. In fact, cracking patterns, failure modes and failure loads are almost identical; only a small reduction of cracks opening and a slight increase of deflection can be found in some cases. In this framework, conventional design predictive models and equations proposed for VCs can be applied also to SCCs. In terms of shear capacity, there is a quite general agreement that the smaller amount of coarse aggregates of SCCs leads to a weakening of the aggregates interlock mechanisms; as a consequence, the shear strength reduces as well with respect to beams made of VC. This difference is not very large (10–15 %) and almost disappears when considering beams with stirrups. As before, also shear models can be applied to elements made of SCC.

The compressive membrane action on slabs has been also investigated, founding out that the positive effect of this type of mechanism applies also to elements cast with SCC.

References

1. Khayat, K.H.: Use of viscosity-modifying admixture to reduce top-bar effect of anchored bars cast with fluid concrete. ACI. Mater. J. **95**(2), 158–167 (1998)
2. Lachemi, M., Hossaini, K.M.A., Lambros, V.: Shear resistance of self-consolidating concrete beams—experimental investigations. Can. J. Civ. Eng. **32**(6), 1103–1113 (2005)
3. Hassan, A.A.A., Lachemi, M., Hossain, K.M.A.: Strength, cracking and deflection performance of large-scale self-consolidating concrete beams subjected to shear failure. Eng. Struct. **32**(5), 1262–1271 (2010)
4. Hassan, A.A.A.: Performance of full scale self-consolidating concrete structural elements in shear, bond and under corrosion attack, Ph.D. Thesis, Ryerson University (2008)
5. Lin, C.H., Hwang, C.L., Lin, S.P.: Self-consolidating concrete columns under concentric compression. ACI.Struct. J. **105**(5), 1262–1271 (2008)
6. Paultre, P., Khayat, K.H., Cusson, D., Tremblay, S.: Structural performance of self-consolidating concrete used in confined concrete columns. ACI.Struct. J. **102**(4), 560–568 (2005)
7. Sonebi, M., Bartos, P.J.M.: Performance of reinforced columns cast with self-compacting concrete. In: Malhotra V. M. (ed) Proceedings of the 5th CANMET/ACI International Conference on Recent Advances in Concrete Technology, SP 200–25. pp. 415–431. Singapore (2001)
8. Khayat, K.H., Paultre, P., Tremblay, S.: Structural performance and in-place properties of self-consolidating concrete used for casting highly reinforced columns. ACI. Mater. J. **98**(5), 371–378 (2005)

9. Lachemi, M., Hossaini, K.M.A., Lambros, V.B.: Axial load behaviour of self-consolidating concrete-filled steel tube columns in construction and service stages. ACI.Struct. J. **103**(1), 38–47 (2006)
10. Restrepo, J.I., Seible, F., Stephan, B., Schoettler, M.J.: Seismic testing of bridge columns incorporating high-performance materIals. ACI.Struct. J. **103**(4), 496–504 (2006)
11. Huang, P.F. ; Workability and earthquake resistance behaviour of self-compacting concrete frame. ACI SP 233–10. pp. 155–178 (2006)
12. Galano, L., Vignoli, A.: Strength and ductility of hsc and scc slender columns subjected to short-term eccentric load. ACI.Struct. J. **105**(3), 259–269 (2008)
13. Cetin, A., Carrasquillo, R.L.: High performance concrete: influence of coarse aggregate on mechanical properties. ACI. Mater. J. **95**(3), 252–261 (1998)
14. Amparano, F.E., Xib, Y., Sook, Y.: Experimental study on the effect of aggregate content on fracture behaviour of concrete. Eng. Fract. Mech. **67**, 65–84 (2000)
15. Zanuy, C., Maya, L.F., Albajar, L., de la Fuente, P.: Transverse fatigue behaviour of lightly reinforced concrete bridge decks. Eng. Struct. **33**, 2839–2849 (2011)
16. De Pauw, P., Boel, V., Clemmens, F., De Schutter, G., De Clercq, E. Application of self-compacting concrete in combination with high strength prestressing strands for precast, prestressed beams fib Symposium "Keep concrete attractive", Budapest 2005
17. Hassan, A.A.A., Lachemi, M., Hossain, K.M.A.: Bond strength of deformed bars in large reinforced concrete members cast with industrial self-consolidating concrete mixture. Constr. Build. Mater. **24**(4), 520–530 (2010)
18. Maghsoudi, A.A.: Theoretical and experimental serviceability performance of SCCs connections. Struct. Eng. Mech. **39**(2), 241–266 (2011)
19. Sonebi, M., Tamimi, A.D., Bartos, P.J.M.: Performance and cracking behaviour of reinforced beams cast with self consolidating-concrete. ACI. Mater. J. **100**(6), 492–500 (2003)
20. Luo, S.: Research on the Bending and Shearing Properties of Self-Compacting Concrete Beams. In: Proceedings of the 1st International Symposium on Design, Performance and use of SCC, Changsha, pp. 641–648. RILEM Publications SARL, Bagneux (2005)
21. De Corte, W., Boel, V.: Crack width analysis of reinforced self-compacting concrete beams. Eng. Mater. **452–453**, 629–632 (2011)
22. Kumar, R., Kumar, R., Kumar, N.: In situ performance of self-compacting concrete in T-beams. J. Mater. Civ. Eng. ASCE. **21**(3), 103–109 (2009)
23. Cuenca, E., Serna, P., Pelufo, M.J.: Structural behaviour of self-compacting and reinforced concrete under shear loading. In: Domingo, A., Lazaro, C. (eds) Proceedings of IASS Symposium 2009, pp. 2920-2931. Valencia, Spain (2009)
24. Collins, M.P., Mitchell, D., Adebar, A., Vecchio, F.J.A.: General shear design method. ACI.Struct. J. **93**(1), 36–45 (1996)
25. Tompos, E.J., Frosch, R.J.: Influence of beam size, longitudinal reinforcement, and stirrup effectiveness on concrete shear strength. ACI.Struct. J. **99**(5), 559–567 (2002)
26. Lachemi, M., Hossain, K.M.A., Patel, R., Shehata, M., Bouzoubaâ, N.: Prediction of flow behaviour of high volume fly ash self-consolidating concrete from the rheology of paste and mortar. Mag. Concr. Res. **7**(59), 517–528 (2007)
27. Kim, J.K., Trejo, D., Hueste, M.D.: Shear characteristics of self-consolidating concrete for precast prestressed concrete members, SP 247–5, pp. 53–63. ACI Publication, Detroit (2007)
28. Cattaneo, S., Mola, F., Rosati, G. Shear strength in self-compacting reinforced concrete beams 4th Int. Conf. on the conceptual approach to structural design, pp.28–29 Venice (2007)
29. Hassan, A.A.A., Lachemi, M., Hossain, K.M.A.: Behaviour of full-scale self-consolidating concrete beams in shear. Cement Concr. Compos. **30**(7), 588–596 (2008)
30. Choulli, Y., Marı′, A., Cladera, A.: Shear behaviour of full-scale prestressed i-beams made with self-compacting concrete. Mater. Struct. **41**, 131–141 (2008)
31. Hassan, A.A.A., Lachemi, M., Hossain, K.M.A.: Structural assessment of corroded self-consolidating concrete beams. Eng. Struct. **32**(3), 874–885 (2010)
32. Rankin, G.I.B., Long, A.E.: Arching action strength enhancement in laterally restrained slabs. ICE. Proc. Struc. Build. **122**, 461–467 (1997)

33. Taylor, S.E., Rankin, G.I.B., Cleland, D.J.: Arching action in high strength concrete slabs. ICE. Proc. Struct. Build. **146**(4), 353–362 (2001)
34. Taylor, S., Mullin, B.: Arching action in FRP-reinforced concrete slabs. Constr. Build. Mater. **20**(1–2), 71–80 (2006)
35. Taylor, S., Sonebi, M., Kidd, D., McCord, W. Arching action in steel reinforced self-compacting concrete slabs, 2nd International Symposium on Design, performance, and use of on self-compacting concrete,pp. 676-686. Beijing (2009)
36. Taylor, S., Sonebi, M. : Compressive membrane action in GFRP reinforced self-compacting concrete bridge deck slabs. In: Khayat K. H., Feys V (eds) Proceedings of SCC 2010, 6th International RILEM Symposium on Self-Compacting Concrete/4th North American Conf. On Design, Placement and Use of SCC 2010. pp. 1119–1128. Springer (2010)
37. Taylor, S., Robinson, D., Sonebi, M.: Basalt-fibre-reinforced polymer reinforcement. Concrete **45**(4), 48–50 (2011)

Chapter 6
Fiber Reinforced SCC

Liberato Ferrara

6.1 Introduction

In the last decade FR-SCC has been used in several partially and fully structural applications, including slabs on grade [1], overlays [2], precast prestressed beams [3], roof elements [4], sheet piles [5], tunnel segments [5], precast post-tensioned girders for slope stabilization [6], panel and slab housing units [7, 8], refractory linings in industrial equipment [9] and façade panels [10, 11].

So far, the largest number of applications has dealt with steel fibers (steel fiber reinforced self-compacting concrete—SFR-SCC). The use of different types of steel fibers (straight, hooked-end, crimped, etc.), having circular, rectangular or even elliptic cross-section has been reported. The fiber length lf varied from 5 to 60 mm and aspect ratios (l_f/d_f), where d_f is the diameter of the fiber cross-section, from 30 to more than 100. The employed dosage of fibers varied, depending on the application, from 25 to 30 kg/m^3 (0.32–0.38 % by volume) for slabs on grade and façade panels, to 40–50 and 80 kg/m^3 (0.64–1.02 % by volume) for housing units, tunnel segments and precast prestressed elements, and up to more than 100 kg/m^3 (1.27 % by volume) in some structural applications (125 kg/m^3—1.59 % by volume for sheet piles) and lab-scale studies (up to 157 kg/m^3—2 % by volume). Examples of dosages even lower than 0.3 % by volume can be found: in this case the sole aim of the fiber addition was the control of shrinkage cracking.

The use of polymeric (mainly polypropylene) macro-fibers (lengths between 25 and 54 mm), either alone or in combination with steel fibers, in volume percentages ranging between 0.25 and 1 % (2.3–9 kg/m^3) has been also reported with reference to lab-scale investigations [12–14]. The only structural application so far found in the literature dealt with façade panels [11]. The use of shorter polypropylene fibers for purposes other than avoiding explosive spalling under fire and

L. Ferrara (✉)
Politecnico di Milano, Milan, Italy
e-mail: liberato.ferrara@polimi.it

K. H. Khayat and G. De Schutter (eds.), *Mechanical Properties of Self-Compacting Concrete*, RILEM State-of-the-Art Reports 14,
DOI: 10.1007/978-3-319-03245-0_6, © RILEM 2014

high temperature has been investigated in recent studies, even in combination with lightweight aggregate [15, 16].

The use, in a self-compacting cementitious matrix of other kind of fibers than steel and polypropylene has been reported in a number of surveyed studies and dealt with either inorganic (glass fibers, [17]) or natural fibers, such as coconut ([18], up to 3 % by weight of fibers, with lengths varying between 3 and 83 mm), sisal ([19], up to 1 % by volume = 1 kg/m^3) or cellulose ([20], up to 3.6 kg/m^3 = 0.12 % by volume). Both structural [16, 18, 19] and non-structural ([17]—for thermo-acoustical insulation) applications were considered.

In all the aforementioned applications, one of the main advantages of using fibers is represented by the possibility of partially or even completely substituting the traditional welded wire mesh reinforcement, such as shear reinforcement in beams and roof elements. This reduces the need for manufacturing, detailing and placing the reinforcement cages and results in improved production efficiency. Furthermore, the element thickness and the structure self-weight can be reduced, since minimum cover requirements do not hold any more.

The major advantage of incorporating fibers in SCC is the elimination of vibration to consolidate the concrete and the enhancement of the stability of the SCC matrix. This results in a randomly uniform dispersion of the fibers within structural elements that is not affected by the downward settlement and segregation of the fibers [4, 21, 22]. This requisite is of paramount importance for a reliable structural performance of elements made with fiber reinforced cementitious composites. Improper compaction and placement, furthermore complicated by the negative effect of fibers on workability [23], may hinder the random dispersion of fibers within structural elements. Areas with a reduced fiber dosage or no fibers, act as flaws, triggering early failure and activating unforeseen mechanisms, thus affecting the load bearing capacity and the structural performance.

It has been recently recognized that, through a suitably balanced performance of the fluid mixture, fibers can be effectively aligned along the casting-flow direction [24–32]. By suitably tailoring the casting process to the intended application, the flow direction of the fresh concrete, along which fibers tend to be aligned, may be made to coincide, as closely as possible, with the anticipated stress pattern (i.e. the direction of principal tensile stresses) within the structural element when in service. This would lead to a superior mechanical and structural performance [29–33] which can also result into optimized structure size and reduced self-weight. Benefits may also follow in terms of time- and cost-effectiveness of the construction process as a whole, including, in the case of precast elements, costs for transportation, lifting equipment, etc.

It is evident that the synergy between the technologies of self-compacting concrete (SCC) and fiber reinforced concrete (FRC) can improve the economic efficiency of the construction process given the reduced construction time, reduced work force, reduced energy consumption, enhanced working environment with reduced noise and health hazards, and enhanced automation of the quality-control process, thanks, among other factors, to the reduction of the ordinary reinforcement and to the simplification of reinforcement detailing and placing. This clearly highlights that FR-SCC favours sustainability as well.

As a matter of fact, the influence of the flow-driven fiber dispersion and orientation on the mechanical performance represents a key distinctive feature of FR-SCC as compared to traditional vibrated FRC, which cannot be disregarded when analysing engineering and mechanical properties of the material. To this aim, a thorough understanding is required of the mechanisms underlying the connection between mix-design and fresh state performance, on one hand, and the dispersion and orientation of the fibers on the other hand, also in the context with monitoring and prediction to achieve the anticipated performance in the hardened state. In this framework the state of the art review of engineering and mechanical properties of FR-SCC will be prefaced by a literature survey focusing on the aforementioned topics. Structural performance related issues will be finally addressed with reference to significant case studies.

6.2 Mix-Design Methods for FR-SCC

The common rationale underlying mix-design methods proposed for plain SCC [34] is the concept of SCC as a "suspension" of a granular aggregate skeleton into a fluid phase, which may be either the paste or the mortar, depending on the "scale of observation" and hence on the fractions of aggregates considered in the former. On one hand the grading of the solid skeleton is optimized to achieve an optimum packing density [35–39]. On the other, the composition of the cement paste/mortar has to be selected to achieve a performance in the fluid state characterized by suitable ranges of yield stress and viscosity, able to guarantee adequate fluidity of the concrete mixture, as well as resistance to static and dynamic segregation [40]. Rheological properties have to be optimized as a function of the particle size distribution and relative volume fraction of the solid skeleton [41]. A denser suspension of the aggregates into the paste requires a lower yield stress and a lower viscosity of the suspending fluid.

In the case of FR-SCC, this approach has to be adapted to take into account the influence of the fibers, as a function of their type and dosage, on the packing density of the solid skeleton.

Experimental assessment of the effects mentioned above has been reported [5, 42, 43], e.g. to determine the optimal sand to total aggregate ratio (for a given type and dosage of fibers). Grunewald [5] reported that, whereas the addition of fibers was detrimental to the packing density of a solid skeleton consisting of the coarse aggregate fraction alone (4–16 mm), optimum packing density (75–80 %) could be obtained, almost independently of the fiber type and dosage (up to 3 % by volume) in mixtures with a sand (0.125–4 mm) to total aggregate ratio of 50 %.

Theoretical approaches have also been formulated, either through the concept of the perturbed volume [35] around a fiber, to take into account the modification of solid skeleton packing, or regarding fibers as a fictitious aggregate fraction, characterised by an equivalent packing diameter [44, 45]. Grünewald [5] applied both approaches obtaining good predictions for a wide range of steel fiber types and dosages (up to 3 % by volume).

Ferrara et al. [43] proposed to include fibers into the optimization of the granular skeleton as a fictitious aggregate fraction with 100 % passing at an "equivalent fiber diameter" ($d_{eq,fiber}$) calculated through specific surface equivalence:

$$d_{eq,fiber} = \frac{3\, l_f}{1 + 2\left(l_f / d_f\right)} \frac{\gamma_{fiber}}{\gamma_{aggregate}} \tag{6.1}$$

where γ_{fiber} and $\gamma_{aggregate}$ denote the specific gravity of the fiber and aggregate respectively. This definition is instrumental to the extension to the mix design of FR-SCC of the so-called "rheology of paste model" [40, 41] formulated for plain SCC and results in a modification of the coarse aggregate fraction to compensate for the addition of fibers.

6.3 Fresh State Performance

The addition of fibers to a concrete matrix affects the fresh state performance of the latter due to both the large surface area of fibers, which requires a higher volume of fluid phase (either paste or mortar) to be properly enveloped and lubricated, and the significant inter-particle friction and interlocking among the fibers as well as between the fibers and aggregates [23, 46]. In order to achieve the desired fresh state performance of a FR-SCC, the importance has been highlighted of both an effective mix-design and a suitably calibrated mixing protocol, which would require fibers to be finally added to a matrix which has already achieved its self-compacting consistence [47, 48].

With reference to plain SCC, extensive research on this topic has been done, including definition of testing protocols [49–52], correlation of field test measurements [53–56] to fundamental rheological properties [57–64] and round robin validation of commercially available or custom-built rheometers [65–68].

In order to transfer and apply the aforementioned achievements to FR-SCC, besides quantifying the effects of fibers on rheological properties and fresh state performance, and transfer related research findings into predictive modelling tools [45], the measurement of fiber dispersing ability has to be addressed as a key feature which distinguishes FR-SCC from conventional vibrated FRC.

Table 6.1 Minimum clear gap spacing (c) to fiber length (l_f) ratio to avoid blockage of FR-SCC as a function of fiber dosage [70]

c/l_f	l_f/d_f	Maximum fiber dosage (kg/m^3)
≥3	80	30
	65	60
≥2	65	30
	45	60
≥1.5	45	30

6.3.1 Test Methods and Equipment

Standard tests employed for SCC [53–56] have been extensively applied also to FR-SCC. Standard or slightly modified tools have been employed, such as the "fiber funnel" [5], which is twice longer and has a nozzle with four times larger cross section area than the standard one. The same target values as for plain SCC proved to be adequate for a quite broad range of fiber types and dosages. Acceptance criteria of FR-SCC based only on the results of slump-flow and V-funnel tests have been recently questioned and more comprehensive ones, also accounting for the filling capacity have been proposed [69].

With reference to the filling capacity as well as to the passing ability of FR-SCC [70–73] guidelines have been proposed for the minimum ratio of bar spacing to fiber length in order to avoid blocking, also as a function of fiber content (Table 6.1) Grunewald [5] also highlighted the dependence of the blocking bar spacing on the mix-composition and on the fundamental rheological properties of the FR-SCC: for the same amount of same type of fibers, the non-blocking bar spacing decreased with increasing the yield stress, which corresponded to a decrease of the paste volume and hence of the thickness of paste/mortar layer enveloping both fibers and coarser aggregates.

Reduced congestion of reinforcement due to the use of fibers may make the L-box, U-box and J-ring tools unnecessarily demanding for the passing ability required by the intended FR-SCC application. Mock-ups of the most "congested" structural detail have been employed in a number of surveyed studies [3, 6]. Plexiglas moulds may allow the flow to be visually inspected (Fig. 6.1). Fiber washing-out and counting/weighing has been also reported in [6] to evaluate the fiber dispersing ability of the mixture and correlate it to the filling capacity (see also Fig. 6.1d, e).

The fiber dispersing ability, which stands as the distinctive feature of FR-SCC with respect to a conventional vibrated FRC: different methods have been proposed to evaluate it, based on the resistance of the mixture to static and dynamic segregation of fibers [74–77] (Fig. 6.2), adapting approaches proposed for plain concrete [78–81]. Correlation between different indicators of fresh state performance (slump-flow diameter, time to final spread, V-funnel flow time) and indices representative of the mixture resistance to static and dynamic segregation of fibers have also been established (Fig. 6.3), in an attempt to assess their acceptable ranges and significance in terms of the fiber dispersing ability [77].

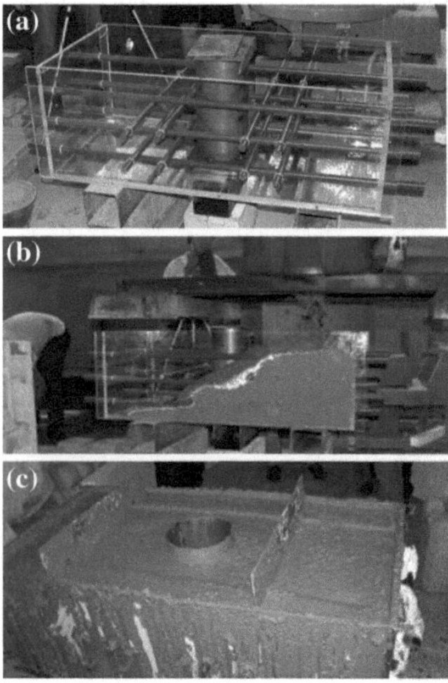

(d)

	Fiber content (kg/m³)		
Sector	Test 1	Test 2	Test 3
1	40.01	60.33	44.43
2	54.48	63.36	69.33
3	54.16	68.12	53.62
4	47.25	61.15	64.66
5	45.34	64.12	57.68
6	50.35	76.00	61.94
7	46.58	70.63	62.41
8	57.56	66.95	62.87
9	55.71	57.82	58.37
Av.	50.16	65.39	59.48

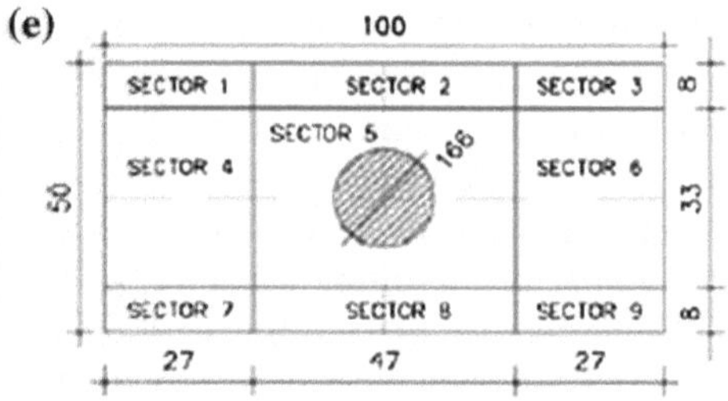

Fig. 6.1 Plexiglas replica of a SCSFRC element, with reinforcement detail for passing ability and filling capacity evaluation (**a**); filling sequence (**b, c**); fiber dispersion evaluation (**d, e**) [6]

6.3.2 Effects of the Fibers on the Fresh State Performance and Rheology of FR-SCC: Experiments and Physical Modelling

Effects of fibers on the fresh state performance of FR-SCC have been effectively quantified and modelled through the so-called fiber factor $V_f\ l_f/d_f$ [1, 5, 23]. The higher the fiber content V_f and the higher the fiber aspect ratio l_f/d_f, the worse the fresh state performance of the FR-SCC mix, i.e. the lower the slump-flow diameter and the higher both the yield stress and the plastic viscosity (Fig. 6.4). Furthermore, the higher the fiber factor, the larger the dispersion of experimental results. The rate of variation of the aforementioned properties with the fiber factor depends also on the mixture composition [5]. This makes the comparison difficult among results obtained by different authors, even in terms of relative performance to the reference plain SCC, because of the different adopted mix-design approaches.

A unifying interpretation has been proposed by Martinie et al. [46], through the concept of relative packing fraction of the granular skeleton of fibers and aggregates, defined as the ratio between the actual and the maximum theoretical packing density (Fig. 6.5).

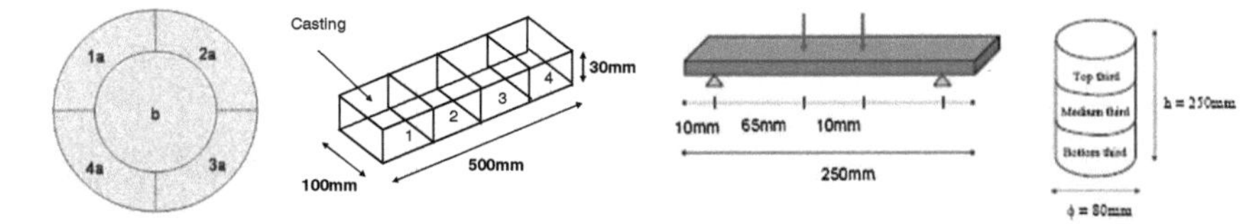

Mini-slump test

Objective: to assess the ability of the mix to disperse fibers in free flow

Procedure: evaluate fiber content in different parts of the flow area and calculate Fiber Dynamic Segregation Index (free flow).

Channel flow test

Objective: to evaluate material's ability to "drive" fibers along a constrained flow

Procedure: pour material from one end and allow to flow to the other. Measure fiber contents along the flow and calculate Fiber Dynamic Segregation Index (channel flow).

Bending test

Objective: to evaluate the effect of static segregation of fibers on the mechanical performance

Procedure: cast two nominally identical specimens and test either downside or upside down to casting sense. Modulus of rupture difference is evaluated.

Cylinder test

Objective: to evaluate resistance to static segregation of fibers

Procedure: measure fiber content in top-medium-bot-tom sections and calculate Fiber Static Segregation Index

Fig. 6.2 Test methods used to assess static and dynamic segregation of fibers of FR-SCC [77]

It appears that a threshold value exists for the relative packing fraction below which the flow behaviour of FR-SCC is dominated by the rheology of the fluid matrix and the influence of the fibers is low; on the other hand, once this threshold has been overcome, the flow behaviour becomes strongly dominated by the contact network between the fibers, whose stiffening effect with reference to the rheology of the unreinforced matrix appears to be strongly nonlinear. This has been also confirmed by recent experiences dealing with shear and adhesive rheology of self-compacting mortars reinforced with cellulose fibers [82].

A similar approach based on the "number of fiber-fiber contact per fiber" has been proposed by Chalencon et al. [83] and yields to similar results. In the same framework, Ghanbari and Karihaloo [84] recently proposed a predictive approach to the plastic viscosity of SFR-SCC, which provides likewise reliable predictions.

The criterion of excess paste/mortar thickness has been also employed to interpret fresh state performance of FR-SCC (Fig. 6.6): the fresh state performance of the fiber reinforced composite depends on the rheological properties and amount of the fluid phase (either paste or mortar) exceeding that strictly necessary to fill the interparticle voids and hence enveloping and lubricating fibers and coarser aggregate particles [43, 85]. The correlation with the approaches reviewed above is physically evident and has been recently supported by micro-tomography X-ray analyses [83], which showed that the "microstructure and local rheology" of the mortar phase in a fiber reinforced cementitious composite "is not affected by the presence of fibers … and that its rheology within the fiber suspensions plays a key role on the rheology" of fiber reinforced composites.

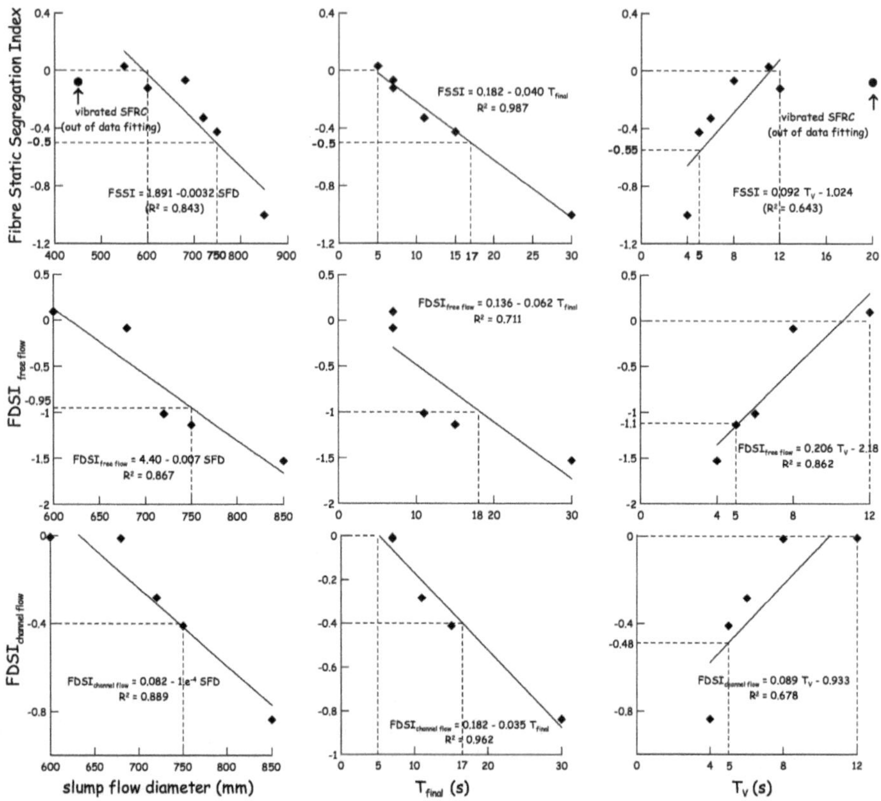

Fig. 6.3 Correlation between fiber static and dynamic segregation indices and indicators of fresh state performance (slump flow diameter, time to final spread, V-funnel time [77])

The role of the excess paste/mortar layer thickness on modelling the performance of SCFRCs in the hardened state, initially addressed by Voigt et al. [86] with reference to drying shrinkage of conventional SFRC.

6.4 Dispersion and Flow-Induced Orientation of Fibers: Experimental Evidence, Monitoring and Predictive Modelling

In an engineering perspective, the foremost outcome of the synergy between SCC and FRC technologies is represented by the randomly uniform dispersion and, in case, tailored orientation of fibers which can be both obtained thanks to the superior fresh state performance of the material. Sound evidence has been provided by dedicated experimental investigation, performed both at the lab-scale [5, 25–32, 87–90] and on full scale applications, including precast beams [4, 5, 24], tunnel segments [5, 91] and underwater slabs [91].

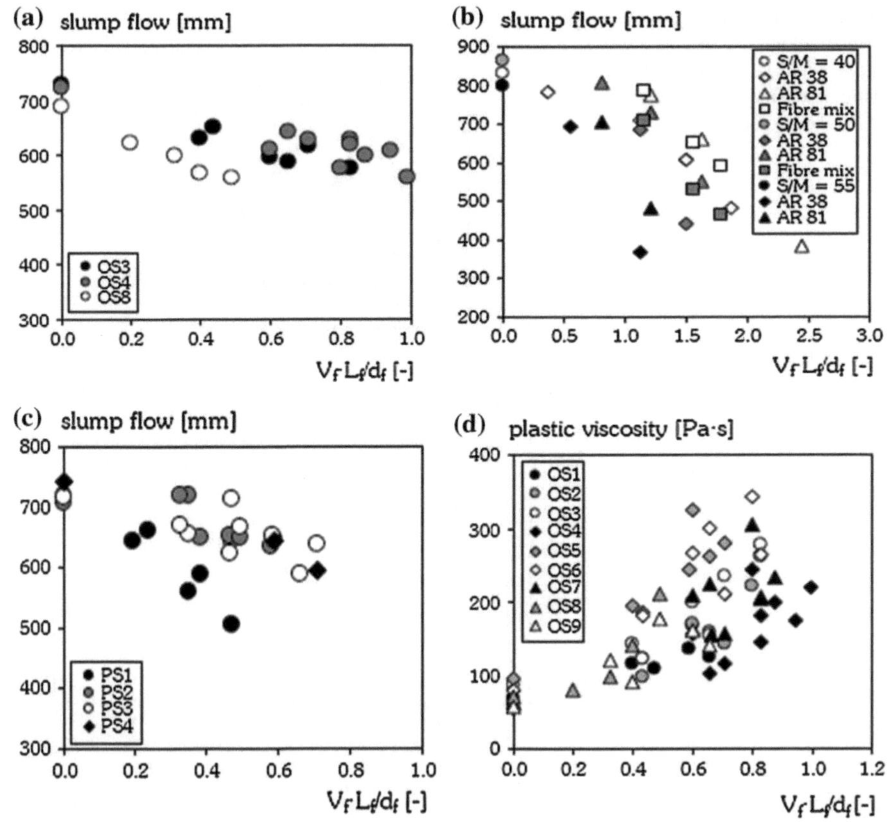

Fig. 6.4 Slump flow (**a–c**) and plastic viscosity versus fiber factor (**d**) for different FR-SCCs [5]

Fig. 6.5 Relative yield stress as a function of the total relative packing fraction. The *dashed line* corresponds to the theoretical random loose packing threshold above which the influence of fiber network starts dominating [46]

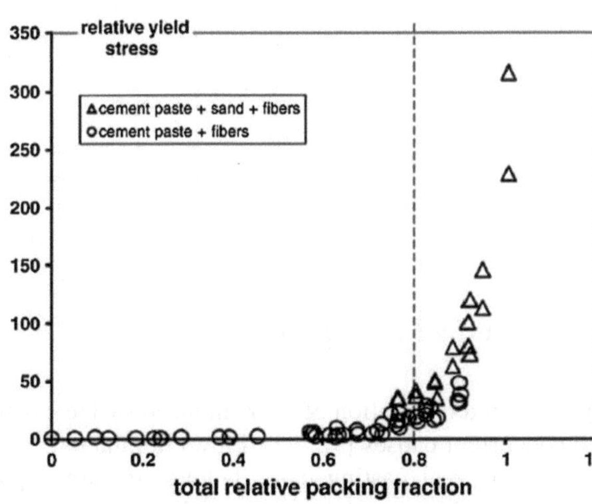

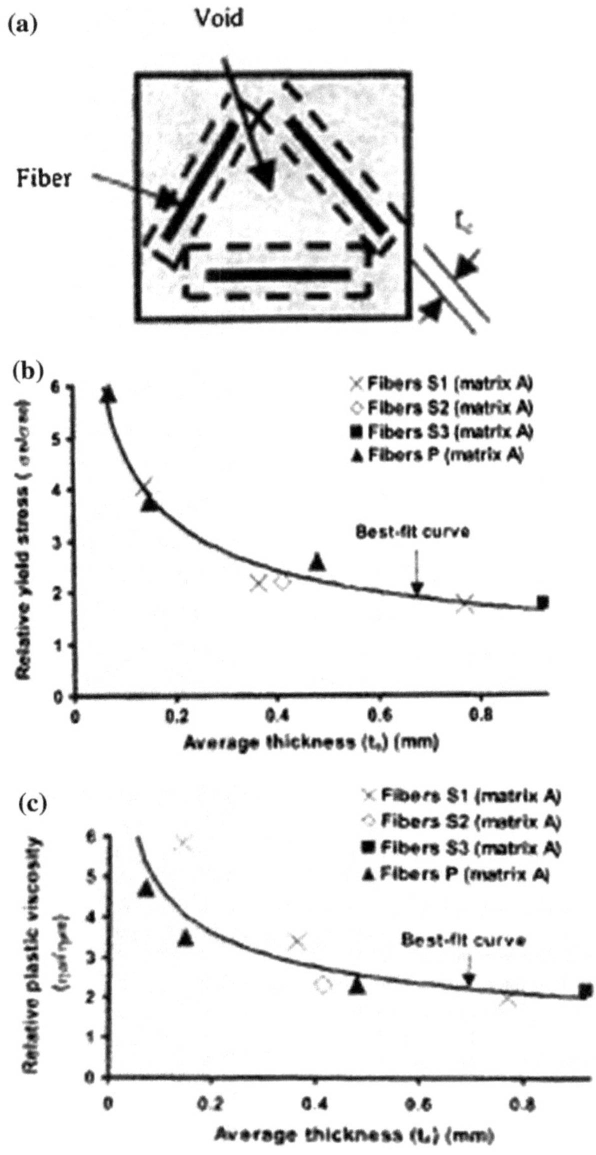

Fig. 6.6 Concept of excess mortar thickness t_c (**a**); relative yield stress (**b**) and plastic viscosity (**c**) of FR-SCC mixes versus t_c [85]

6.4.1 Physical Background

The preferred orientation of fibers induced by the casting flow of FR-SCC is due to two concurrent causes [92, 93]. Flows dominated by shear stresses, to which category many real casting cases belong, feature a parabolic flow-velocity cross profiles and a likewise distribution of drag forces. This, acting transverse to the axis

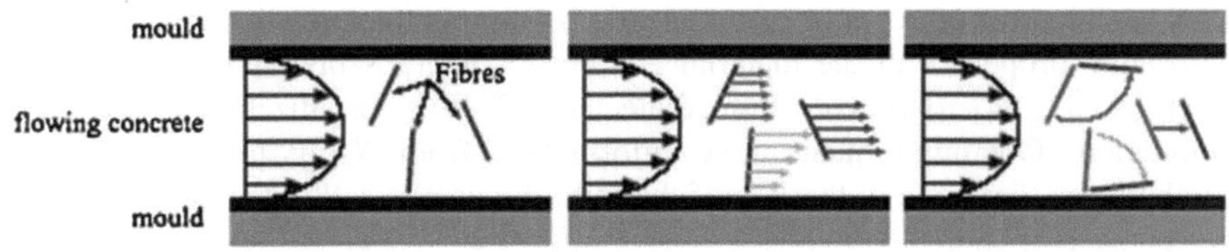

Fig. 6.7 Mechanisms of fiber orientation in a shear-flow [28]

of a fiber immersed in the flow, is likely to induce a torque which makes the fiber itself to align parallel to the flow direction, along which this torque becomes minimum (Fig. 6.7). In the case of a yield stress fluid, the cross profile of flow-velocity, is characterized by a plug-flow zone, i.e. a zone across which, being the shear stress lower than the yield strength, the flow features a constant velocity profile, which yields no or negligible torque on the fiber and this is not able to trigger any preferred alignment [92]. The through-flow extension of the plug-flow zone depends on the value of the yield stress, the higher the latter the thicker the former. The viscosity of the fluid, on its hand, governs the gradient of the flow velocity profile and hence of the drag forces acting transverse to the fiber axis and the magnitude of the resulting "fiber-orientating torque". It has been furthermore shown [93] that this alignment can be practically considered, at the scale of industrial casting processes, as an almost instantaneous phenomenon, since it occurs in a "maximum fiber orientation characteristic time" which is of the order of a couple seconds, which is far shorter than the duration of any real casting process.

The second cause for preferred alignment of fibers is the so called "wall effect", since "it is indeed not possible to find a fiber perpendicular to a wall at a distance lower than half the length of the fiber" [93].

6.4.2 Fiber Dispersion Monitoring

In most of the surveyed investigations, fiber dispersion and orientation were (destructively) monitored by manually counting the fibers on selected "specimen locations" (e.g. over fracture cross-sections of specimens employed for toughness fracture tests). Manual counting is rather easy when low dosages of longer fibers with average aspect ratios are employed [25, 87].Image analysis techniques are required for higher dosages of shorter fibers with higher aspect ratios or whenever quantitative information on the orientation of fibers with respect to the plane of the cut surface is needed [5, 29, 88, 90].

The need for a time- and cost- effective, non-destructive method for the assessment of issues related to fiber dispersion and orientation has hence become crucial, also in the sight of developing "ad-hoc" quality control and acceptance procedures, which can be instrumental to anticipate undesired scattering of mechanical properties due to poor fiber dispersion [77].

X-ray pictures of cores or of thin sections (sample thickness is dictated by the absorption properties of the material and power of the X-ray equipment) can be taken [4, 87, 94] to provide immediate and effective visualization of the fiber orientation. Obtaining quantitative information from X-ray pictures may be cumbersome, both due to the loss of the third dimension and in the case of higher percentages of short fibers.

Electrical methods, based on the effects of the fibers on the resistivity/conductivity of the composite material, has received lots of attention recently [95]. Ozyurt et al. [96] and Woo et al. [97] employed the Alternate Current Impedance Spectroscopy (AC-IS) for the detection of fiber dispersion and orientation. The technique demonstrated its reliability and sensitivity to fiber orientation, clumping, and segregation by means of extensive comparison with results obtained from destructive methods. Attempts were also made to address the application of the aforementioned method to industrial scale problems [24]. The method is based on the frequency-dependent behaviour of cementitious composites reinforced with conductive fibers, such as steel and carbon ones. These were in fact shown to be practically insulating under Direct Current (DC) and low frequencies Alternate Current (AC) while they are conductive under high frequencies AC. The method consists of applying to the specimen a voltage excitation over a range of frequencies (e.g. 10 MHz–1 Hz—[97]) and measuring the amplitude and phase of the flowing current. When the real and imaginary parts of the calculated impedance Z are plotted on a Nyquist diagram, fiber reinforced cementitious composites exhibit the so called dual-arc behaviour (Fig. 6.8a). It features a low-frequency cusp (fibers act insulating), which gives the (higher) resistance of the matrix, R_m, and a high frequency cusp (fibers are conductive), which corresponds to the (lower) resistance of the fiber reinforced composite R. In order to overcome the drawback of the sensitivity of the resistivity of the concrete matrix to moisture conditions, the so-called matrix normalized conductivity is used, from which information about local fiber concentration should be easily obtained by means of a simple mixture rule approach:

$$\frac{R_m}{R} = \frac{\sigma}{\sigma_m} = 1 + [\sigma_{fibers}]\, V_f \tag{6.2}$$

where σ and σ_m are the conductivity of the fiber reinforced composite and of the matrix respectively, $[\sigma_{fibers}]$ is the intrinsic conductivity of the fibers which, in the case of highly conductive fibers, only depends on their geometrical properties (fiber aspect ratio) [98] and V_f is the fiber volume fraction.

The method has been extensively employed to assess the influence of the fresh state performance on the dispersion of the fibers ([28]—Fig. 6.7) and their correlation with the mechanical properties (e.g. fracture toughness) [22]. An application to industrial scale problems has also been addressed [24]. By the way, any direct quantitative comparison between, e.g. the concentration of fibers, as evaluated from Eq. (6.2) and data obtainable from destructive tests (e.g. crushing samples, separating and weighing fibers) could hardly be found in the literature.

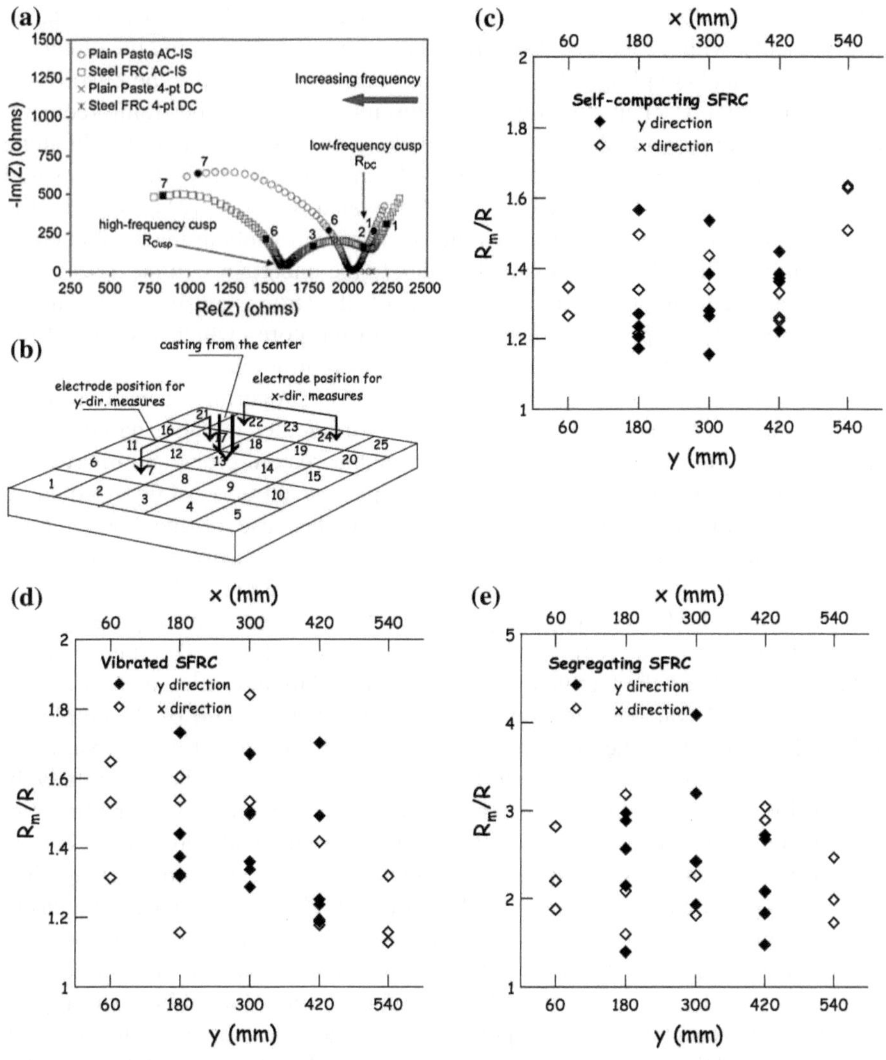

Fig. 6.8 Principle of AC-IS method for fiber dispersion monitoring—example of a Nyquist plot for plain and fiber reinforced cement pastes (**a**—[96]); scheme for AC-IS measurements on 600 mm square slabs (**b**) and matrix normalized conductivities for FR-SCC (**c**), vibrated (**d**) and segregation consolidating SFRC (**e**) plates [28]

The need of a dedicated expensive instrumentation, as required by the extension of the employed frequency range, and the sensitivity of the method itself to the contact impedance between the surface electrodes and the specimen/structure surface, has led several researchers to investigate different approaches and techniques in the mainstream of electrical methods.

Lataste et al. [99] employed a method based on low frequency resistance measurements, with a four electrode arrangement, aimed at reducing the effects of the poor electrical coupling. The method has been demonstrated to be effective in detecting orientation characteristics of the discrete dispersed fiber reinforcement, because of the different resistance measured along the two directions at right angle to each other; qualitative correlation with mechanical properties measured according to the same directions as above was also provided [31]. Anyway the method is not able to provide any quantitative information about the local average concentration of the fibers. This is mainly due to the uncertainty in the assessment of the concrete matrix resistivity, because of its strong sensitivity to aging, moisture content and presence of electrolytes in the pores which also affect the measured resistivity beside the effects of fiber concentration.

The effects of conductive fibers on the capacitive properties of the fiber reinforced composite have led to the development of another method, based on the measurement of the effective permittivity through a coaxial probe and microwave reflectometry techniques [100]. The local average concentration of fibers could be assessed by assuming a random orientation and because of the known fiber geometry (aspect ratio) which governs their capacitive behaviour: as a matter of fact the method is unsuitable for FR-SCC featuring a preferential alignment.

A new method which employs a probe sensitive to the magnetic properties of the steel fibers has been recently proposed and validated [101]. The fundamentals of the method rely on the fact that the presence and relative position of the fibers in a fiber reinforced concrete element modify the magnetic circuit of the employed probe, when placed on the element/structure surface, thus resulting in a variation of the measured inductance. Both fiber concentrations can be quantitatively assessed by suitable calibrating the method and fiber preferential orientation can be estimated. The method, besides its good sensitivity and robustness, is also characterized by an intriguing ease of use, due to the simple equipment which just needs to be positioned of the surface of a structural element, which can be easily done even on vertical elements or slabs accessible only from the bottom, without any dedicated care about the electrical coupling. The method has been applied in a recent study [102, 103] to quantitatively demonstrate the ability of a self-consolidating mortar matrix to disperse and orientate steel fibers along the flow direction in a slab-casting experience. Extensive calibration (Fig. 6.9) was also performed by means of comparison with destructive evaluation of both fiber orientation, through image analysis of slab cross sections at selected locations, and fiber dispersion, evaluated by crushing different specimens extracted from the cast slab and separating fibers with a magnet.

Methods based on the study of heat transients and hence on the effect of fibers on the thermal properties of concrete have been also tentatively applied for the non-destructive assessment of fiber concentration, thanks to the influence of fiber volume fraction on the thermal diffusivity of the composite [104]. Temperature fields within a structural element may be easily surveyed through IR

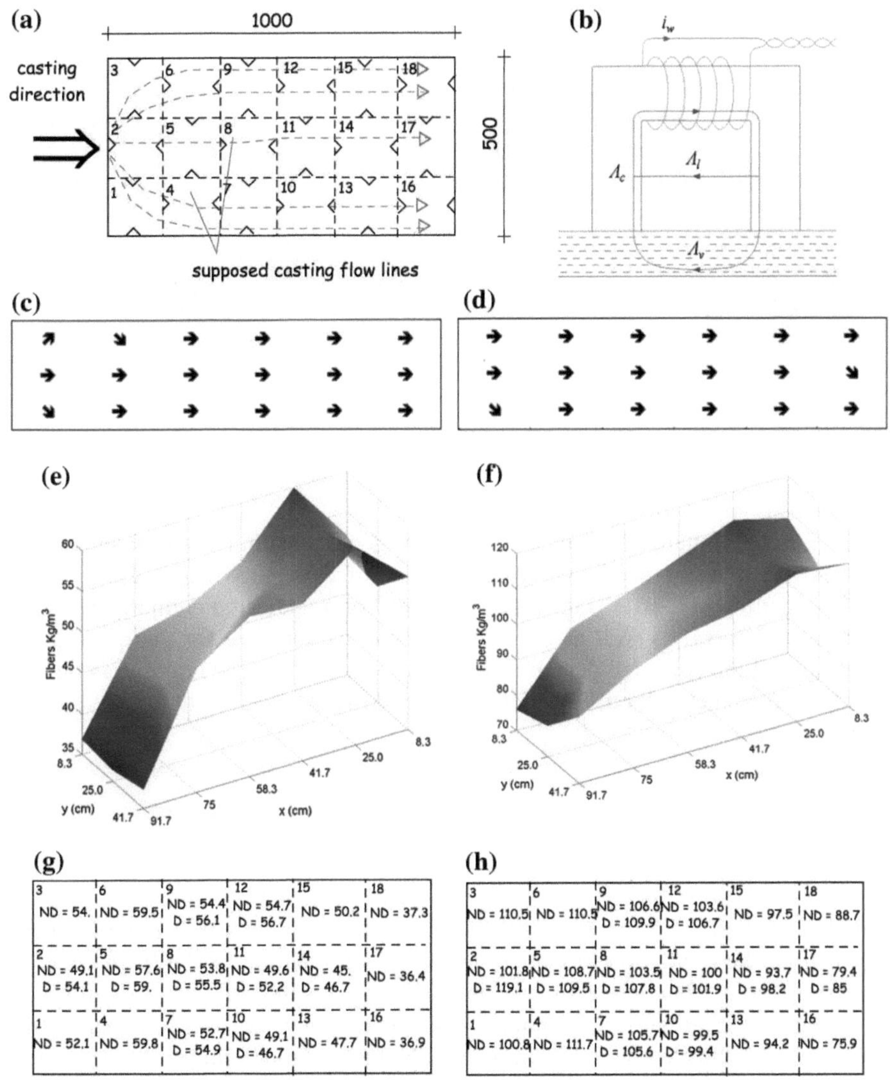

Fig. 6.9 Scheme of slab casting for fiber orientation testing and calibration of magnetic method for not-destructive monitoring of fiber dispersion (**a**); scheme of the measuring set-up (**b**); directions of maximum magnetic inductance for 50 (**c**) and 100 (**d**) kg/m^3 fibers; estimated dispersion of fiber content for 50 (**e**) and 100 (**f**) kg/m^3 fibers and comparison between ND inferred concentration and destructively measured one for 50 (**g**) and 100 (**h**) kg/m^3 fibers [101–103]

thermography, but the slow propagation of temperature variations in thick members, which makes it difficult to provide a controlled input excitation over large areas, may limit the applicability of the method.

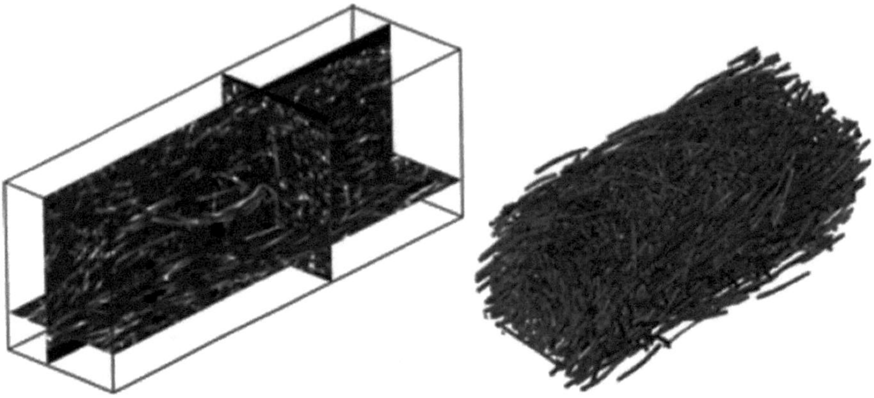

Fig. 6.10 Three different orthogonal slices through a reconstructed FRC specimen from a CT-scan and 3D reconstruction of the steel fibers only [27]

A nice 3D visualization of the fiber arrangement within a specimen can be obtained, as it was recently shown, by means of Computed Axial Tomography (CAT) scanning [27, 105] (Fig. 6.10). The need of a dedicated equipment (like in the case of X-rays) as well as of an image analysis software for the quantitative processing of the collected data, may be singled out, so far, as a major drawback to a wider use of the method, especially at an industrial scale.

6.4.3 *Numerical Prediction of Flow-Induced Fiber Orientation*

With the aim of tailoring the casting process to the requisites of the intended application, the prediction of fiber orientation as a function of the fresh concrete properties and casting flow geometry and process plays the major role, mainly with reference to the alignment of fibers with the direction of the principal tensile stresses within the element when in service. Numerical modelling of fresh concrete flow has been recently recognized as an effective tool to simulate casting processes and optimize them with reference to specific application [106]. Computational Fluid Dynamic approaches, based on the Galerkin FEM formulation of the Navier–Stokes equations and including moving boundaries and free surfaces, have been successfully applied to model the flow of fresh concrete in simple tests [107–110] as well as to a few examples of full scale castings [111].

The inclusion of a dispersed fiber network in these computational tools requires peculiar issues to be taken into account, concerning, e.g. the interaction of the wire-like particles with the suspending fluid and the other aggregates, as a function of the grading and size of these, and also of the fiber geometry and content. Theoretical studies and reviews [92, 112–118] can be found in the literature. The problem has been tackled also from the numerical point of view [119, 120], even if distance still has to be travelled before such an approach could be applied to real casting flow modelling. It is in fact evident that the inclusion of solid elongated particles into the fluid phase represents the most comprehensive and reliable way to deal with the flow of FR-SCC, but computational efforts required to deal with a large number of particles, as may be the case when a true scale casting has to be modeled, may become huge and for this reason make the approach unpractical.

Martinie [92] and Martinie and Roussel [121] have recently implemented a tensor approach, based on the classical Jeffery orbit theory [122] for the orientation process of a rotating ellipsoid immersed in an incompressible fluid. They have interestingly applied it to study the effect of the matrix rheology on the flow induced orientation of discrete (two) wire-like inclusions in a confined shear flow. Because of the computational costs of this kind of approach when applied to full scale casting processes, the following semi-empirical equation has been proposed by the same authors to predict the average fiber orientation factor (α) over a cross section perpendicular to the casting flow, as a function of the concrete yield stress (τ_0), fiber length (l_f) and characteristic dimension of the casting flow (e) (Fig. 6.11):

$$\alpha = 0.7 - \frac{\tau_0}{\rho g e} + 0.05\frac{l_f}{\mathrm{e}} \tag{6.3}$$

The same authors have further demonstrated that, since "at the scale of industrial casting processes, shear induced fiber orientation is almost instantaneous phenomenon …simple computational fluid mechanics simulations only aiming at predicting stream lines give good overviews of fiber orientation in a structural element … and the cartography of the stream lines through the last seconds of the casting process correlates very well with the cartography of the final fiber orientation no matter the flow history". In this framework an attempt has been recently made [123] to correlate the orientation of the fibers, as detected through AC-IS, to the direction of the shear rate vectors, as computed through a single fluid casting flow simulation. With reference to the case study shown in Fig. 6.8, the orientation of the fibers, as inferred from fractional conductivities measured through AC-IS Eq. (6.2), has been then compared (Fig. 6.12) to fractional orientation of the shear rate vectors, as from the CFD modelling of the casting:

$$(\text{Fractional fiber conductivity})_{\mathrm{x,y}} = \frac{[\sigma]_{\mathrm{fibersx,y}}}{[\sigma]_{\mathrm{fibersx}} + [\sigma]_{\mathrm{fibersy}}} \tag{6.4}$$

$$(\text{Fractional Orientation})_{\mathrm{x,y}} = \frac{\sin\alpha_{\mathrm{x,y}}}{\sin\alpha_{\mathrm{x}} + \sin\alpha_{\mathrm{y}}} \tag{6.5}$$

An alternative modelling approach relies on the often made analogy between the flow of liquids and that of granular media, even though the physical properties of the two are quite different. Concrete by nature is dominated by its fluid-like behaviour or by its granular media-like behaviour according to its mix design. In the case of SCC, the amount of coarse particles in the mixture is low and this

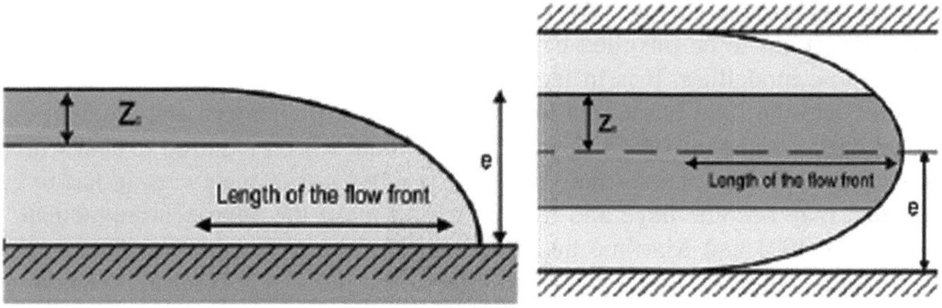

Fig. 6.11 Illustration of characteristic dimension of casting flow *e* to be included in Eq. (6.3) for the case of free surface slow (e.g. slab casting) and confined flow (e.g. wall casting) [93]

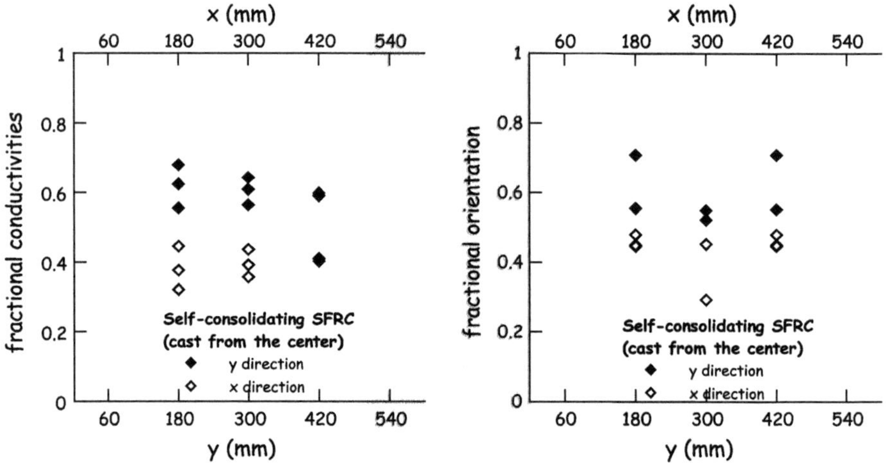

Fig. 6.12 Fractional conductivities of FR-SCC (AC-IS measures) versus fractional orientation angles of shear rate vectors (CFD modelling) [123]

modern concrete behaves as a fluid suspension, whereas, in the case of ordinary concrete with greater amount of coarse particles, the behaviour is dominated by the granular nature of the material. Recently, it has been shown that the distinct element methods (DEM) used to simulate dry granular media flows allows for the simulation the behaviour of fresh concrete with different consistency during transport, placement and compaction [124–128]. The correlation between mix design and rheology was also investigated through the effect of the aggregate particle size distribution. Constitutive relations based on the Bingham formula were developed in order to describe the interactions between two neighbouring particles for simulating fresh concrete. In the case of the DEM fibers can be modelled as clusters of solid spheres (Fig. 6.13): the number of needed particles might be very large but calculations are much faster than if fibers have to be embedded into continuum fluid modelling.

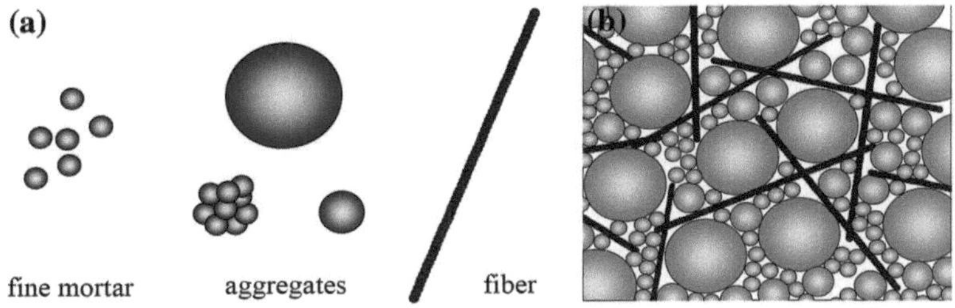

Fig. 6.13 Schematic view of: **a** Modelling the basic constituents of concrete meso-structure by spherical particles, **b** discretisation of fiber reinforced concrete (from [127])

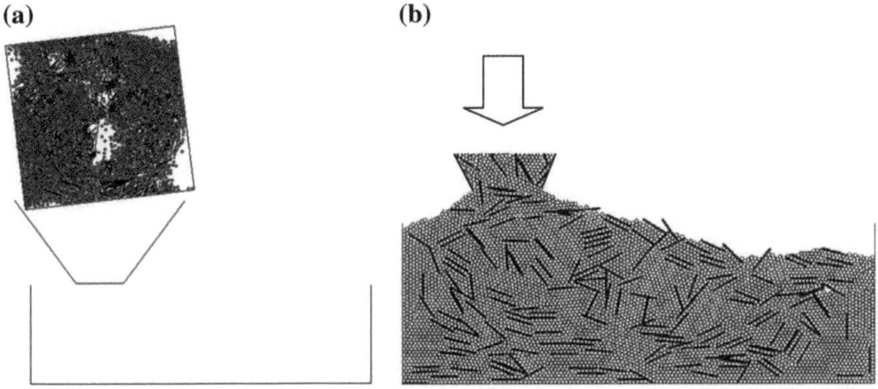

Fig. 6.14 Example of form filling with a fiber reinforced SCC (1 % fiber volume ratio) [128]

Besides the simulation of a study example of FR-SCC mould filling (Fig. 6.14), a prediction has been recently proposed of the flow induced dispersion and orientation of fibers in slump-flow patties for fiber reinforced concretes featuring three different levels of performance in the fresh state (self-compacting, segregating and vibrated) [129]: the experimentally detected trends [77] were captured with an acceptable qualitative agreement (Fig. 6.15). From the quantitative point of view, some discrepancies have been detected in some of the investigated cases, which may be attributed to the employed particle generation techniques, highlighting the need for further investigation in order to fully and reliably profit of the predictive skills of the approach. Detailed information could also be obtained on the distribution of fiber orientation (Fig. 6.16), confirming, as detected in experiments, that in a free-surface free-front flow fibers tend to align, also as a function of concrete rheology, transverse to the flow itself [31, 32].

A "heuristic" approach has been recently proposed by Laranjeira et al. [130] in an attempt to "create an engineering toolbox for the prediction of fiber orientation in practical applications of fiber reinforced concrete". The approach implies a stepwise calculation of the fiber orientation factor, the concept of which is

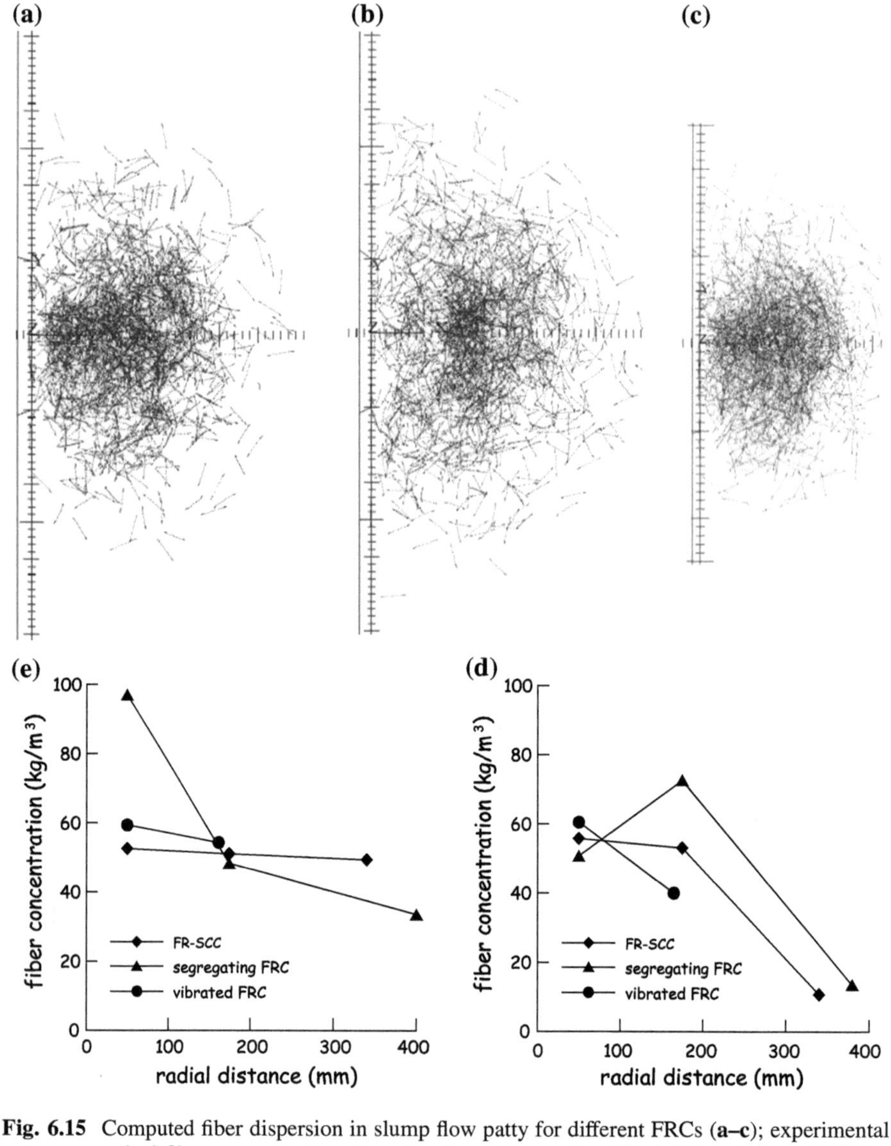

Fig. 6.15 Computed fiber dispersion in slump flow patty for different FRCs (**a–c**); experimental versus numerical fiber concentration along the slump flow radius (**d, e**) (adapted from [129])

introduced through a solid probabilistic 3d stereological analysis, as the additive result of the effects of material properties and issues related to production processes, from casting method to, in case, either vibration or flow and formwork wall effect

The steps of the approach are summarized in detail in Fig. 6.17a, together with the prediction versus experimental comparison for a series of experimental cases

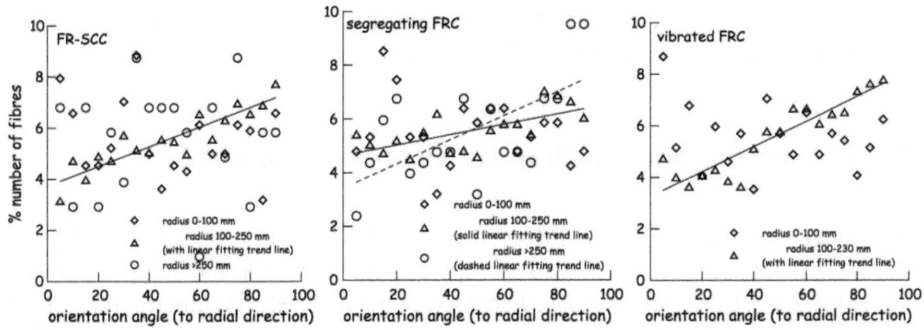

Fig. 6.16 Computed orientation angles of fibers in different sectors of slump flow patties [129]

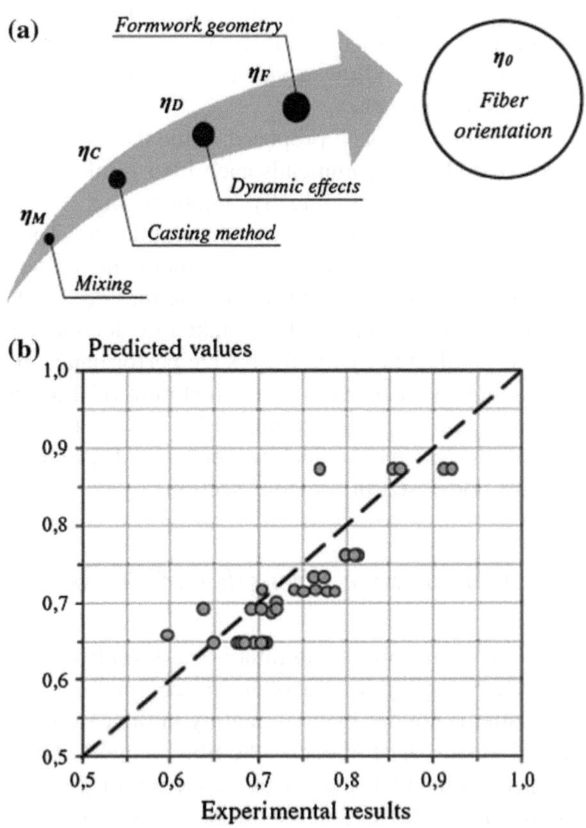

Fig. 6.17 Step-wise procedure for the proposed predictive approach of fiber orientation (**a**) and overall prediction versus experimental comparison (**b**) [130]

detailed in [130] (Fig. 6.17b). The interesting reliability of the approach highlights once again the need of integrating, into the concept and design of structures made with advanced concretes and cementitious composites, such as FR-SCC, material properties, production processes and the geometry and function of the structure itself, in order to pursue a cost-effective tailoring of both the material and the production processes themselves to the intended structural application.

6.5 Engineering and Mechanical Properties of FR-SCC

Mechanical properties of fiber reinforced cementitious composites are actually the outcome of a synergy between the properties of the concrete matrix and the toughening effect exerted by the dispersed wire-like reinforcement: this effect depends on the geometry and dosage of the fibers, strength of the filament, fiber dispersion and orientation with respect to the applied stress and on the fiber-matrix bond. Approaches aimed at predicting the mechanical response of FRCs from pull-out behaviour of single fibers and fiber distribution data [131–137] also because of the previously recalled advances in non-destructive fiber dispersion monitoring techniques, which allowed valuable database to be effectively garnered for validating such approaches.

The design approach proposed in the recently issued final version of the Model Code 2010 [138] recommends each fiber reinforced cementitious composite to be regarded as a macroscopically homogeneous material, with its own properties. These have to be identified through suitable test procedures which not only have to "reproduce" the aforementioned synergistic effect between matrix and fibers, consistently with the performance requested to the material by the intended structural application, but also to take into account issues related to the production process, as affected by the fresh state performance and the structure geometry and as affecting the dispersion and orientation of the fibers with respect to the stress state to be applied. This section will review in the aforementioned framework, the engineering and mechanical properties of FR-SCC, also taking advantage of the concepts addressed in previous section.

6.5.1 Compressive Strength and Stress–Strain Behaviour

As well-known from the literature on conventional vibrated SFRC (see e.g. [139]), fibers have negligible effects on the compressive strength of the fiber reinforced composite, if not detrimental, mainly in the case of higher dosages, because of the "steric" disturbance effect by the fiber network. On the other hand, they affect the post-peak behaviour, mainly as a function of their dosage, improving the residual load bearing capacity and the ultimate strain. The negligible effect on strength has been confirmed also for SFR-SCC [3, 140], whereas any investigation on the toughening effect of the compression softening is so far lacking. Interestingly, the compressive strength gain with time was found to be faster for SFR-SCC than for conventional vibrated SFRC having the same 28-day compressive strength (e.g. same strength class) [3] or than what obtainable through predictive models, also with reference to same strength class [141, 142] (Fig. 6.18).

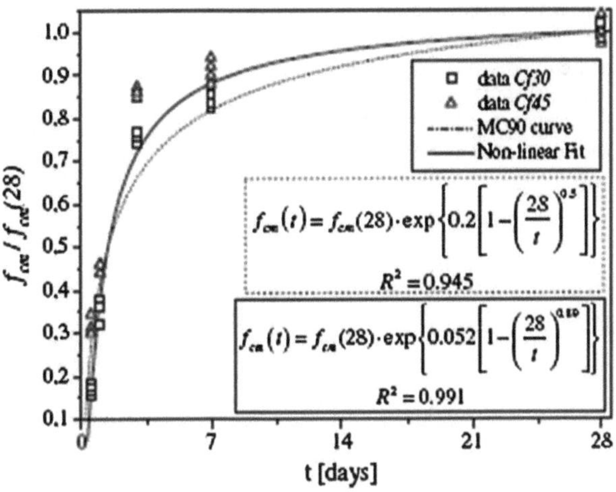

Fig. 6.18 Influence of aging on the compressive strength of FR-SCC [141]

Similar findings were obtained also for plain SCC as compared to VC. This can be attributed to a reduced W/CM [143, 144], to the higher content of cementitious materials [3], or, when applicable, to the presence of limestone filler, whose fine particles act as nucleation sites for the formation of calcium silicate hydrates [143, 144].

The Young modulus of FR-SCC was found to be lower than what is predicted by models calibrated on conventional VC for the same strength class [141]. This also agrees with related findings for plain SCC and VC, because of the higher paste content and lower maximum aggregate diameter which feature the mix-design composition of the former [143] (Fig. 6.19).

Dhonde et al. [3] also found that fibers slightly enhance the modulus of elasticity, in the case of SCSFRC mixtures, pointing out this is an additional advantage of using fibers in SCC, due to its intrinsic higher deformability. Similarly to what happens for conventional concrete and, to a larger extent, for plain SCC, actual "in structure" properties of FR-SCC are likely to exhibit a "spatial" variation which, among the other factors, is governed by the concrete rheology, the geometry of the structural element and the adopted casting procedure.

Torrijos et al. [145] investigated the so-called "top column" effect on compressive strength and modulus of elasticity, i.e. the variation of the aforementioned properties along columns, 2.5 m high and cast from the top. They found a reasonably uniform dispersion of both the compressive strength and the Young modulus along the column height, except for the top 50 cm (20 % of the total height), where the material featured a somewhat 20 % lower compressive strength and a 10 % lower Young modulus. The measured trend of both the mechanical properties along the column height was also interestingly compared to the likewise measured material density and fiber content, once again highlighting the correlation among the micro- and meso-scale structure of the material, its macroscopical mechanical properties and the casting technique (Fig. 6.20).

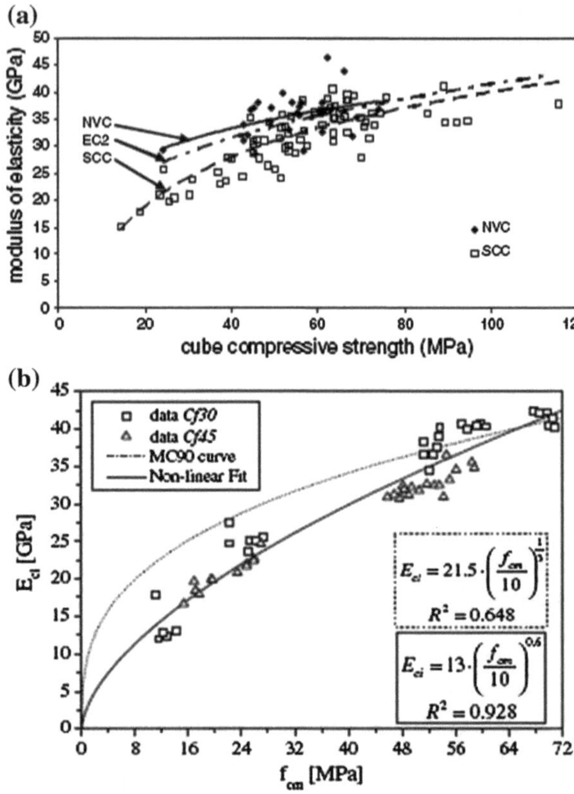

Fig. 6.19 Relationship between modulus of elasticity and compressive strength for plain (**a**—[143]) and fiber reinforced (**b**—[142]) SCC and vibrated concrete (NVC)

6.5.2 Biaxial and Multi-Axial Compression

Mohammed and Eliot [146] performed a study on the behaviour of SFR-SCC under biaxial compressive stresses, employing 100 × 100 × 50 mm specimens, obtained from larger 100 mm side cubes. Two fiber volume percentages (0.5 and 1 %) were considered. Hooked end fibers 35 mm long and with an aspect ratio equal to 65 were employed. They found the ultimate biaxial strength increase, with respect to the reference plain concrete matrix, by 24 and 55 % respectively for $V_f = 0.5$ and 1 % (stress ratio = 0.5). No significant increment was measured for uniaxial strength with $V_f = 0.5$ % and a 25 % increment for $V_f = 1$ %). Fibers also positively affect the biaxial to uniaxial strength ratio, which was equal to 1.39 and 1.51, respectively for $V_f = 0.5$ and 1 %, as compared to 1.12 for the reference plain SCC matrix (Fig. 6.21). Finally, as expected, fibers increase the stiffness and ductility along the major principal compressive stress direction as well as changed the mode of failure from tensile splitting to shear-type. Results were in good agreement with previous studies on vibrated SFRC under biaxial stress states

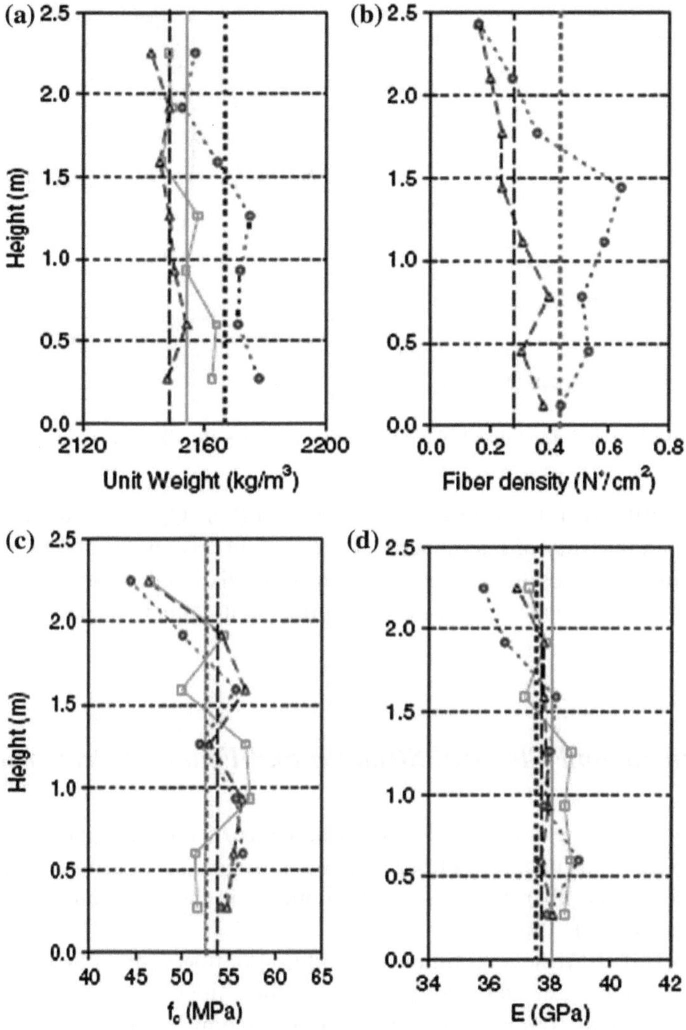

Fig. 6.20 Top-column effect on material density (**a**), fiber content (**b**), compressive strength (**c**) and modulus of elasticity (**d**) for SCSFRC (*squares* plain SCC; *triangles* SCC with 25 kg/m^3 steel fibers; *circles* SCC with 50 kg/m^3 steel fibers) [145]

[147, 148]—where it was moreover found that the biaxial to uniaxial compressive strength ratio may be also affected by the fiber aspect ratio.

With reference to compressive behaviour under lateral confinement, an experimental investigation has been recently performed on SCC, both plain and reinforced with either 35 or 70 kg/m^3 hooked end steel fibers (l_f 35 mm, $d_f = 0.55$ mm), and under three different levels of lateral confining pressure (1, 3 and 10 MPa, compared to an average uniaxial compressive strength equal to

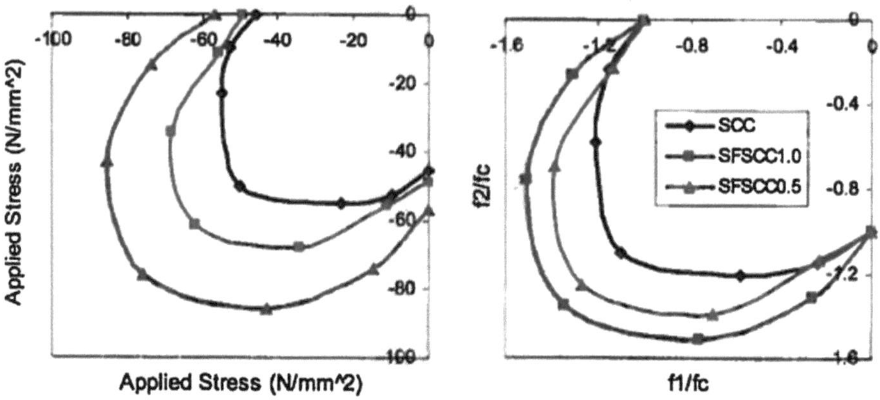

Fig. 6.21 Absolute and normalised biaxial failure envelopes for FR-SCC [145]

30 MPa) [149]. Results confirmed, also for SFR-SCC, the positive effect on toughness and ductility of lateral active confinement, also highlighting, mainly for lower confinement pressure, some beneficial synergistic effect between the active confinement itself and the passive confinement effect provided by fibers (Fig. 6.22).

6.5.3 Tensile and Flexural Strength and Fracture Properties

Fibers improve the tensile and flexural behaviour of concrete due to the crack-bridging effect, as a function of fiber type and dosage, aspect ratio l_f/d_f, bond with the concrete matrix, fiber dispersion and alignment in the direction of the principal tensile stresses [130, 150, 151].

As remarked above several times, the addition of fibers into a SCC may benefit from its superior fresh state performance, which does not require mechanical vibration or manual compaction, to achieve a randomly uniform dispersion, with no fiber-free spots and controlled downward settlement [4]. Comparative investigation on 300 mm high cylinders cast with either vibrated SFRC or SFR-SCC highlighted the importance of the rheological stability on the dispersion not only of the fiber content along the cylinder height, but also of the splitting tensile strength, measured on 75 mm thick disks cut along the height of the same aforementioned cylinders [22] (Fig. 6.23).

Similarly, Cunha et al. [152] reported comparable fiber contents in the bottom and top halves of 300 mm high cylinders, cast with SFR-SCCs containing different amounts of steel macro-fibers; the two 150 mm high half-cylinders were further notched and tested in direct tension, providing similar values of residual stresses at different levels of crack openings, coherent with the number of fibers on the fracture cross section (Fig. 6.24).

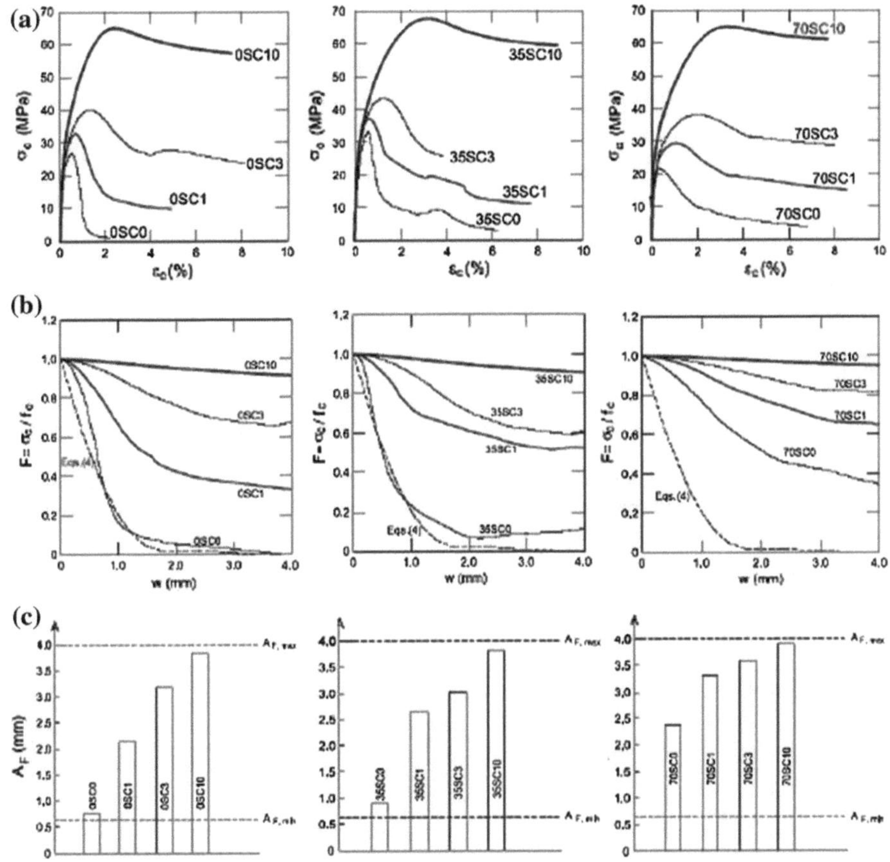

Fig. 6.22 Results of experimental investigation on multi-axial behaviour of SFR-SCC [149]: (**a**) stress–strain curves (specimen nomenclature: ex 35SC10 SCC with 35 kg/m^3 fibers and 10 MPa lateral confining pressure); dimensionless post-peak stress versus inelastic displacement (**b**) and material post-peak toughness (**c**—area subtended by the dimensionless stress versus inelastic displacement curves) for plain SCC and SCC reinforced with 35 and 70 kg/m^3 hooked-end steel fibers

In a similar framework, the role of a "flow-driven" orientation of the fibers, due to the "tailored" specimen casting procedure, can be called to justify the results obtained by Grunewald [5], SFR-SCC featuring superior flexural toughness than a same strength class vibrated SFRC having the same fiber content (Fig. 6.25), irrespective also of the similar fiber pull-out strength. Also Pons et al. [153] reported, for VC and SCC with different kinds of hybrid fiber reinforcement, better flexural performance of the latter, attributing it to better fiber matrix-bond, even if neither single fiber pull-out tests were performed nor fiber dispersion/orientation information, as related to casting procedure, was provided.

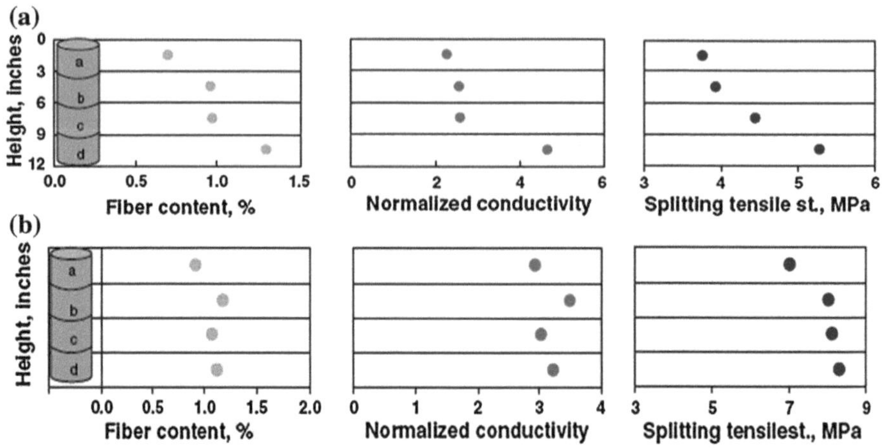

Fig. 6.23 Variation along 300 mm high cylinder of fiber content, matrix-normalized conductivity (AC-IS method) and splitting tensile strength for vibrated SFRC (**a**) and SFR-SCC (**b**) (1 % by volume of 40 mm long steel fibers) [22]

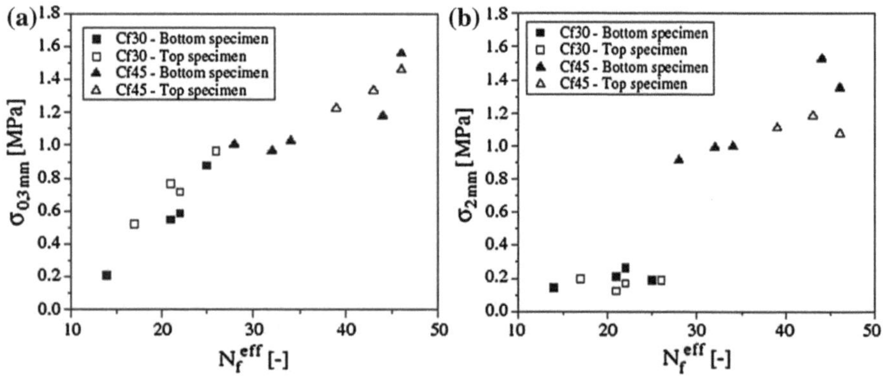

Fig. 6.24 Correlation between tensile post-cracking residual stresses at 0.3 mm (**a**) and 2 mm (**b**) crack opening versus number of fibers on fracture surfaces; direct tension tests on notched cylinders made with SC-SFRC containing either 30 or 45 kg/m^3 hooked end 60/80 steel fibers; specimens were obtained as top and bottom part of twice as long cast cylinders [152]

Velasco et al. [154] also reported positive effects of the orientation of fibers driven by the fresh concrete flow on either the tensile strength and peak strain in direct tension tests (dumbbell specimens) on self-consolidating concretes reinforced with up to 2.5 % by volume 65/35 steel fibers.

Correlation among the performance in the fresh state, the dispersion and flow-induced orientation of fibers and the resulting mechanical performance in bending was investigated by Ferrara et al. [28]. Figure 6.26 shows the results of 4-point bending tests performed on beam specimens cut from the same slabs shown in

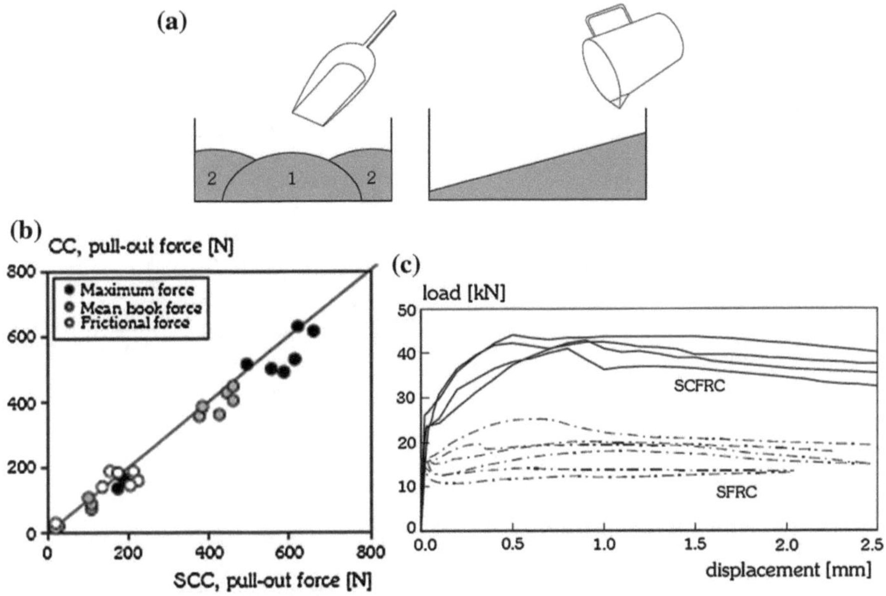

Fig. 6.25 Mould filling method for SFRC (*left*) and SFR-SCSC (*right*) beam specimens (**a**); results of fiber pull-out tests in conventional concrete (CC) and SCC matrices (**b**) and results of 3-point bending tests (**c**) (beams 150 × 150 × 550 mm with a 25 mm notch at mid-span) [5]

Fig. 6.8 (reported also in Fig. 6.26a). Beams were cut in such a way that their axis resulted, when applicable, either parallel or orthogonal to the main flow direction during casting. A higher dispersion characterizes the results obtained from either vibrated SFRC or poorly designed SFR-SCC (prone to segregation), whereas randomly uniform dispersion and a tendency to flow driven alignment of fibers can be highlighted for SFR-SCC beams. Sensitiveness to fiber downward settlement due to either mechanical vibration or poor performance in the fresh state also appears from the results of beams tested upside down to casting, together with the effectiveness of properly designed SCC.

Quantitative correlation between flexural toughness and fiber orientation on the fracture surface has been provided by several other authors [25, 27, 87–89, 155]. Specimen casting was always carefully designed in order to trigger the orientation of fibers along the direction of flow of the fresh concrete. Special investigations were performed by Grunewald [5], with reference to splitting tensile strength, by Ferrara et al. [29, 30], with reference to bending behaviour of a self-consolidating high performance fiber reinforced cementitious composite (SC-HPFRCC—Fig. 6.27), and by Ferrara et al. [102, 103], Pansuk and Sato [156], and by Kang and Kim [157], with reference to tensile behaviour of a SC-HPFRCC. Fiber orientation affected tensile strength and residual stresses as well as the strain capacity of the material (e.g. crack opening at peak load).

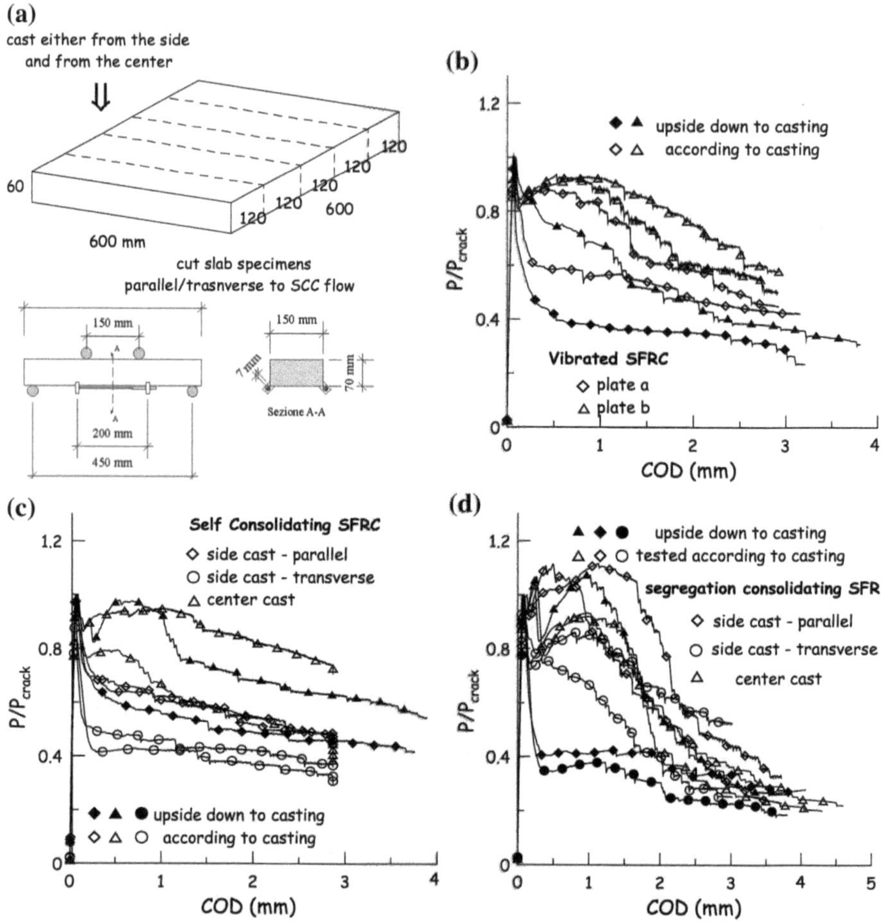

Fig. 6.26 Scheme for beam cutting from slabs and 4pb test set-up (**a**); normalized load vs. COD for: (**b**) vibrated, (**c**) self-consolidating and (**d**) segregation prone SFRC [28]

More recently, Ferrara et al. [77] also correlated the "spatial scattering" of flexural toughness of SFRC-SCCs, as occurring in a real scale casting mock-up, to several indicators of fresh state performance, as affecting the effectiveness of the fiber orientation along the casting flow.

All the aforementioned results highlight the "feasibility" of a casting process tailored to the intended structural application that aims at obtaining a flow-induced orientation of the fibers which matches, as close as possible, with the direction of the principal tensile stresses in structural elements when in service, thus resulting in superior performance at both the material and structural levels.

Moreover, the following open questions have been highlighted: the manufacturing of the specimens to be employed for the identification of material properties

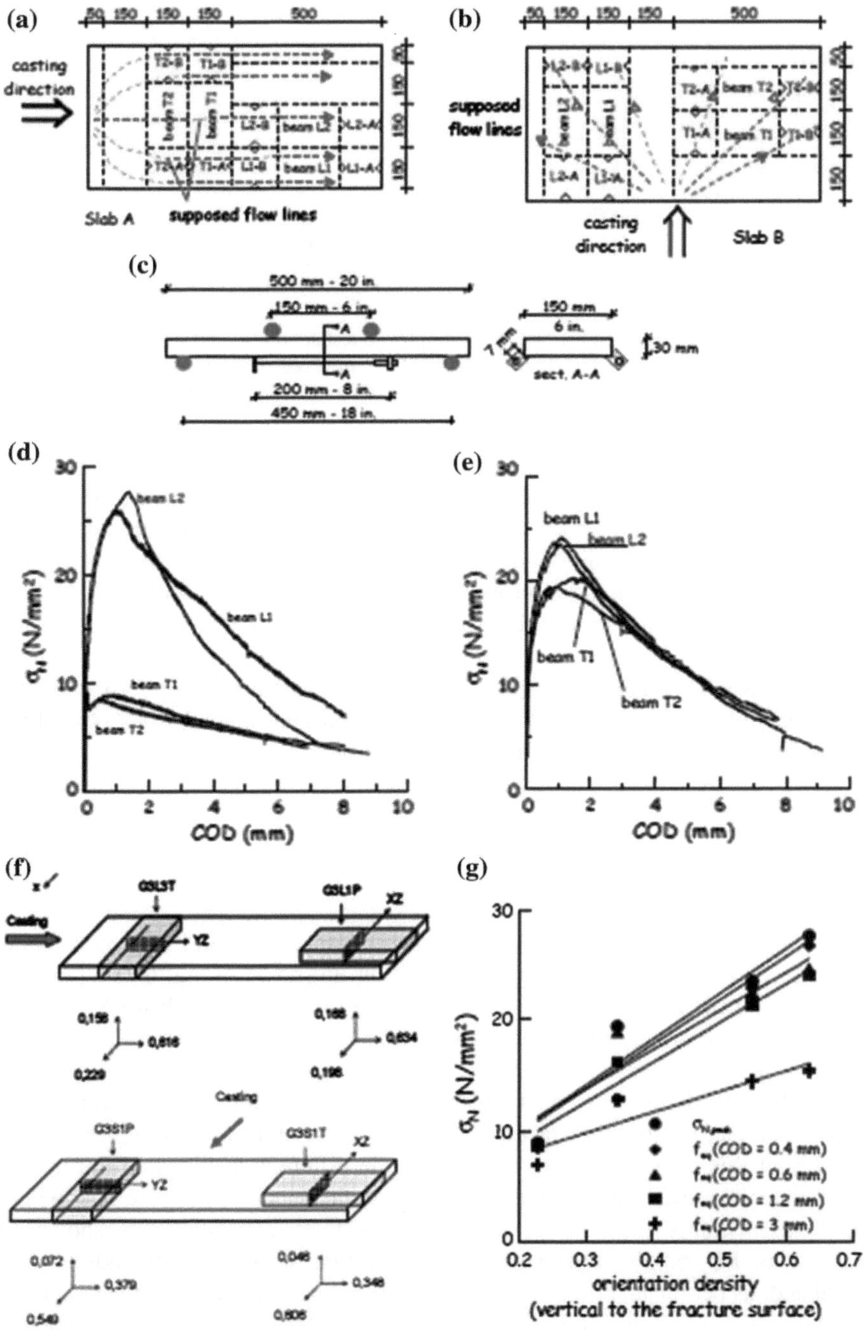

Fig. 6.27 Influence of fiber orientation on toughness of FR-SCC: specimen manufacturing (**a, b**); test set-up for 4pb tests (**c**); nominal stress versus crack opening curves (**d, e**); fiber orientation factors—from image analysis—obtained through tailored casting at selected locations in the slabs (**f**) and correlation of peak and residual tensile stresses with fiber orientation factors (**g**) [29]

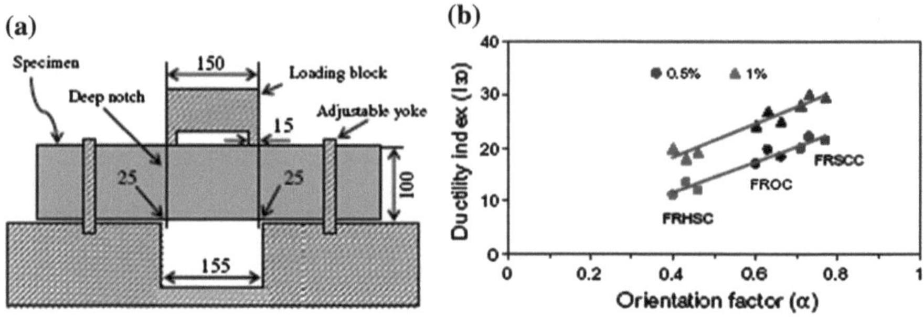

Fig. 6.28 Direct shear tests on FR-SCC [33]: employed test set-up (**a**) and shear ductility indices for different investigated mixes (**b**) - legend: FROC fiber reinforced ordinary concrete–f_c 30 N/mm^2; FRSCC fiber reinforced SCC–f_c 55 N/mm^2; FRHSC fiber reinforced high strength concrete–f_c 80 N/mm^2

has to "mimic" as close as possible the casting of structural elements. In this way, the flow-governed fiber dispersion and orientation can be replicated at the lab-specimen scale [158] as they will most likely occur in the structure to be. This allows identifying the material properties relevant to the intended application.

The material anisotropy resulting from the flow-induced fiber orientation should be suitably taken into account in the design-oriented identification of material properties as well as in structural design calculations. Whereas the former issue is addressed, though in a general way, by codes and recommendations [138, 155], provisions addressing the latter are so far lacking.

6.5.4 Shear Behaviour

The shear behaviour of both vibrated and self-compacting SFRC was investigated by Boulebache et al. [33], employing the test set-up shown in Fig. 6.28a. Different matrices (normal and high strength for VC and medium strength for SCC) were considered, with either 0.5 or 1 % by volume of either 35 or 60 mm long hooked end steel fibers. The results obtained confirmed an expectable strong dependence of the ultimate shear stress on both the compressive strength and, even though more moderate, on the fiber factor V_f l_f/d_f. What is more interesting, in the perspective of this chapter, is that, because of the different rheology of the investigated concretes, different orientation factors on the shear plane were obtained, to which significantly different shear ductility indices correspond, as shown for example in Fig. 6.28b. It is worth remarking that shear ductility indices were calculated, as per an adaption of ASTM C1010, as the ratio between the area under the load-slip curves up to a prescribed multiple of the shear slip at first cracking and the area under the same curve up to first cracking.

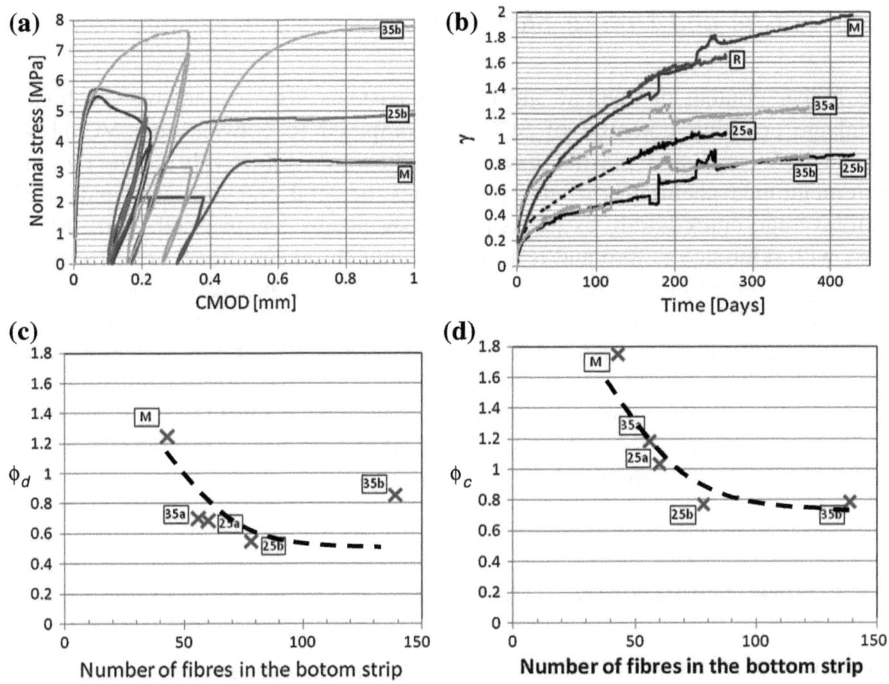

Fig. 6.29 Creep bending tests on FR-SCC [159]: recorded sustained stress versus CMOD evolutions (**a**), creep coefficient (**b**); correlation of the creep coefficient in terms of deflections and CMOD with the number of fibers in the tension chord of the beams (**c, d**). (*Legend 25 a–b* SC-SFRC with 25 kg/m^3 steel fibers $l_f/d_f = 50/0.75$; *35 a–b* SC-SFRC with 35 kg/m^3 steel fibers $l_f/d_f = 33/0.6$; *R* reference plain SCC; *M* SC-SFRC with a cocktail of aforementioned fiber types)

6.5.5 Creep

The increase of long term deformations due to the higher paste content is a major concern with regard to the structural performance of SCC [159]. According to a recent study [160], even a limited amount of steel fibers added to a normal strength SCC ($f_{c,cube} = 40$ N/mm^2) can effectively reduce the effects of creep on both deflections and crack openings (published results refer up to about 9 months—Fig. 6.29a, b). Tests were performed on beam specimens made with SFR-SCC containing either 25 kg/m^3 steel fibers $l_f/d_f = 50/0.75$ or 35 kg/m^3 steel fibers $l_f/d_f = 33/0.6$ as well as a cocktail of both fiber types. A sustained stress level was applied equal to 50 % of the residual stress measured, on a monotonic reference test, at a 0.2 mm crack opening. Interestingly, the number of fibers active on the fracture cross section was found to effectively govern the evolution of the creep deformation both in terms of deflection and crack opening (Fig. 6.29c, d).

6.5.6 Plastic and Drying Shrinkage

The higher content of cement and cement paste is also responsible for higher shrinkage cracking potential of SCC, as compared to conventional VC [144]. Fibers, both steel and plastic, have been shown to effectively counteract the aforementioned phenomenon, either reducing the total cracking area and maximum/average crack width, as far as plastic shrinkage is concerned [160]. Moreover, with reference to free and restrained drying shrinkage, fibers are effective in reducing both the time for the first crack to appear and its progressive opening with time [162–164]. Steel fibers (hooked end macro fibers) were found to be more effective, in this respect, than either polypropylene or PVA fibers, when the same volume fraction was added to the same matrix [161–163]. Results quantitatively agreed with similar experiences on conventional vibrated SFRC as far as the effect of fibers is concerned on both plastic [164, 165] and drying shrinkage (see e.g. [86]), as a function of their volume fraction, type, aspect ratio, etc.

6.5.7 High Temperature Behaviour

A comprehensive one of a kind investigation on thermal and mechanical properties of fiber reinforced SCC at elevated temperatures has been recently published by Khaliq and Kodur [166]. Results of tests performed on a high performance SCC (f_c = 60 N/mm^2), plain and reinforced with either steel fibers (l_f/d_f = 38/1.14—dosage equal to 0.5 % by volume) and polypropylene fibers (dosage equal to 1.1 % by volume, or even to a cocktail of both, have highlighted the following main issues, which also confirm the results of previous investigation on the high temperature behaviour of either plain SCC [167] or vibrated SFRC [168]:

- addition of fibers, either steel or polypropylene or hybrid, does not significantly alter, in the temperature range 20–800 °C, the thermal conductivity of the SCC matrix, which, at room temperature, is anyway higher than that of a comparable vibrated high strength concrete, most likely because of the higher content of cement paste and admixtures (higher amount of ions dissolved in water);
- steel fibers increase the specific heat of the SCC matrix in the temperature range beyond 400 °C; polypropylene fibers, because of their melting which results in a pervious concrete, temper this effect and may even yield, beyond 650 °C, to a specific heat lower than that of the reference matrix;
- fibers increase the thermal expansion of the SCC matrix, the higher the temperature range, the higher the increase, steel fibers always providing the highest increment; this may be due to the role of crack arresters played by fibers;
- addition of steel, polypropylene or hybrid fibers does not significantly alter the high temperature compressive strength of the FR-SCC. On the contrary fibers help to keep the splitting tensile strength of FR-SCC almost constant up to 400 °C, whereas polypropylene fibers, even when combined with a significant

amount of steel fibers, reduce the splitting tensile strength at high temperatures, once again because of the increased porosity due to their melting;

- addition of fibers improves the elastic modulus of the SCC matrix at high temperatures, steel fibers, as above, providing the best performance.

Similar results were also obtained by Alonso et al. [169], who tested up to 700 °C a SC-SFRC reinforced with 40 kg/m^3 steel fibers $l_f/d_f = 30/0.48$, also in combination with polypropylene fibers $l_f/d_f = 54/0.05$, and by Lourenço et al. [170], who tested up to 800 °C a SC-SFRC reinforced with 45 kg/m3 steel fibers $l_f/d_f = 30/0.38$.

Romano et al. [171] investigated the behaviour of a self-compacting refractory concrete (produced with calcium aluminate cement thermally treated to 588 °C) reinforced with either 0.7 and 1 % by volume of steel fibers ($l_f/d_f = 25/0.51$): two temperatures were investigated (110 and 650 °C) but no room temperature data were provided for comparison.

6.6 Structural Performance

6.6.1 Bond with Reinforcement

Fibers are known to positively affect the steel to concrete bond, in case of splitting bond failure, thanks to the crack bridging effect provided by fibers, which may be as effective as the confinement due to transverse reinforcement [172, 173]. A specific study [174], focused on the bond between steel reinforcement and a self-consolidating high performance FRC provided results coherent with the afore-mentioned statement. No comparative investigation with vibrated SFRC was performed to highlight peculiarities, if any, of SFR-SCC.

6.6.2 Tension Stiffening

The crack bridging effect of fibers positively influences the tension stiffening behaviour of tie members, leading to higher initial stiffness and a more ductile cracking behaviour, characterized by reduced crack spacing and crack width and making it possible even to further increase the tie bearing capacity beyond the yielding of steel reinforcing bars [175]. It has been interestingly pointed out [173] that, "depending on the combination of steel fibers (amount, geometry, orientation, bond properties) and reinforcing bars (amount, hardening properties) localization of deformations in one large crack can occur", thus reducing the deformation capacity of the member. Studies on SFR-SCC ties [176–178] confirmed previous findings related to vibrated SFRC. Fiber effect was well accounted considering uniform dispersion and orientation across cracks.

6.6.3 Columns in Compression

Fibers have been shown to be able to effectively (at least partially) replace transverse reinforcement, e.g. closely spaced hoops which are needed to provide confinement in critical regions of columns in buildings and structures susceptible of high seismic excitation [179]. A reduction of congested reinforcement which would be required in such cases may lead to improved constructability, which is likely to furthermore benefit from the use of a SCC matrix. Moreover, fibers delay cover spalling, thereby resulting in a greater material integrity at larger strains (even if the effect was not such to prevent the buckling of longitudinal reinforcement). This was confirmed by a recent study [180] in which SFR-SCC columns were tested, characterized by different arrangement of transverse hoop reinforcement and different fiber contents (1, 1.5 and 2 % by volume). It has to be remarked that in the referred study, the fresh state behaviour of the concrete containing the highest tests amount of fibers was hardly self-compacting/consolidating (slump flow diameter 360 mm). Anyway, this lower flowability may not have affected the reliability of the results and related conclusions. An explanation for this could be given considering that the addition of higher and higher amounts of fibers to the plain SCC matrix was not at all compensated by a reduction of the (coarser) aggregate content, e.g. in the framework of the equivalence of solid particle lateral surface.

6.6.4 Beams in Shear

A significant number of experimental investigation on VC structural elements has shown that the inclusion of randomly distributed and orientated, short discrete steel fibers results in improved shear resistance (see e.g. [181] for a review). This is due to either the crack growth controlling effect provided by the fibers, as well as by a greater effectiveness in crack arresting mechanisms and better distribution of tensile cracks due to the smaller spacing existing between fibers than between stirrups [182]. In the post-cracking regime, fibers, bridging shear cracks, may not only increase the ductility in shear failure but also potentially change the failure mode from diagonal tension to shear/tension or shear/compression [183] or even to flexure [184] (Fig. 6.30). The addition of fibers is hence effective to supplement or replacing conventional shear reinforcement (stirrups), also offering substantial time- and cost savings over it, which tends to be labour intensive. Moreover, due to the resulting reduction in reinforcement congestions, the use of fibers may be extremely useful in thin webbed elements or thin slabs, where stirrups are both uneconomical and difficult to implement. Further advantages in terms of constructability due to the use of SCC matrices makes it extremely attractive to add fibers into SCCs to improve shear resistance.

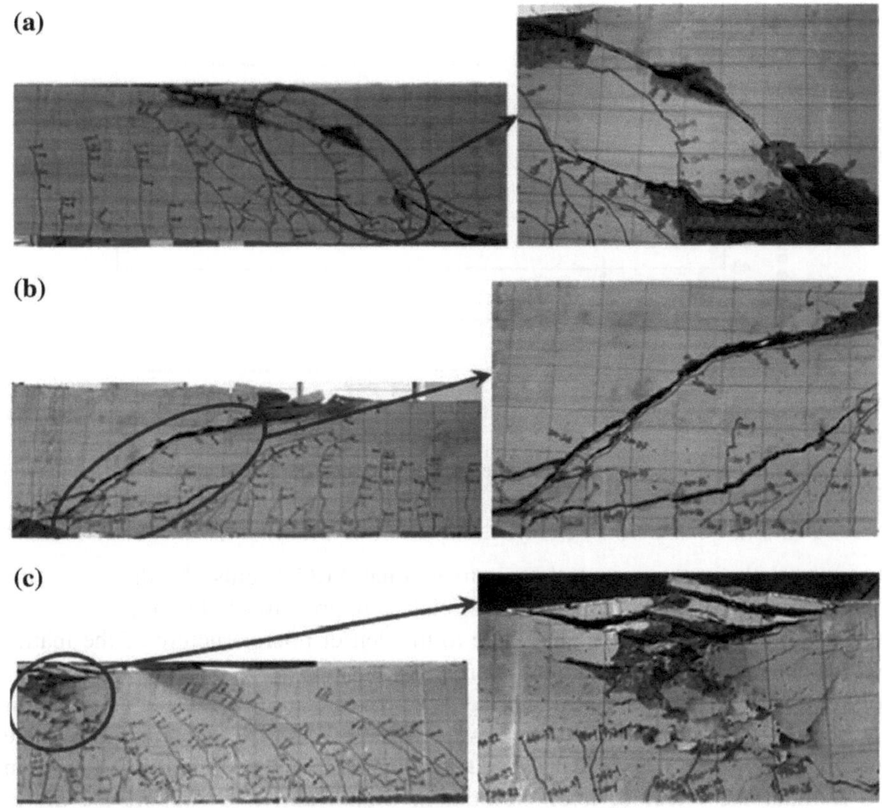

Fig. 6.30 Diagonal tension failure in SCC beam (**a**); shear-compression failure in a SC-SFRC bam (25 kg/m^3 fibers—**b**) and bending failure in a SC-SFRC beam (50 kg/m^3 fibers—**c**); all beams also contained ϕ 6.5@150 stirrups [184]

Greenough and Nehdi [185] performed an extensive investigation on the shear behaviour of SFR-SCC slender beams, studying the influence of fiber type (steel, polypropylene), length (30 and 50 mm) and aspect ratio (respectively 43 and 50), fiber anchorage (flat- or hooked-end) and fiber content (0.5, 0.75 and 1 % by volume). The addition of higher and higher amounts of fibers resulted in a reduction of the coarse aggregate content as well as in adjustments of HRWRA and VMA dosages to achieve the same target slump flow diameter (625 and 650 mm).

As expected, the addition of fibers, with respect to the plain concrete unreinforced reference beam, resulted in higher cracking and maximum loads and (mainly in the case of steel fibers) in increased peak and ultimate deflections, hence in increased ductility. What is even more interesting is the fact, clearly shown by the authors with reference to a wide database (31 cases), that SFR-SCC beams always have, as compared to the plain reference case, a more pronounced

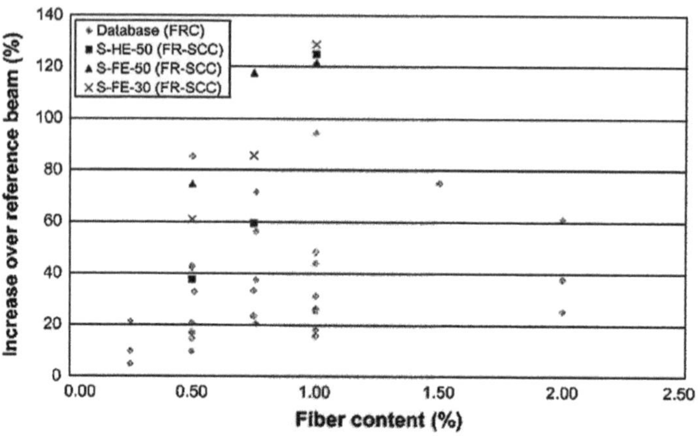

Fig. 6.31 Comparison between the relative increase in shear strength of self-compacting and vibrated FRC beams [185]

shear strength improvement than conventional FRC beams. Furthermore, the higher the fiber content, the higher the relative improvement (Fig. 6.31).

The authors claim this is likely due to the denser microstructure of the matrix and the enhanced properties in the interfacial transition zone, but also to the "self-compacting ability of SCC", which may have led to better fiber dispersion (even if not explicitly stated). The authors also provided a suitable modification to the current ACI approach for shear capacity of slender beams, explicitly including in the formula the effect of fiber orientation. Significantly, to support the aforementioned conclusion, assuming a random orientation of fibers, predictions for FRSCC beams always underestimate the measured shear capacity, which is not the case for either vibrated plain and fiber reinforced concrete beams. With the authors: "it is believed that by optimizing the mixture design (and the casting process), substantial improvements in the ultimate shear strength of FR-SCC beams over that of traditional FRC beams can be achieved, and this aspect deserves further research".

6.6.5 Impact Resistance

The higher energy dissipation capacity imparted by fibers to a fiber reinforced cementitious composite leads to improvements in impact resistance as well as guarantees material and structural integrity [186]. The impact resistance of SC-FRC elements (beams and slabs) has been recently investigated in a few studies.

He and Yang [15] employed SCC reinforced with PP fibers (up to 1.2 kg/m^3—1.25 % by volume), 12 mm long and 0.031 mm in diameter, whereas Jiang et al. [20] used cellulose sheeted fibers, 6 mm long and 0.02 mm in diameter, up to

Table 6.2 Mixture composition and properties of SFR-SCC mix employed for sheet piles [5]

Mix design	
Constituent	Dosage (kg/m^3)
Cement I 52.5 R	358
Cement III/A 52.5	555
Silica fume	61
Sand 0.125–0.5	549
Sand 0.5–1.0	549
Steel fibers OL 13/0.16	125
Water	226
Superplasticiser	21
Slump flow diameter	718 mm
t_{50}	3.2 s
Cube compressive strength (N/mm^2)	75.4 (24 h—prestressing release) 134.3 (28 days)

3.6 kg/m^3 (2.4 % by volume). Drop weight and impact pendulum tests confirmed that the presence of fibers increased the number of blows to first crack and to failure, enhancing energy dissipation capacity.

6.7 Applications and Case Studies

A number of projects and case studies with FR-SCC has highlighted its potential and was instrumental to spread its use for both cast-in-place and precast structural applications.

6.7.1 Prestressed Sheet Piles

Grunewald [5] reported the use of a high performance SCSFRC for the production of precast sheet piles. The composition of the employed mix and its main fresh state properties are summarized in Table 6.2. The study included the optimization of the mix composition and of the sheet pile geometry, considering either structural or economic aspects, thorough characterization of the mechanical performance of the optimized mix and full scale trial tests of sheet piles at the precast factory. A "demonstrative" installation of a 100 m front by the Dutch Ministry of Transport, Public Works and Water Management is under discussion. A careful optimization of the casting procedure was performed in order to achieve flow driven orientation of the fibers along the longitudinal axis of the sheet pile, along which the tensile strength demand is the highest.

Prestressed sheet piles produced with SCFRC have several advantages compared with standard concrete ones (Fig. 6.32): the placement of the concrete is

Fig. 6.32 Prestressed sheet piles cast with self-compacting high performance fiber reinforced mortar (*left*) and conventional concrete (*right* with bar reinforcement) [5]

easier (no bar reinforcement has to be arranged), the storage of the segments requires less space, more sheet piles can be transported with one truck (they are lighter and occupy less space) and the placement in the ground is easier. Without an increase of the total production costs, the load bearing capacity should be at least that of a comparable concrete sheet pile of the same cross section height.

6.7.2 Tunnel Segments

Steel fiber reinforced tunnel segments have been successfully applied in several infrastructural projects. In most cases the bar reinforcement could be completely left out, which simplified the manufacturing process. Fibers can also avoid spalling of concrete that is caused by tolerance differences and inaccurate placement of the segments. Furthermore, during their service life, tunnel linings may be subjected to high temperatures caused by fire. Fibers are able to prevent the consequent deterioration and spalling of concrete.

The reported case study [5—Fig. 6.33] focused on determining how the flow affects the orientation of the fibers and how the orientation relates to mechanical performance. The employed mixture composition is summarized in Table 6.3 together with results of fresh state tests. Plain SCC matrix was prepared at a ready mix plant and fibers were added manually from paper bags into the inlet gap of the truck mixer.

Fig. 6.33 Tunnel segment fabricated with SFR-SCC [5]

Table 6.3 Mixture composition and properties of SFR-SCC employed for tunnel segments [5]

Mix design	
Constituent	Dosage (kg/m^3)
Cement III/A 42.5	382
Fly ash	179
Sand 0.125–4	1044
Gravel 4–16	489
Steel fibers Dramix 60/80 or 45/30	60
Water	183
Superplasticiser	4
Slump flow diameter	620 mm (60/80) 590 mm (45/30)

Four segments 400 mm thick, 1.47 m wide and 4.23 m long (inner length) were cast (Fig. 6.33), two per each type of fiber. From each segment, 15 cores were drilled, as shown in Fig. 6.31a, and deformation-controlled splitting tensile tests were performed. Slices were then cut from cores and X-ray photographs were taken (Fig. 6.34b, c) to determine fiber orientation, to be correlated to the splitting tensile strength, also as a function of the flow distance (Fig. 6.35).

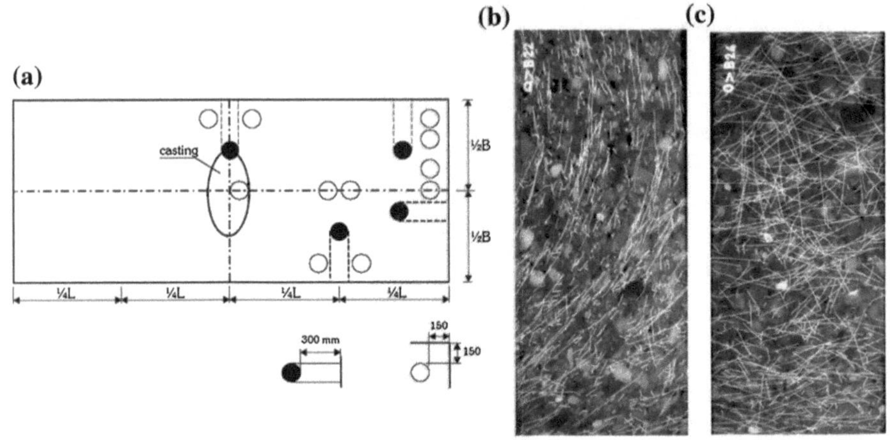

Fig. 6.34 Scheme of the cores drilled for splitting tensile tests (**a**) and X-ray images of cores taken at mid-length of the segment parallel and perpendicular to the flow (**b, c**) [5]

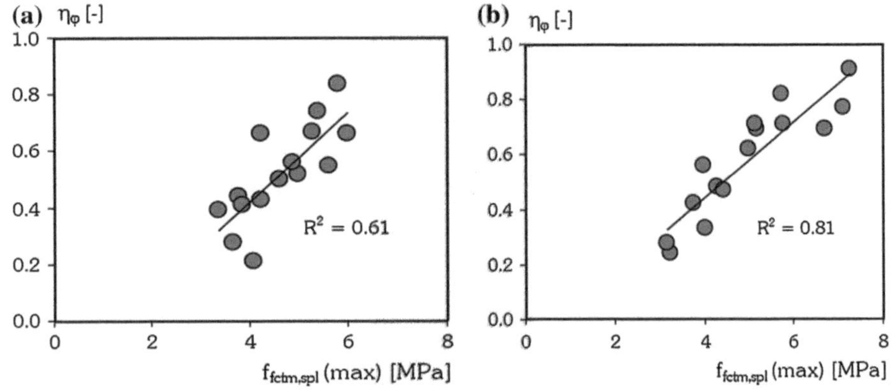

Fig. 6.35 Splitting tensile strength versus fiber orientation number: fibers 45/30 (**a**) and 60/80 (**b**) [5]

6.7.3 Prestressed Roof Elements

Ferrara and Meda [4] reported about a series production of 40 precast prestressed roof elements (Fig. 6.36) made with SCSFRC.

Mixture optimization studies led to employ the mix-design in Table 6.4. All along the scheduled period of manufacturing (about one month June–July 2003) besides thorough fresh and hardened state characterization of the material, including checks on the fiber content in samples taken from each batch, cores were drilled from seven roof elements (cast on different days) and crushed, to measure the fiber content. This was meant as a "semi-destructive" quality control to

Fig. 6.36 Demoulding of a SFR-SCC element [4]

Table 6.4 Mixture composition of SFR-SCC mix employed for prestressed roof elements [4]

Constituent	Dosage (kg/m^3)
Cement I 52.5/R	500
Limestone filler	50
Sand 0 – 3	825
Sand 0–12	190
Gravel 8–15	735
Steel fibers Dramix 45/30	50
Water	160
Superplasticiser	6
Viscosity modifier	2

determine the homogeneity of fiber dispersion within the cast elements and as a robustness check of the precast production. Results are summarized in Fig. 6.37; larger deviations were traced back to fresh state performance and physical properties and explained by high water demand in hotter days, due to a rough evaluation of moisture content of aggregates (supposedly drier in hotter days).

6.7.4 Post-Tensioned Girders for Slope Stabilization

Di Prisco et al. [6] reported the use of a SCSFRC (Table 6.5) for precast post-tensioned girders to be used for stabilization of sub-vertical slopes (Fig. 6.38). During production of the elements (10 non consecutive days Sept.–Oct. 2005), a

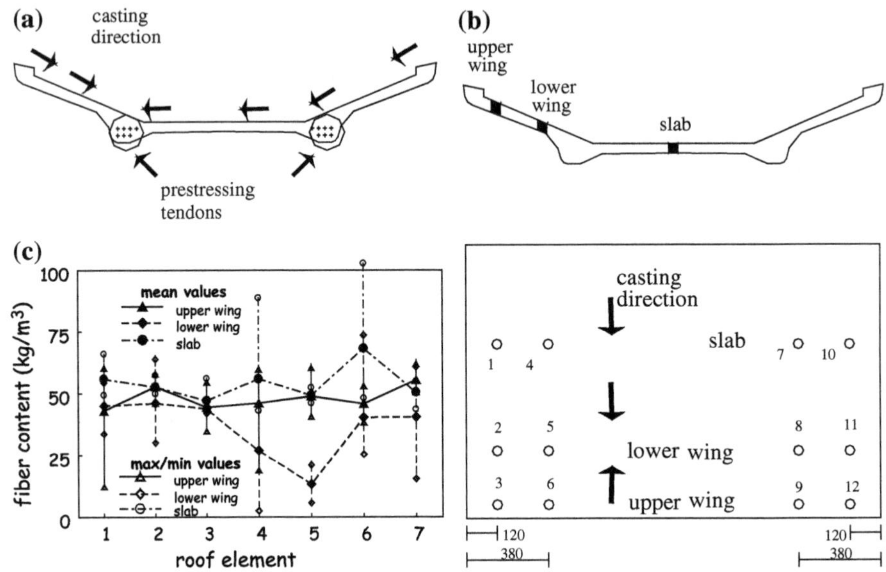

Fig. 6.37 Scheme of roof element cross section with directions of casting flow (**a**); scheme of core drilling (**b**) and dispersion of fiber content inside elements (**c**) [4]

Table 6.5 Mixture composition of SFR-SCC for post-tensioned girders for slope stabilization [6]

Constituent	Dosage (kg/m^3)
Cement I 42.5/R	400
Limestone filler	96
Sand 0–3	676
Sand 0–12	569
Gravel 8–15	403
Steel fibers Dramix 60/80	50
Water	195
Superplasticiser	8.8

check on the filling and fiber dispersing ability with reference to the most congested structural detail was performed (see also Fig. 6.1). Robustness of material fresh and hardened state performance was checked along the production of precast units in the factory [6].

6.7.5 *Façade Panels*

Pereira et al. [141] proposed the use of SFR-SCC (Table 6.6) for façade panels. Prototype panels were tested in bending and shear (Fig. 6.39—the prototype tested in shear was actually a mock-up of the central part of the panel tested in bending).

Fig. 6.38 Precast SFR-SCC post-tensioned girders for slope stabilization installed

Table 6.6 Mixture composition of SFR-SCC employed for façade panels [141]

Constituent	Dosage (kg/m^3)
Cement I 42.5/R	359.4
Limestone filler	308.1
Fine sand	172.2
Coarse sand	859.2
Gravel	698.4
Steel fibers Dramix 60/80	30
Water	97
Superplasticiser	7.1

The use of fibers dispersed in a SCC matrix (and consequent elimination of conventional reinforcement) resulted in thin slabs and ribs, and in an easier construction, as well as in a tremendously larger ductility and energy dissipation capacity in punching failure.

6.7.6 SC-SFRC Precast Roof System

Gonçalves et al. [187] designed a precast roof-deck system (Fig. 6.40) consisting of prestressed main beams and T-section secondary beams, all made with SC-SFRC and no conventional reinforcement (except for pre-stressing tendons in the main beams). The employed mixture composition is detailed in Table 6.7.

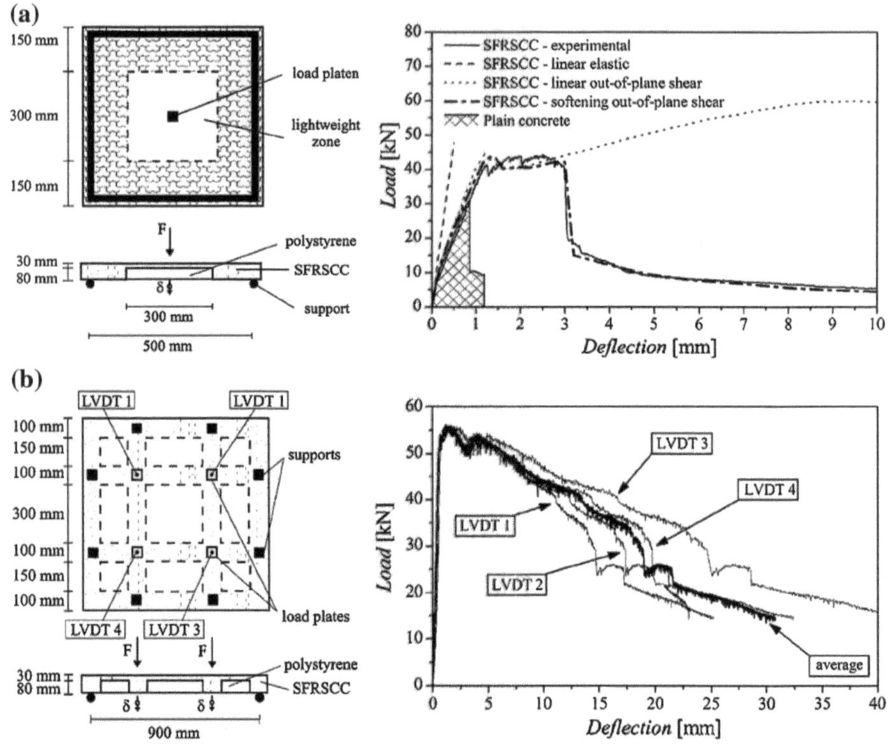

Fig. 6.39 Test setup and load–deflection experimental and numerical results for punching (**a**) and flexural tests (**b**) on panel prototype (**b**) [141]

The use of a SC-SFRC mixture, besides allowing optimized cross-section dimensions because of the elimination of conventional reinforcement (e.g. of stirrups in both the main pre-stressed and the T-section secondary beams) was also functional to minimize construction time, better durability and aesthetical appearance (surface finishing). The study included mechanical characterization of the material in compression and bending, full-scale testing up to failure of a T beam mock-up (with manual measurement of deflections along the beam axis) and an example of design employing the prescriptions for FRC structures by the recently issued *fib* Model Code 2010 [138].

6.7.7 Repairing

Mesbah et al. [188] proposed the use of fiber reinforced SCC to repair damaged or deteriorated beams. Self-consolidating concrete and mortars reinforced with either steel or polypropylene fibers were used, proving their efficacy in restoring the load bearing capacity as well as the ductility of the undamaged beams.

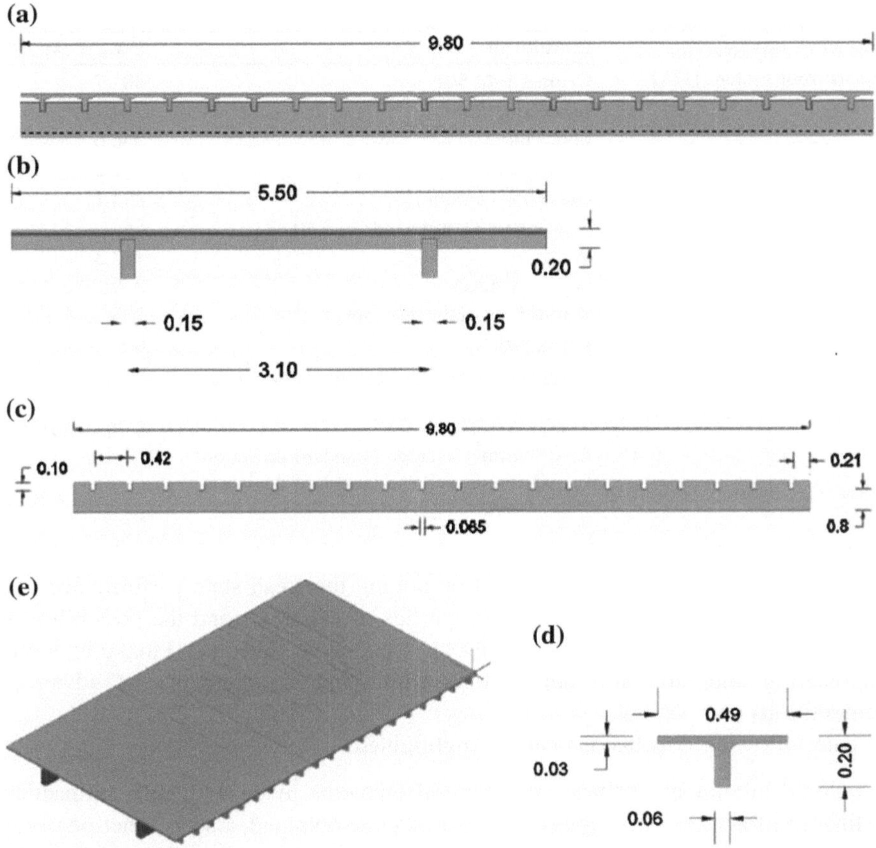

Fig. 6.40 Scheme of the designed precast SC-SFRC roof system [187]: front (**a**) and side (**b**) view of the roof system; geometry of the main prestressed beam (**c**); perspective view of the roof system (**d**) and geometry of the secondary T element (**e**)

Other examples of repairing employing self-consolidating high performance fiber reinforced cementitious composites (SC-HPFRCC) have been reported [189], the superior fresh state performance allowing for an easier successful accomplishment of the repairing task, even in difficult accessible conditions [190].

6.8 Concluding Remarks

In an engineering perspective the foremost outcome of the synergy between SC-FRC technologies is represented by the randomly uniform dispersion and, in case, tailored orientation of fibers which can be both obtained thanks to the superior

Table 6.7 Mixture composition and properties: SFR-SCC employed for precast roof system [187]

Mix design	
Constituent	Dosage (kg/m^3)
Cement I 42.5/R	401.7
Limestone filler	344.3
Fine sand	356.6
Coarse sand	550.2
Gravel 5–12 mm	494.1
Steel fibers Dramix $l_f/d_f = 30/0.38$	45
Water	117.3
Superplasticiser	7.7
Slump flow diameter (mm)	700
L-box ratio	0.8
Compressive strength (28 d—N/mm^2)	60
Residual flexural strength (RILEM TC 162-TDF)	
$f_{eq,2}$ (N/mm^2) average (standard deviation)	9.88 (1.46)
$f_{eq,3}$ (N/mm^2) average (standard deviation)	9.79 (0.89)

fresh state performance of the material. This chapter has reviewed the main findings with reference to the correlation among the fresh state performance, the fiber dispersion and the mechanical properties of FR-SCCs and the possibility of governing the aforementioned correlations to conceive, design and make high end engineering and structural applications employing this category of advanced cementitious composites.

The following conclusions can be highlighted:

- tailored alignment of fibers in structural elements made with self-compacting fiber reinforced cementitious composites can be obtained, also as function of the geometry of the element, due to the superior fresh state performance of the material and as the outcome of a suitably designed casting process;
- methods for non-destructive monitoring of fiber dispersion in full scale castings have been developed by several researchers worldwide and are, in the author's opinion, ready to be exported from university labs into the construction practice;
- tools to design the casting process and predict through it the most likely alignment of the fibers are still at an embryonic stage, most likely because of the high computational time still required to have meaningful information for a real scale casting flow simulation. Interesting results, also from the quantitative point of view in terms of predicted versus measured dispersion and orientation of the fibers, have been obtained for small volume castings. Furthermore simple empirical formulations and heuristic approached have been developed and calibrated on wide and sound experimental databases;
- fiber orientation dependent procedures for the identification of design material properties of FR-SCCs, coherently with current design approaches, represent the next step of the knowledge transfer into engineering construction practice, through an integrated design framework for structural elements made with FR-SCCs;

- the experimental characterization and modelling of the "flow induced" anisotropy of the material behaviour, under both static and cyclic loadings, have to be urgently tackled also in the sight of emerging applications of FR-SCCs.

The possibility of governing the correlation among fresh state performance and fiber dispersion and orientation in FR-SCCs would lead to the concept of a "holistic" design approach which tailors both the material composition and the casting process to the anticipated structural performance. This would require the orientation of fibers to match as close as possible with the direction of the principal tensile stress within the element when in service, so to achieve a more efficient structural use of the material. A closer correspondence between the shape of an element and the function it performs in a structure assembly could thus be achieved: a suitably balanced fresh-state performance would allow moulding the shape of an element and, thanks to a tailored casting process, to align the fibers along the direction of the principal tensile stresses resulting from its structural function.

References

1. Groth, P., Nemegeer, D.: The use of steel fibers in self-compacting concrete. In: Proceedings of the 1st International Rilem Symposium on Self-Compacting Concrete, pp. 497–507
2. Carlsward, C., Emborg, M.: Shrinkage cracking of steel fiber reinforced self-compacting concrete overlays – tests methods and theoretical modelling. In: De Schutter, G., Boel, V. (eds.) Proceedings of the SCC2007, 5th International RILEM Symposium on Self-Compacting Concrete, Gent, Belgium, pp. 793–798, 3–5 Sept 2007. RILEM Publications, Gent (2007)
3. Dhonde, H.B., Mo, Y.L., Hsu, T.T.C., Vogel, J.: Fresh and hardened state properties of self-consolidating fiber-reinforced concrete. ACI Mater. J. **104**, 491–500 (2007)
4. Ferrara, L., Meda, A.: Relationships between fiber distribution, workability and the mechanical properties of SFRC applied to precast roof elements. Mater. Struct. **39**, 411–420 (2006)
5. Grunewald, S., Performance based design of self-compacting steel fiber reinforced concrete. Ph.D. Thesis, Delft University of Technology (2004)
6. di Prisco, C., di Prisco, M., Mauri, M., Scola, M.: A new design for stabilizing round slopes. In: Proceedings of the Fib Conference, Naples, June 2006 (paper 823-CD-Rom)
7. Barragàn, B., Zerbino, R., Gettu, R., Soriano, M., de la Cruz, C., Giaccio, G., Bravo, M.: Development and application of steel fiber reinforced self-compacting concrete. In: Di Prisco, M. et al. (eds.) BEFIB 2004, Proceedings of the 6th International RILEM Symposium, Varenna, Italy, pp. 455–464, 20–22 Sept 2004. RILEM Publications, Varenna (2004)
8. Borralleras, P., Barragàn, B., Gettu, R.: Comparison for durability parameters between conventional concrete standard SCC and steel fiber reinforced SCC for construction of thin elements. Part 2: thin concrete walls application. In: De Schutter, G., Boel, V. (eds.) Proceedings of the SCC2007, 5th International RILEM Symposium on Self-Compacting Concrete, Gent, Belgium, pp. 1092–1098, 3–5 Sept 2007. RILEM Publications, Gent (2007)

9. Romano, G.Q., Silva, F.A., Toledo Filho, R.D., Fairbairn, E.M.R., Battista, R.C.: Mechanical characterization of steel fiber reinforced self-compacting refractory concrete. In: De Schutter, G., Boel, V. (eds.) Proceedings of the SCC2007, 5th International RILEM Symposium on Self-Compacting Concrete, Gent, Belgium, pp. 881–886, 3–5 Sept 2007. RILEM Publications, Gent (2007)
10. Pereira, E.N.B, Barros, J.A.O., Ribeiro, A.F., Camoes, A.: Post-cracking behaviour of self-compacting steel fiber reinforced concrete. In: Di Prisco, M., et al. (eds.) BEFIB 2004, Proceedings of the 6th International RILEM Symposium, Varenna, Italy, pp. 1371–1380, 20–22 Sept 2004. RILEM Publications, Varenna (2004)
11. Bigas, J.P., Pellerin, B., Deschryver, F., Massinari, P., Plizzari, G.: Synthetic macro-fibers reinforced self-compacting concrete for lightweight precast elements. A case study. In: De Schutter, G., Boel, V. (eds.) Proceedings of the SCC2007, 5th International RILEM Symposium on Self-Compacting Concrete, Gent, Belgium, pp. 1013–1018, 3–5 Sept 2007. RILEM Publications, Gent (2007)
12. Nakamura, S., van Mier, J.G.M., Masuda, Y.: Self compactability of hybrid fiber concrete containing PVA fibers. In: Di Prisco, M. et al. (eds.) BEFIB 2004, Proceedings of the 6th International RILEM Symposium, Varenna, Italy, pp. 527–538, 20–22 Sept 2004. RILEM Publications, Varenna (2004)
13. Mantegazza, G., Gatti, A.: Role of polycarboxilate superplasticizer and hybrid polymeric fibers in self-compacting fiber-reinforced concrete. ACI SP, pp. 69–86
14. Krage, G., Wallevik, O.H.: Rheology of synthetic fiber reinforced SCC. In: De Schutter, G., Boel, V. (eds.) Proceedings of the SCC2007, 5th International RILEM Symposium on Self-Compacting Concrete, Gent, Belgium, pp. 347–352, 3–5 Sept 2007. RELIM Publications, Gent (2007)
15. He, X., Yang, Q.: Research on impact properties of flexible fiber reinforced self-compacting concrete in middle-low intensity. In: Shi, C., et al. (eds.) Proceedings 2nd International Symposium on the Design Performance and Use of SCC, SCC2009, Beijing, pp. 577–585, 5–7 June 2009. RILEM Publications, Beijing (2009)
16. Mazaheripour, H., Ghanbarpour, S., Mirmoradi, S.H., Hosseinpour, I.: The effect of polypropylene fibers on the properties of fresh and hardened lightweight self-compacting concrete. Constr. Build. Mater. **25**, 351–358 (2011)
17. Suresh Babu, T., Seshagiri Rao, M.V., Rama Seshu, D.: A study on the flexural behaviour of glass fiber reinforced self-compacting concrete. In: Gettu, R. (ed.) Fiber Reinforced Concrete: Design and Applications, Proceedings of the 7th International RILEM Symposium BEFIB 2008, Chennai, India, pp. 793–802, 17–19 Sept 2008. RILEM Publications, Chennai (2008)
18. Carneiro, V.L., Lima, P.R.L., Leite, M.B.: "Avaliaçao das propriedades fisicas e mecânicas de pastas compósitas auto-compactavéis reforçadas com fibras curtas de coco", in BAC 2010, Proceedings 2° Congreso Iberico Betão Auto-Compactável – Hormigón AutoCompactante, J. Barros et al. eds., 1–2 July 2010, Guimarães, Portugal, Multicomp, 10 pp. (CD-ROM), abstract p. 83 (in Portuguese)
19. Hernandex, F.J.R., Navarro, J.F.V., del Valle, A.J.R. and Cosp. J.P..: "Hormigón Autocompactante renforzado con fibras de sisal", in BAC 2010, Proceedings 2° Congreso Iberico Betão Auto-Compactável – Hormigón AutoCompactante, J. Barros et al. eds., 1-2 July 2010, Guimarães, Portugal, Multicomp, 9 pp. (CD-ROM), abstract p. 115 (in Spanish)
20. Jiang, Z., Banthia, N., Delbar, S.: Effect of cellulose fibers on properties of self-compacting concrete with high volume mineral admixtures. In: Shi, C., et al. (eds.) Proceedings 2nd International Symposium on the Design Performance and Use of SCC, SCC2009, Beijing, pp. 495–505, 5–7 June 2009. RILEM Publications, Beijing (2009)
21. Ferrara, L.: On the use of viscosity enhancing admixtures to improve the homogeneity of fiber distribution in fiber reinforced concretes. In: Brandt et al. (eds.) Proceedings of the International Symposium "Brittle Matrix Composites 7", A.M., Warsaw, pp. 287–300, 13–15 Oct 2003. Zturek RSI and Woodhead Publishers, Warsaw (2003)

22. Ozyurt, N., Mason, T.O., Shah, S.P.: Correlation of fiber dispersion, rheology and mechanical performance of FRCs. Cement Concr. Compos. **29**, 70–79 (2007)
23. Bayasi, M.Z., Soroushian, P.: Effect of steel fiber reinforcement on fresh mix properties of concrete. ACI Mater. J. **89**(4), 369–374 (1992)
24. Ozyurt, N., Mason, T.O., Shah, S.P.: Non-destructive monitoring of fiber orientation using AC-IS: an industrial scale application. Cem. Concr. Res. **36**, 1653–1660 (2006)
25. Ferrara, L., Dozio, D., Di Prisco, M.: On the connections between fresh state behavior, fiber dispersion and toughness properties of steel fiber reinforced concrete. In: Naaman, A., Reinhardt, H.W. (eds.) Proceedings of HPFRCC5, Mainz (Germany), pp. 249–258, 13–15 July 2007, RILEM Publications, Mainz (2007)
26. Stahli, P., van Mier, J.G.M.: Manufacturing, fiber anisotropy and fracture of hybrid fiber concrete. Eng. Fract. Mech. **74**, 223–242 (2007)
27. Stahli, P., Custer, R., van Mier, J.G.M.: On flow properties, fiber distribution, fiber orientation and flexural behaviour of FRC. Mater. Struct. **41**, 189–196 (2008)
28. Ferrara, L., Park, Y.D., Shah, S.P.: Correlation among fresh state behaviour, fiber dispersion and toughness properties of SFRCs. ASCE J. Mater. Civ. Eng. **20**, 493–501 (2008)
29. Ferrara, L., Ozyurt, N., di Prisco, M.: High mechanical performance of fiber reinforced cementitious composites: the role of "casting-flow" induced fiber orientation. Mater. Struct. **44**, 109–128 (2011)
30. Ferrara, L., Di Prisco, M., Lamperti, M.G.L.: Identification of the stress-crack opening behavior of HPFRCC: the role of flow-induced fiber orientation. In: Oh, B.H., et al. (eds.) Proceedings FraMCoS 7, 7th International Conference of Fracture Mechanics of Concrete and Concrete Structures, Jiejiu, South Korea, pp. 1541–1550, 23–28 May 2010
31. Barnett, S.J., Lataste, J.F., Parry, T., Millard, S.G., Soutsos, M.N.: Assessment of fiber orientation in ultra high performance fiber reinforced concrete and its effect on flexural strength. Mater. Struct. **43**, 1009–1023 (2010)
32. Boulekbache, B., Hamrat, M., Chemrouk, M., Amziane, S.: Flowability of fiber-reoinforced concrete and its effect on the mechanical properties of the material. Constr. Build. Mater. **24**, 1664–1671 (2010)
33. Boulekbache, B., Hamrat, M., Chemrouk, M., Amziane, S.: Influence of yield stress and compressive strength on direct shear behaviour of steel fiber-reinforced concrete. Constr. Build. Mater. **27**, 6–14 (2012)
34. RILEM TC 174-SCC: "Self-Compacting concrete. State of the art report", RILEM Report 23, Å. Skarendahl and Ö Petersson eds., 2000, pp. 154 + XI
35. De Larrard, F.: Concrete mixture proportioning. A scientific approach. E&FN Spoon (1999)
36. Su, N., Hsu, K.C., Chai, H.W.: A simple mix-design method for self-compacting concrete. Cem. Concr. Res. **31**, 1799–1807 (2001)
37. Brouwers, H.J.H., Radix, H.J.: Self compacting concrete: theoretical and experimental study. Cem. Concr. Res. **35**, 2116–2136 (2005)
38. Domone, P.: Self-compacting concrete: an analysis of 11 years of case studies. Cement Concr. Compos. **28**(2006), 197–208 (2006)
39. Naaman, A.E., Wille, K.: Some correlation between high packing density, ultra-high performance, flow-ability and fiber reinforcement of a concrete matrix. In: BAC 2010, Proceedings 2° Congreso Iberico Betão Auto-Compactável – Hormigón AutoCompactante, J. Barros et al. eds., 1–2 July 2010, Guimarães, Portugal, Multicomp, pp. 3–18
40. Saak, A.W., Jennings, H.M., Shah, S.P.: New methodology for designing self-compacting concrete. ACI Mater. J. **98**, 429–439 (2001)
41. Bui, V.K., Akkaya, J., Shah, S.P.: Rheological model for self-consolidating concrete. ACI Mater. J. **99**, 549–559 (2002)
42. Pereira, E.N.B., Barros, J.A.O., Ribeiro, A.F., Camoes, A.: Post cracking behavior of self-compacting steel fiber reinforced concrete. In: Di Prisco, M., et al. (eds.) Proceedings BEFIB04, Paris, 2004, pp. 1371–1380. RILEM Publications, Paris (2004)
43. Ferrara, L., Park, Y.D., Shah, S.P.: A method for mix-design of fiber reinforced self-compacting concrete. Cem. Concr. Res. **37**, 957–971 (2007)

44. Yu, A.B., Zou, R.P., Standish, N.: Packing of ternary mixtures of non-spherical particles. J. Am. Ceram. Soc. **75**, 265–272 (1992)
45. Yu, A.B., Standish, N., McLean, A.: Porosity calculation of binary mixtures of non-spherical particles. J. Am. Ceram. Soc. **76**, 2813–2816 (1993)
46. Martinie, L., Rossi, P., Roussel, N.: Rheology of fiber reinforced cementitious materials: classification and prediction. Cem. Concr. Res. **40**, 226–234 (2010)
47. Liao, W.C., Chao, S.H., Naaman, A.E.: Experience with self-consolidating high-performance fiber-reinforced mortar and concrete. In: Aldea, C.M., Ferrara, L. (eds.) Fiber Reinforced Self Consolidating Concrete: Research and Application, ACI-SP 274, 2010, pp. 79–94
48. Camacho, E., Serna, P.: "Optimización de dosificaciones de hormigón autocompactante de muy alto rendimiento renforzado con fibras híbrido", in BAC 2010, Proceedings 2° Congreso Iberico Betão Auto-Compactável – Hormigón AutoCompactante, J. Barros et al. eds., 1–2 July 2010, Guimarães, Portugal, Multicomp, 10 pp. (CD-ROM), abstract p. 41 (in Spanish)
49. Tattersall, G.H., Banfill, P.F.G.: The rheology of fresh concrete, pp. 356 + XII. Pitman Books Ltd., Boston (1983)
50. Tattersall, G.H.: Workability and quality control of concrete. E&FN Spon, London (1991)
51. Assaad, J., Khayat, K., Daczko, J.: Evaluation of static stability of self-consolidating concrete. ACI Mater. J. **101**, 207–215 (2004)
52. Khayat, K., Assaad, J., Daczko, J.: Comparison of field-oriented test method to assess dynamic stability of self-consolidating concrete. ACI Mater. J. **101**, 168–176 (2004)
53. prEN12350-8, Testing fresh concrete – part 8 – Self compacting concrete – slump flow test
54. prEN12350-9, Testing fresh concrete – part 9 – Self compacting concrete – V-funnel test
55. prEN12350-10, Testing fresh concrete – part 10 – Self compacting concrete – L-box test
56. prEN12350-12, Testing fresh concrete – part 12 – Self compacting concrete – J-ring test
57. Saak, A., Jennings, H., Shah, S.P.: A generalized approach for the determination of yield stress by slump and slump flow. Cem. Concr. Res. **34**, 363–371 (2004)
58. Roussel, N., Coussot, P.: Fifty-cent rheometer for yield stress measurements: from slump to spreading flow. J. Rheol. **49**, 705–718 (2005)
59. Roussel, N.: Correlation between yield stress and slump: comparison between numerical simulations and concrete rheometer results. Mater. Struct. **39**, 501–509 (2006)
60. Wallevik, J.E.: Relationships between Bingham parameters and slump. Cem. Concr. Res. **36**, 1214–1221 (2006)
61. Roussel, N., LeRoy, R.: The Marsh cone: a test or a rheological apparatus. Cem. Concr. Res. **35**, 823–830 (2005)
62. Tregger, N., Ferrara, L., Shah, S.P.: Identifying cement rheological properties from the mini-slump-flow test. ACI Mater. J. **105**, 558–566 (2008)
63. Cremonesi, M., Ferrara, L., Frangi, A.: "Rheological properties of cementitious composites: identification through field tests and Computational Fluid Dynamics modelling", n BAC 2010, Proceedings 2° Congreso Iberico Betão Auto-Compactável – Hormigón AutoCompactante, J. Barros et al. eds., 1-2 July 2010, Guimarães, Portugal, Multicomp, 10 pp. (CD-ROM), abstract p. 65
64. Bouvet, A., Chorbel, E., Bennacer, R.: The mini-conical slump flow test: analysis and numerical study. Cem. Concr. Res. **40**, 1517–1523 (2010)
65. Hu, C., de Larrard, F., Sedran, T., Boulay, C., Bosc, F., Deflorenne, F.: Validation of BTRHEOM, the new rheometer for soft to fluid concrete. Mater. Struct. **29**, 620–631 (1996)
66. Domone, P., Yongmo, X., Banfill, P.F.G.: Developments of the two-point workability test for high performance concrete. Mag. Concr. Res. **51**, 171–179 (1999)
67. Wallevik, J.E.: Development of parallel plate based measuring system for the CONTEC viscometer. In: Wallevik, O.H. et al. (eds.) Proceedings of the 3rd International RILEM Symposium on Rheology of Fresh Cementitious Suspensions such as Fresh Concrete, Reykjiavik, pp. 18–24, 19–21 Aug 2009. RILEM Publications, Reykjiavik

68. Kuder, K.G., Ozyurt, N., Mu, E.B., Shah, S.P.: Rheology of fiber reinforced cementitious materials. Cem. Concr. Res. **37**, 191–199 (2007)
69. El-Dieb, A.S., Reda Taha, M.M.: Flow characteristics and acceptance criteria of fiber-reinforced self-compacted concrete (FR-SCC). Constr. Build. Mater. **27**, 585–596 (2012)
70. Groth, P.: Steel fiber reinforced SCC, Final report of task 6, Brite Euram project (BE 96-3801) – rational production and improved working environment through using SCC
71. Borralleras, P., Barragàn, B., Gettu, R.: Comparison for durability parameters between conventional concrete standard SCC and Steel Fiber Reinforced SCC for construction of thin elements. Part 1: durability tests. In: De Schutter, G., Boel, V. (eds.) Proceedings of the SCC2007, 5th International RILEM Symposium on Self-Compacting Concrete, Gent, Belgium, pp. 1079–1091, 3–5 Sept 2007. RILEM Publications, Gent (2007)
72. Cheung, A.K.F., Leung, C.K.Y.: Experimental study of the bond between steel reinforcement and self-compacting high strength fiber reinforced cementitious composites. In: Gettu, R. (ed.) Proceedings of the Befib2008, 7th International RILEM Symposium on FRC, Chennai, India, pp. 6673–6678, 17–19 Sept 2008. RILEM Publications, Chennai
73. Khayat, K.H., Roussel, Y.: Testing and performance of fiber-reinforced, self-consolidating concrete. Mater. Struct. **33**, 391–397 (2000)
74. Ferrara, L., Ozyurt, N.: Mix-design optimization of steel fiber reinforced SCC. In: Shah, S.P. (ed.) Proceedings 3rd North American Conference on SCC, Chicago, 10–12 Nov 2008, CD_Rom, paper n° 1124
75. Ozyurt, N., Tregger, N., Ferrara, L., Sedan, I., Shah, S.P.: Adapting fresh state properties of fiber reinforced cementitious material for high performance thin-section elements. In: Wallevik, O.H., et al. (eds.) Proceedings Rheo-Iceland 2009, 3rd International RILEM Symposium on Rheology of Cement Suspensions such as Fresh Concrete, pp. 313–321. RILEM Publications (2009)
76. Marangon, E., Filho, R.D.T.: "Comportamento reológico de concretos auto-adensáveis reforçados comelevadas frações volumétricas de fibras de aço", in BAC 2010, Proceedings 2° Congreso Iberico Betão Auto-Compactável – Hormigón AutoCompactante, J. Barros et al. eds., 1–2 July 2010, Guimarães, Portugal, Multicomp, 10 pp. (CD-ROM), abstract p. 109 (in Portuguese)
77. Ferrara, L., Bamonte, P., Caverzan, A., Musa, A.M., Sanal, I.: An experimental methodology to assess the performance of steel Fiber-Reinforced Self-Compacting Concrete (FR-SCC). In: Studies and Researches - Annual Review of Structural Concrete/ Politecnico di Milano and Italcementi, 31, 2011, pp. 155–184. Starrylink publications (2011)
78. Shen, L., Struble, L., Lange, D.: Modeling static segregation of Self-Consolidating Concrete. ACI Mater. J. **106**, 367–374 (2009)
79. Shen, L., Struble, L., Lange, D.: Modeling dynamic segregation of Self-Consolidating Concrete. ACI Mater. J. **106**, 375–380 (2009)
80. Tregger, N., Ferrara, L., Shah, S.P.: Predicting dynamic segregation resistance of self-consolidating concrete from the slump-flow test. ASTM Int. **7**, 1–7 (2010)
81. Tregger, N., Gregori, A., Ferrara, L., Shah, S.P.: Correlating dynamic segregation of SCC to the slump-flow test. Constr. Build. Mater. **28**, 499–505
82. Kaci, A., Bouras, R., Phan, V.T., Andreani, P.A., Chaouche, M., Brossas, H.: Adhesive and rheological properties of fresh fiber-reinforced mortars. Cem. Concr. Compos. **33**, 218–224 (2011)
83. Chalencon, F, Orgeas, L., Dumont, P.J.J., Foray, G., Cavaillé, J.Y., Maire, E., Rolland du Roscoat, S: Lubricated compression and X-ray microtomography to analyse the rheology of a fiber-reinforced mortar. Rheol. Acta **49**, 221–235 (2010)
84. Ghanbari, A., Karihaloo, B.L.: Prediction of the plastic viscosity of self-compacting steel fiber reinforced concrete. Cem. Concr. Res. **39**, 1209–1216 (2009)

85. van Bui, K., Geiker, M., Shah, S.P.: Rheology of fiber reinforced cementitious materials. In: Naaman, A., Rheinhardt, H.W. (eds.) Proceedings of the HPFRCC4, Paris, 2003, pp. 221–231. RILEM Publications, Paris (2003)
86. Voigt, T., Van Bui, K., Shah, S.P.: Drying shrinkage of concrete reinforced with fibers and welded-wire fabric. ACI Mater. J. **101**, 233–241 (2004)
87. Vandewalle, L., Heriman, G., van Rickstal F.: Fiber orientation in self-compacting fiber reinforced concrete. In: Gettu, R. (ed.) Fiber Reinforced Concrete: Design and Applications, Proceedings of the 7th International RILEM Symposium BEFIB 2008, Chennai, India, pp. 719–728, 17–19 Sept 2008. RILEM Publications, Chennai (2008)
88. Torrijos, M.C., Barragan, B.E., Zerbino, R.L.: Placing conditions, mesostructural characteristics and post-cracking response of fiber reinforced self-compacting concretes. Constr. Build. Mater. **24**, 1078–1085 (2010)
89. Yardimci, M.Y., Baradan, B., Tasdemir, M.A.: Studies on the relation between fiber orientation and flexural performance of SFRSCC. In: Gettu, R. (ed.) Fiber Reinforced Concrete: Design and Applications, Proceedings of the 7th International RILEM Symposium BEFIB 2008, Chennai, India, pp. 711–718, 17–19 Sept 2008. RILEM Publications, Chennai (2008)
90. Sanal, I., Ozyurt, N.: Effects of formwork dimensions on the performance of fiber-reinforced cement based materials. In: Proceedings of the 9th International Conference on Advances in Civil Engineering, Trabzon, Turkey, 27–30 Sept 2010
91. Kooiman, A.G: Modelling steel fiber reinforced concrete for structural design. Ph.D. thesis. Department of Structural and Building Engineering, Delft University of Technology, 2000
92. Martinie, L.: Comportement rhéologique et mise en œuvre des matériaux cimentaires fibrés. Ph.D. thesis, Universitè Paris Est, 2010, 193 + vi pp
93. Martinie, L., Roussel, N.: Simple tools for fiber orientation prediction in industrial practice. Cem. Concr. Res. **41**, 993–1000 (2011)
94. Stroeven, P., Shah, S.P.: Use of radiography-image analysis for steel fiber reinforced concrete. In: Swamy, R.N. (ed.) Testing and Test Methods of Fiber Cement Composites, pp. 345–353. Construction Press, Lancaster (1978)
95. Torrents, J.M., Mason, T.O., Peled, A., Shah, S.P., Garboczi, E.J.: Analysis of the impedance spectra of short conductive fiber-reinforced composites. J. Mater. Sci. **36**, 4003–4012 (2001)
96. Ozyurt, N., Woo, L.Y., Mason, T.O., Shah, S.P.: Monitoring fiber dispersion in fiber reinforced cementitious materials: comparison of AC impedance spectroscopy and image analysis. ACI Mater. J. **103**, 340–347 (2006)
97. Woo, L.Y., Wansom, S., Ozyurt, N., Mu, B., Shah, S.P., Mason, T.O.: Characterizing fiber dispersion in cement composites using AC Impedance Spectrometry. Cement Concr. Compos. **27**, 627–636 (2005)
98. Douglas, J.F., Garboczi, E.J.: Intrinsic viscosity and the polarizability of particles having a wide range of shapes, pp. 2265–2270. Advances in Chemical Physics, XCI, Wiley, New York (1995)
99. Lataste, J.F., Behloul, M., Breysse, D.: Characterisation of fibers distribution in a steel fiber reinforced concrete with electrical resistivity measurements. NDT and E Int. **41**(8), 638–647 (2008)
100. Van Damme, S., Franchois, A., De Zutter, D., Taerwe, L.: Nondestructive determination of the steel fiber content in concrete slabs with an open-ended coaxial probe. IEEE Trans. Geosci. Remote Sens. **42**, 2511–2521 (2004)
101. Faifer, M., Ottoboni, R., Toscani, S., Ferrara, L.: Non-destructive testing of steel fiber reinforced concrete using a magnetic approach. IEEE Trans. Instrum. Meas. **60**(5), 1709–1717 (2011)
102. Ferrara, L., Faifer, M., Toscani, S.: A magnetic method for non-destructive monitoring of fiber dispersion and orientation in Steel Fiber Reinforced Cementitious Composites – part 1: method calibration. Mater. Struct. **45**(4), 575–589 (2012)

103. Ferrara, L., Faifer, M., Muhaxheri, M., Toscani, S.: A magnetic method for non-destructive monitoring of fiber dispersion and orientation in steel fiber reinforced cementitious composites – part 2: correlation to tensile fracture toughness. Mater. Struct. **45**(4), 591–598 (2012)
104. Felicetti, R., Ferrara L.: The effect of steel fiber on concrete conductivity and its connection to on-site material assessment. In: Gettu, R. (ed.) Proceedings of Befib 2008, 7th International RILEM Symposium on Fiber Reinforced Concrete, Chennai, India, pp. 525–536, 17–19 Sept 2008, RILEM Publications, Chennai
105. Molins Borrel, C., Martinez Martinez, J.A., Arnaiz Alvaro, N.: "Distribucion de fibras de acero en probetas prismaticas de hormigon", Proc. IV Congreso Internacional de Estructuras de la Asociaci6n Científico-tecnica del Hormig6n Estructural, 2008, Valencia, Espana
106. Roussel, N., Geiker, M., Dufour, F., Thrane, L.N., Szabo, P.: Computational modeling of concrete flow: general overview. Cem. Concr. Res. **37**, 1298–1307 (2007)
107. Dufour, F., Pijaudier-Cabot, G.: Numerical modeling of concrete flow: homogeneous approach. Int. J. Numer. Anal. Methods Geomech. **29**, 395–416 (2005)
108. Cremonesi, M., Ferrara, L., Frangi, A., Perego, U.: Simulation of the flow of fresh cementitious suspensions by a Lagrangian Finite Element approach. J. Non-Netwonian Fluid Mech. **165**, 1555–1563 (2010)
109. Gram, A., Silfwerbrand, J.: Numerical simulation of fresh SCC flow: applications. Mater. Struct. 9 (2010) doi:10.1617/s11527-010-9666-9
110. Kulesagaram, S., Karihaloo, B., Ghanbari, A.: Modelling the flow of self-compacting concrete. Int. J. Numer. Anal. Methods Geomech. **35**, 713–723 (2011)
111. Roussel, N., Staquet, S., D'Aloia SCchwartzentruber, L., Le Roy, R., Toutlemonde, F.: SCC casting prediction for the realization of prototype VHPC-precambered composite beams. Mater. Struct. **40**, 877–887 (2007)
112. Petrie, C.J.S.: The rheology of fiber suspensions. J. Non-Netwonian Fluid Mech. **87**, 369–402 (1999)
113. Moses, K.B., Advani, S.G., Rheinhardt, A.: Investigation of fiber motion near solid boundaries in single shear flow. Rheol. Acta **40**, 296–306 (2001)
114. Holm, R., Soderberg, D.: Shear influence on fiber orientation. Rheol. Acta **46**, 721–729 (2007)
115. Gunez, D.Z., Scirocco, R., Mewis, J., Vermart, J.: Flow induiced orientationj of non-spherical particles: effect of aspect ratio and medium rheology. J. Non-Newtonian Fluid Mech. **155**, 39–50 (2008)
116. Ferec, J., Ausias, G., Heuzey, M.C., Carreau, P.J.: Modeling fiber interactions in semiconcentrated fiber suspensions. J. Rheol. **53**, 49–72 (2009)
117. Keshtkar, M., Heuzey, M.C., Carreau, P.J.: Rheological behaviour of fiber filled model suspensions: effect of fiber flexibility. J. Rheol. **53**, 631–650 (2009)
118. Ma, W.K.A., Chinesta, F., Ammar, A., Mackley, M.R.: Rheological modeling of carbon nanotube aggregate suspensions. J. Rheol. **52**, 1311–1330 (2008)
119. Patankar, N.A., Joseph, D.D.: Modelling and numerical simulation of particulate flow by Eulerian Lagrangian approach. Int. J. Multiph. Flow **27**, 1659–1684 (2001)
120. Patankar, N.A., Joseph, D.D.: Lagrangian numerical simulation of particulate flow. Int. J. Multiph. Flow **27**, 1685–1706 (2001)
121. Martinie, L., Roussel, N.: Fiber reinforced cementitious materials: from intrinsic isotropic behavior to fiber alignment. In: Khayat, K.H., Feys, D., (eds.) Design, Production and Placement of Self-Consolidating Concrete, Proceedings of SCC2010, 6th International RILEM Symposium on Self-Compacting Concrete and 4th North American Conference on the Design and Use of SCC, Montreal, Canada, pp. 407–416, 26–29 Sept 2010
122. Jeffery, G.B.: The motion of ellipsoidal particles immersed in a viscous fluid. Proc. R. Soc. Lond. Ser. A **102**, 161–179 (1922)
123. Ferrara, L., Tregger, N., Shah, S.P.: Flow-induced fiber orientation in SCSFRC: monitoring and prediction. In: Khayat, K.H., Feys, D. (eds.) Design, Production and Placement of Self-

Consolidating Concrete, Proceedings of SCC2010, 6th International RILEM Symposium on Self-Compacting Concrete and 4th North American Conference on the Design and Use of SCC, Montreal, Canada, pp. 417–428, 26–29 Sept 2010

124. Mechtcherine, V, Shyshko, S., Simulating the behaviour of fresh concrete using distinct element method. In: De Schutter, G. (ed.) Proceedings 5th International RILEM Symposium on Self Compacting Concrete, Ghent, Belgium, pp. 467–472, 3–5 Sept 2007. RILEM Publications, Ghent (2007)
125. Mechtcherine, V, Shyshko, S.: Self-compacting conçrete simulation using distinct element method. In: Wallevik, O.H., et al. (eds.) Proceedings 3rd International RILEM Symposium of Rheology of Cement suspensions, Reykjavik, Iceland, pp. 171–179, 19–21 Aug 2009. RILEM Publications, sarl, Reykjavik (2009)
126. Mechtcherine, V, Shyshko, S.: Continuous numerical modelling of concrete from fresh state to hardened state. In: Proceedings of the 16th Building Material Congress Ibausil, Weimar, vol. 2, pp. 165–172 (2006)
127. Mechtcherine, V, Shyshko, S., Simulating the workability of fresh concrete. In: Proceedings of the International RILEM Symposium on Concrete Modelling, CONMOD 08, Delft, pp. 173–181. RILEM Publications, s.a.r.l. (2008)
128. Mechtcherine, V., Shyshko, S.: Virtual concrete laboratory – Continuous numerical simulation of concrete behaviour from fresh to hardened state. In: Grosse, U. (ed.) Advances in Construction Materials, Berlin-Heidelberg, pp. 479–488. Springer-Verlag, Berlin (2007)
129. Ferrara, L., Shyshko, S., Mechtcherine, V.: Predicting the flow-induced fiber dispersion and orientation in self-consolidating concrete by distinct element method. Submitted for presentation to BEFIB 2012
130. Laranjeira, F. Aguado, A., Molins, C., Grunewald, S, Walraven, J and Cavalaro, S.: Framework to predict the orientation of fibers in FRC: a novel philosophy. Cem. Concr. Res. **42**, 752–768 (2012)
131. Armelin. H.S., Banthia, N.: Predicting the flexural post-cracking performance of steel fiber reinforced concrete from the pullout of single fibers. ACI Mater. J. **94**, 18–31 (1997)
132. Jones, P.A., Austin, S.A., Robis, P.J.: Predicting the flexural load-deflection response of steel fiber reinforced concrete from strain, crack-width, fiber pull-out and distribution data. Mater. Struct. **41**, 449–463 (2008)
133. Cunha, V.M.C.F., Barros, J.A.O., Sena-Cruz, J.M.: Pull.out behavior of steel fibers in self-compacting concrete. ASCE J. Mater. Civ. Eng. **22**, 1–9 (2010)
134. Laranjeira de Oliveira, F.: Design-oriented constitutive model for steel fiber reinforced concrete. Ph.D. thesis, Universitat Politecnica de Catalunya, 318 + VI pp (2010)
135. Laranjeira, F., Molins, C., Aguado, A.: Predicting the pullout response of inclined hooked steel fibers. Cem. Concr. Res. **40**, 1471–1487 (2010)
136. Laranjeira, F., Aguado, A., Molins, C.: Predicting the pullout response of inclined straight steel fibers. Mater. Struct. **43**, 875–895 (2010)
137. Laranjeira, F., Grunewald, S., Walraven, J., Blom, C., Molins, C., Aguado, A.: Characterization of the orientation profile of steel fiber reinforced concrete. Mater. Struct. **44**, 1093–1111 (2011)
138. *fib* Model Code 2010, 1st draft, March 2010, 1st vol., pp. 292 + xx, 2nd vol., pp. 288 + xviii
139. Balaguru, P. N., Shah. S.P.: Fiber reinforced cement composites, pp. 530. Mc Graw Hill, Singapore New York (1992)
140. Aoude, H., Cook, W.D., Mitchell, D.: Behavior of columns constructed with fibers and self-consolidating concrete. ACI Mater. J. **106**, 349–357 (2009)
141. Cunha, V.M.C.F., Barros, J.A.O., Sena-Cruz, J.: Modelling the influence of age of steel fiber reinforced self-compacting concrete on its compressive behaviour. Mater. Struct. **41**, 465–478 (2008)
142. Pereira, E.N.B., Barros, J.A.O., Camoes, A.: Steel fiber reinforced self-compacting concrete: experimental research and numerical simulation. ASCE J. Struct. Eng. **134**, 1310–1321 (2008)

143. Ferrara, L.: Statistical properties of steam-cured plant-produced SCC for prestressed precast applications. In: Shi, C., et al. (eds.) Proceedings SCC2009, 2nd International Symposium on Design, Performance and use of Self-Consolidating Concrete, Beijing, pp. 483–494, 5–8 June 2009. RILEM Publications, Beijing (2009)
144. Domone, P.L.: A review of the hardened mechanical properties of self-compacting concrete. Cement Concr. Compos. **29**, 1–12 (2007)
145. Torrijos, M.C., Barragan, B.E., Zerbino, R.L.: Physical-mechanical properties, and mesostructure of plain and fiber reinforced self-compacting concrete. Constr. Build. Mater. **22**, 1780–1788 (2008)
146. Mohammed, R.N., Elliot, K. S.: Behaviour of steel fiber self-compacting concrete under biaxial loading. In: De Schutter, G., Boel, V. (eds.) Proceedings of the SCC2007, 5th International RILEM Symposium on Self-Compacting Concrete, Gent, Belgium, pp. 1079–1091, 3–5 Sept 2007. Rilem Publications, Gent (2007)
147. Yin, W.S., Su, E.C.M., Mansur, M.A., Hsu, T.C.H.: Biaxial tests on plain and fiber concrete. ACI Mater. J. **86**, 236–243 (1989)
148. Traina, L.A., Mansour, S.H.: Biaxial strength and deformational behaviour of plain and steel fiber concrete. ACI Mater. J. **88**, 354–362 (1991)
149. Fantilli, A.P., Vallini, P., Chiaia, B.: Ductility of fiber-reinforced self-consolidating concrete under multi-axial compression. Cement Concr. Compos. **33**, 520–527 (2011)
150. Soroushian, P., Lee, C.D.: Distribution and orientation of fibers in steel fiber reinforced concrete. ACI Mater. J. **87**, 433–439 (1990)
151. Stroeven, P.: Stereological principles of spatial modeling applied to steel fiber reinforced concrete in tension. ACI Mater. J. **106**, 213–222 (2009)
152. Cunha, V.M.C.F., Barros, J.A.O., Sena Cruz, J.M.: An integrated approach for modeling the tensile behaviour of steel fiber reinforced self-compacting concrete. Cem. Concr. Res. **41**, 64–76 (2011)
153. Pons, G., Mouret, M., Alcantara, M., Granju, J.L.: Mechanical behaviour of self-compacting concrete with hybrid fiber reinforcement. Mater. Struct. **40**, 201–210 (2007)
154. Velasco, R.V., Toledo Filho, R.D., Fairbairn, E.M.R.: Behavior under direct tension and bending of self-consolidating concrete reinforced with high volume fractions of steel fibers. In: Gettu, R. (ed.) Fiber Reinforced Concrete: Design and Applications, Proceedings of the 7th International RILEM Symposium BEFIB 2008, Chennai, India, pp. 751–759, 17–19 Sept 2008. RILEM Publications, Chennai (2008)
155. Ferrara, L., Caverzan, A., Muhaxheri, M., di Prisco M.: Identification of tensile behaviour of SFR-SCC: direct vs. indirect tests. *Accepted for presentation to* BEFIB2012
156. Pansuk, W, Sato, I., Sato, Y., Shionaga, R.: Tensile behaviors and fiber orientation of UHPC. In: Fehling et al. (eds.) Ultra High Performance Concrete, Proceedings 2nd International Symposium on UHPC, KUP, Kassel, Germany, pp. 161–168, 05–07 March 2008
157. Kang, S.T., Kin, J.K.: The relation between fiber orientation and tensile behavior in ultra high performance fiber reinforced cementitious composites (UHPFRCC). Cem. Concr. Res. **41**, 1001–1014 (2011)
158. Wille, K., Parra Montesinos, G.J.: Effect of beam size, casting method and support conditions on flexural behavior of ultra-high performance fiber-reinforced concrete. ACI Mater. J. **109**, 379–388 (2012)
159. Buratti, N., Mazzotti, C., Savoia, M.: Long term behaviour of self-compacting fiber-reinforced concrete beams. In: Khayat, K.H. Feys, D. (eds.) Design, Production and Placement of Self-Consolidating Concrete, Proceedings of SCC2010, 6th International RILEM Symposium on Self-Compacting Concrete and 4th North American Conference on the Design and Use of SCC, Montreal, Canada, 26–29 Sept 2010, pp. 439–450
160. Forgeron, D., Omer, A.: Flow characteristics of macro-synthetic fiber reinforced self-consolidating concrete. In: Aldea, C.M., Ferrara, L. (eds.) Fiber Reinforced Self Consolidating Concrete: Research and Application, ACI-SP 274, pp. 1–14 (2010)

161. Kwon, S.H., Ferron, R.P., Akkaya, Y., Shah, S.P.: Cracking of fiber reinforced self-compacting concrete due to restrained shrinkage. Int. J. Concr. Str. Mater. **1**(1), 3–9 (2007)
162. Carlsward, J., Emborg, M.: Prediction of stress development and cracking in steel fiber reinforced self-compacting concrete overlays due to restrained shrinkage. In: Aldea, C.M., Ferrara, L. (eds.) Fiber reinforced Self Consolidating Concrete: Research and Application, ACI-SP 274, 2010, pp. 31–50
163. Kassimi, F., Khayat, K.H.: Drying shrinkage model for fiber reinforced SCC. In: Khayat, K.H. Feys, D. (eds.) Design, Production and Placement of Self-Consolidating Concrete, Proceedings of SCC2010 vol II, 6th International RILEM Symposium on Self-Compacting Concrete and 4th North American Conference on the Design and Use of SCC, Montreal, Canada, 26–29 Sept 2010, pp. 1223–1233
164. Naaman, A.E., Wongtanakitcharoen, T., Hauser, G.: Influence of different fibers on plastic shrinkage cracking of concrete. ACI Mater. J. **102**, 49–58 (2005)
165. Banthia, N., Gupta, R.: Influence of polypropylene fiber geometry on plastic shrinkage cracking in concrete. Cem. Concr. Res. **36**, 1263–1267 (2006)
166. Khaliq, W., Kodur, V.: Thermal and mechanical properties of fiber reinforced high performance self-consolidating concrete at elevated temperatures. Cem. Concr. Res. **41**, 1112–1122 (2011)
167. Bamonte, P., Gambarova, P.G.: A study on the mechanical properties of self-compacting concrete at high temperature and after cooling. Mater. Str. **45**, 1375–1387 (2012). doi:10.1617/s11527-012-9839-9
168. Colombo, M., di Prisco, M., Felicetti, R.: Mechanical properties of steel fiber reinforced concrete exposed at high temperatures. Mater. Struct. **43**, 475–491 (2010)
169. Alonso, M.C., Rodriguez, C., Sanchez, M., Barragán, B.: "Respuesta al fuego de HAC con y sin renfuerzo de fibras", in BAC 2010, Proceedings 2° Congreso Iberico Betão Auto-Compactável – Hormigón AutoCompactante, J. Barros et al. eds., 1–2 July 2010, Guimarães, Portugal, Multicomp, 10 pp. (CD-ROM), abstract p. 143 (in Spanish)
170. Lourenço, L., Durães, B., Barros, J.A.O., Gonçalves, D.: "Comportamento mecanico do betão auto-compactável reforçado com fibras de aço após exposição a temperaturas elevadas", in BAC 2010, Proceedings 2° Congreso Iberico Betão Auto-Compactável – Hormigón AutoCompactante, J. Barros et al. eds., 1–2 July 2010, Guimarães, Portugal, Multicomp, 10 pp. (CD-ROM), abstract p. 85 (in Portuguese)
171. Romano, G.Q., Silva, F.A., Toledo Filho, R., Fairbairn, E.M.R., Battista, R.C.: Mechanical characterization of steel fiber reinforced self-compacting refractory concrete. In: De Schutter, G., Boel, V., (eds.) Proceedings of the SCC2007, 5th International RILEM Symposium on Self-Compacting Concrete, Gent, Belgium, pp. 881–886, 3–5 Sept 2007. Rilem Publications, Gent (2007)
172. Ezeldin, A.S., Balaguru, P.N.: Bond behavior of normal and high.strength fiber reinforced concrete. ACI Mater. J. **86**(5), 515–524 (1989)
173. Schumacher, P.: Rotation capacity of self-compacting steel fiber reinforced concrete. Ph.D. thesis, Delft University of Technology (2006)
174. Cheung, A.K., Leung, C.K.Y.: Experimental study on the bond between steel reinforcement and self-compacting high strength fiber reinforced cementitious composites. In: Gettu, R. (ed.) Fiber Reinforced Concrete: Design and Applications, Proceedings of the 7th International RILEM Symposium BEFIB 2008, Chennai, India, pp. 667–678, 17–19 Sept 2008. RILEM Publications, Chennai (2008)
175. Abrishami, H.H., Mitchell. D.: Influence of steel fibers on tension stiffening. ACI Struct. J. **94**, 769–776 (1997)
176. Aoude, H., Cook, W.D., Mitchell, D.: Tensile behaviour of reinforced concrete specimens constructed with steel fibers and SCC. In: Gettu, R. (ed.) Fiber Reinforced Concrete: Design and Applications, Proceedings of the 7th International RILEM Symposium BEFIB 2008, Chennai, India, pp. 689–698, 17–19 Sept 2008. RILEM Publications, Chennai (2008)
177. Shionaga, R., Walraven, J.C., den Ujil, J.A., Sato, Y.: Tension stiffening of high performance fiber reinforced concrete. In: Radic, J. (ed.) Concrete Structures: Stimulators of

Development, Proceedings of the *fib* Symposium, Dubrovnik, Croatia, pp. 259–266, 20–23 May 2007 (Secon HDGK)
178. Jansson, A., Flansbjer, M., Lofgren, I., Lundgren, K., Gylltoft, K.: Experimental investigation of surface crack initiation, propagation and tension stiffening in self-compacting steel-fiber reinforced concrete. Mater. Struct. 45, 1127 (2012). doi:10.1617/s11527-012-9821-6
179. Campione, G.: The effects of fibers on the confinement models for concrete columns. Can. J. Civ. Eng. **29**, 742–750 (2002)
180. Aoude, H., Cook, W.D., Mitchell, D.: Behavior of columns constructed with fibers and self-consolidating concrete. ACI Struct. J. **106**, 349–357 (2009)
181. Minelli, F.: Plain and fiber reinforced concrete b eams under shear: structural behaviour and design aspects, p. 429. Ph.D. thesis, University of Brescia (2005)
182. Batson, G., Jenkins, E., Spatney, R.: Steel fibers as shear reinforcement in beams. ACI Journal **69**, 640–644 (1972)
183. Cuenca, E., Serna, P.: Shear behaviour of self-compacting concrete and fiber reinforced concrete beams. In: Khayat, K.H., Feys, D. (eds.) Design, Production and Placement of Self-Consolidating Concrete, Proceedings of SCC2010 vol II, 6th International RILEM Symposium on Self-Compacting Concrete and 4th North American Conference on the Design and Use of SCC, Montreal, Canada, pp. 1273–1282, 26–29 Sept 2010
184. Ding, Y., You, Z., Jalali, S.: The composite effect of steel fibers and stirrups on the shear-behaviour of beams using self-consolidating concrete. Eng. Struct. **33**, 107–117 (2011)
185. Greenough, T., Nehdi, M.: Shear behaviour of fiber.reinforced self-consolidating concrete slender beams. ACI Struct. J. **105**, 468–477 (2008)
186. Bindiganavile, V., Banthia, N., Aarup, B.: Impact response of ultra-high strength fiber reinforced cement composites. ACI Mater. J. **99**, 543–548 (2002)
187. Gonçalves, D., Barros, J., Lourenço, L., Sampaio, C.: "Cobertura pré-fabricada em betão auto-compactável reforçado com fibras de aço", in BAC 2010, Proceedings 2° Congreso Iberico Betão Auto-Compactável – Hormigón AutoCompactante, J. Barros et al. eds., 1-2 July 2010, Guimarães, Portugal, Multicomp, 10 pp. (CD-ROM), abstract p. (in Portuguese)
188. Mesbah, H.A., Kassimi, F., Yahia, A., Khayat, K.H.: Flexural performance of reinforced concrete beams repaired with fiber-reinforced SCC. In: De Schutter, G., Boel, V. (eds.) Proceedings of the SCC2007, 5th International RILEM Symposium on Self-Compacting Concrete, Gent, Belgium, pp. 637–644, 3–5 Sept 2007. Rilem Publications, Gent (2007) (PRO 54)
189. Martinola, G., Meda, A., Plizzari, G., Rinaldi, Z.: Strengthening and repair of RC beams with fiber reinforced concrete. Cem. Concr. Compos. **32**, 731–739 (2010)
190. Beschi, C., Maringoni, S., Meda, A., Riva, P., Simonelli, F.: Use of HPC jacketing for seismic retrofitting of columns. In: Atti 17° Congresso CTE, Roma, 5-8 novembre 2008, pp. 913–920 (in Italian)

Chapter 7
Specialty SCC

Manuel Vieira, Liberato Ferrara, Mohamed Sonebi and Caijun Shi

7.1 Introduction

The last decades have seen a significant and continuously increasing proportion of concrete used in construction that has become substantially modified from the traditional cement–water-aggregate mixture. The modifications were made to obtain better concrete in more demanding service conditions (e.g. higher strength, higher toughness, higher durability) and using different specialised methods of construction (pumping, underwater placing, spraying). In some cases, the developments simply aimed at increasing one or more of the many well established performance parameters of concrete in either its fresh or hardened state. Such developments lead to concretes which can be broadly classified as 'special', rather than the 'ordinary' ones used for the bulk of concrete construction. Special types of concrete are those with out-of-the-ordinary properties or those produced by unusual techniques or materials [1].

There is no firm boundary between 'ordinary' and 'specialty' concrete mixtures. In some situations, one and the same mixture can show both 'ordinary' behaviour and 'special' behaviour, depending on which of its properties is examined or in which specific type of practical application it is used [2].

M. Vieira (✉)
LNEC, Lissabon, Portugal
e-mail: mvieira@lnec.pt

L. Ferrara
Politecnico di Milano, Milan, Italy

M. Sonebi
Queen's University, Belfast, UK

C. Shi
Hunan University, Changsha, China

K. H. Khayat and G. De Schutter (eds.), *Mechanical Properties of Self-Compacting Concrete*, RILEM State-of-the-Art Reports 14,
DOI: 10.1007/978-3-319-03245-0_7, © RILEM 2014

In the context of ordinary concrete, SCC is classified a special concrete. Meanwhile, out-of-the-ordinary properties or use of unusual techniques or materials lead to the definition of special SCC. Considering those SCC that retain no appreciable amount of entrapped air other than entrained air, after its application, one could point for a large range of special types of SCC. Analogously to ordinary concrete, examples of those types include lightweight or heavyweight concrete; preplaced aggregate concrete; self-compacting slurry infiltrated concrete (SIF-CON); recycled aggregate concrete; fiber reinforced concrete; controlled shrinkage concrete; polymer-modified concrete; and geo-polymer concrete.

In this chapter, specialty SCC that will be considered include SCC with uncommon aggregates, self-compacting mortars, preplaced aggregate SCC, special fiber reinforced SCC, and application oriented SCC will be presented. Chapter 6 focuses on fiber reinforced concrete. Reports on the use of special fibers in SCC for specific applications are presented in the current chapter.

7.2 SCC with Uncommon Aggregates and Self-Compacting Mortars

This section reports about SCC produced mainly with aggregates different from the standard that are normally employed in the production of concrete: lightweight, heavyweight, recycled aggregates, and only fine aggregates.

7.2.1 Lightweight SCC

Lightweight SCC (LSCC) is defined as SCC having an oven-dry unit weight lower than 2000 kg/m^3. It is produced using lightweight aggregate (LWA) for all or part of the total aggregate content.

SCC partially incorporating LWA and presenting a normal-weight, i.e. having an oven-dry unit weight greater than 2000 kg/m^3 but not exceeding 2600 kg/m^3, according to EN 206-1 [3], must be considered as normal SCC. However, the inclusion of lightweight particles can enhance mechanical properties due to the internal curing [4].

Lightweight concrete has been used for a number of applications and is also known for its good performance and durability [5]. In structural applications, the self-weight of the concrete structure is important since it represents a large portion of the total load. The reduced self-weight of lightweight concrete will reduce gravity load and seismic inertial mass, resulting in reduced member size and foundation force. The use of lightweight concrete can be of interest in retrofit applications, for instance, where a concrete column jacket is desired due to architectural reason over other methods, such as steel or composite jackets. In that situation, a normal weight concrete jacket might result in foundation forces that would require an expensive foundation retrofit whereas a lightweight concrete jacket may not require a footing retrofit [6].

Repair applications with lightweight concrete are also advantageous due to the reduced elastic modulus and lower risks of cracking. It was reported that the use of lightweight aggregate concrete in a structure could result in lower overall costs [5]. While lightweight concrete may cost more per unit volume than normal weight concrete, the structure may cost less as a result of reduced dead weight and lower foundation costs. A large body of publications deal with normal weight self-compacting concrete (SCC), but only a limited number of publications can be found on LSCC.

LSCC combines the well-known advantages of lightweight concrete with those of SCC. Due to its favourable physical properties, namely its low unit weight, in combination with an excellent workability and a low noise emission during concreting, LSCC might find a broad application in construction practice, especially in the area of rehabilitation and reconstruction of buildings [7]. However, the high absorption of the LWA may require a prolonged pre-saturation of the aggregates in order to obtain an adequate slump-flow at lower superplasticiser content [8].

An optimized adjustment of unit weight and compressive strength is the key element of mix design of LSCC. A lower unit weight leads to a decrease of compressive strength due to the impact of the higher volume of LWA. Meanwhile the risk of "inverse" segregation—the floating of the lightweight particles, must be mitigated, by either using viscosity modifying agent (VMA) or a higher powder content to increase the paste or mortar viscosity.

7.2.1.1 Mechanical Properties

The strength of LSCC depends on aggregate, W/C, content of cement, and mineral powder used in the mixture. Generally speaking, the relation in strength development between LSCC and vibrated lightweight concrete is similar to that between SCC and current vibrated concrete.

Due to the high range of densities in which lightweight concrete can be produced, it is common to characterise these concretes by the ratio between their compressive strength and unit weight, designated by specific compressive strength. Papanicolaou and Kaffetzakis [9] presented a survey of the specific compressive strength values of a number of LSCC reported in the literature. Fig. 7.1 presents the results from references indicated in that survey (Ref.10 to Ref.22) together with recent studies [9–11].

The average strength to unit weight value (LWSS specific compressive strength) reported in Fig. 7.1 is 26×10^{-3} MPa/(kg/m^3). This means that a concrete with a unit weight of 1400 kg/m^3 and a compressive strength of about 36 MPa, or a concrete with a unit weight of 2000 kg/m^3 and a compressive strength of about 52 MPa. This specific strength value is equivalent to a

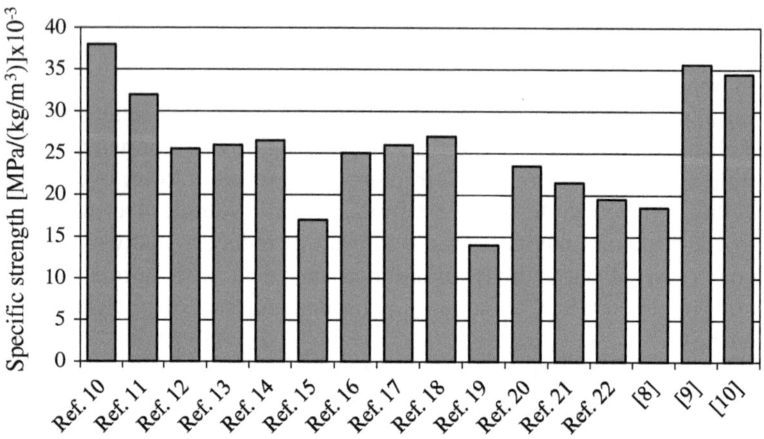

Fig. 7.1 LSCC specific compressive strength results from literature

compressive strength of 60 MPa for a normal unit weight SCC with a unit weight of about 2300 kg/m^3.

Tensile strength has been reported to vary between 6 and 12 % of compressive strength values depending on the type of aggregate and mixture composition [9, 11, 12].

In general, the elastic modulus of LSCC is affected by the type and properties of the aggregate, as well as the compressive strength of the concrete, ranging from 40 to 70 % of the equivalent normal concrete [10, 13]. This is because the elastic modulus of the aggregates itself is relatively low.

Freezing-thawing resistance of LSCC was reported to be excellent [15]. Many studies have confirmed that the use of lightweight concrete can significantly reduce autogenous shrinkage of concrete due to the transport of water from the porous saturated aggregate to the matrix [14]. Actually, a replacement of 25 % normal weight aggregate with water saturated lightweight concrete could eliminate autogenous shrinkage [14]. LSCC also demonstrated negligible autogenous shrinkage regardless of different mineral powders used [15].

7.2.1.2 Applications

LSCC can be designed and used where regular lightweight concrete is used. However, it can also be applied for some special applications. For example, it was demonstrated for use in insulated concrete foam system where both low density and high flowability are required [16]. A lightweight precast steel fiber reinforced SCC panel was also developed for building facades [17]. Precast construction is one of the most prominent applications of LSCC, namely for precast insulating panels [18].

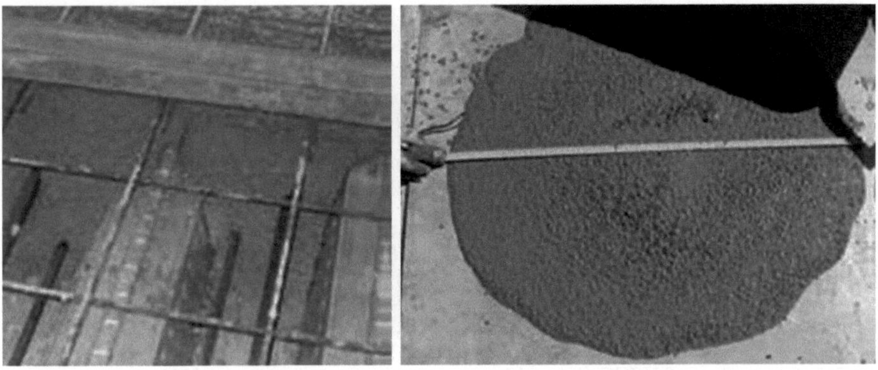

Fig. 7.2 Application of LSCC on a metallic structure [19]

In-situ LSCC was applied in a structure for the 32 edition of 2007 Sailing America Cup [19]. The aim was to cast the concrete slabs, as shown in Fig. 7.2, with a supporting metallic formwork, limiting the weight on the structure. For that, the concrete unit weight was limited to 1900 kg/m^3 on dry conditions and due to environmental class exposure the required characteristic strength was 40 MPa.

7.2.2 Heavyweight SCC

Heavyweight SCC by definition has an oven-dry density greater than 2600 kg/m^3. The manufacturing of heavyweight concrete has been directed only to applications for counterbalance elements or as ballast. It was the development of the nuclear industry that increased the use of this type of concrete since it gives an effective protection against radiation emitted by nuclear or by particle accelerators. It has also been applied in hospitals and other facilities where equipment that emits radiation, such as X-ray and NMR, is used.

The radiation in question may be undermined by elements of high atomic weight. The protective wall thickness can be reduced with the use of concrete made with the incorporation of high density aggregates, such as barite (barium sulphate), with specific gravity values of 4.0–4.4. Other heavyweight aggregates that can be used in concrete include iron minerals, such as hematite, magnetite, limonite, goethite, ilmenite (iron titanium oxide), and also wolframite (iron manganese tungstate), ferrophosphorus, and ferrous wastes. The characteristics of heavyweight concrete should also take into account the action of fast neutrons that must be mitigated by low atomic weight elements, such as hydrogen, which can be obtained by water of crystallization in limonite and goethite. The presence of oxygen also helps to moderate the action of neutrons and therefore the presence of silica is also desirable [20].

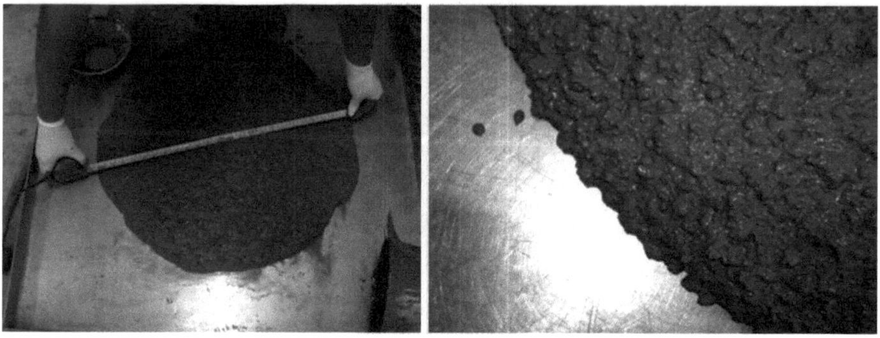

Fig. 7.3 Slump flow of HSCC (*left*) with no signs of segregation (*right*) [21]

The application of heavyweight concrete requires the use of formwork that can withstand high lateral pressure. Moreover, segregation due to the high density of the aggregate should be controlled. Heavyweight concrete used for containment should be compact and its cracking due to thermal expansion or shrinkage is prohibited. In applications where structural elements are densely reinforced, SCC technology is required. In spite of its potential for application, there are a limited number of publications on heavyweight SCC (HSCC) [21, 22]. In these publications, the results of a study to produce a SCC with heavyweight barite aggregate were reported. Starting from a conventional concrete mix design, a SCC (Fig. 7.3) was achieved with a fresh unit weight of 3300 kg/m^3 and a compressive strength of 24 MPa and 31 MPa at 7 and 28 days, respectively [21].

7.2.3 Recycled Aggregate SCC

Construction activities generate a huge amount of construction and demolition (C&D) materials each year. Deposit landfills have been the most common destination of those materials. However, to accommodate C&D materials in this way is not sustainable. In addition, there is the tendency to reduce the environmental impact of the quarries decreasing the supply of natural aggregates. Recycling is one of the measures to reduce the deposit of C&D materials and to complement the supply of virgin aggregates for the concrete production.

While the use of recycled aggregate in concrete production is still limited by its cost and standard restrictions, the use of powders from that source has a wide field of application in SCC, namely in powder-type SCC [23–26].

For the same W/C, no significant strength loss was observed when a fine or coarse recycled aggregate fraction was used instead of the natural ones, mainly for low W/B [24, 25]. However, this conclusion is not consensual; other researchers had found decrease in strength, mainly due to the increase in W/C needed to achieve the same flowability [23, 26].

7.2.4 Self-Compacting Mortars

Self-compacting mortars are suitable for the construction of thin elements or for repair applications, which will be focused on in Sect. 7.5.3. Self-compacting mortars can be used for the manufacturing of aesthetically appealing kitchen work surfaces and sinks, uniquely designed cupboards, sinks, and other interior or exterior design objects [27].

Self-compacting, fiber reinforced high-strength mortars have been developed and used to cast sheet piles [28]. In that study, it was investigated if sand with a maximum diameter of 2 mm could be used instead of coarse aggregate of separate fractions. The modified mixtures performed equally well compared to the reference one, showing that as long as the maximum diameter is reduced to 2 mm and the total volume of the sand plus steel fibers does not exceed 43 %, a mixture fulfilling the structural needs can be produced even with different aggregate size distributions.

In a study performed to evaluate the effectiveness of various mineral additives and chemical admixtures in producing self-compacting mortar [29], a good correlation between the ultrasonic pulse velocity (UPV) and compressive strength of mortars was demonstrated. Consequently, UPV could be used to non-destructively assess the internal structure of such mortar, namely on repair applications.

7.3 Preplaced Aggregate SCC

Preplaced aggregate concrete (PAC), as defined in ACI 116R, is "Concrete produced by placing coarse aggregate in a form and later injecting a Portland cement-sand grout, usually with admixtures, to fill the voids" [30]. As it is defined, there is a need for specific equipment and formworks to assure proper injection.

Meanwhile, the development of SCC technology allowed achieving cementitious mixtures that flow and compact by their own weight. Following, two variants of preplaced aggregate concrete were already developed and applied in practical conditions: rock-filled concrete and preplaced aggregate filled with self-compacting mortar.

7.3.1 Rock-Filled Concrete

Rock-Filled Concrete (RFC) is produced by filling the working space with large-scale blocks of rock to form a rock-block mass. Then, SCC is either pumped into place or poured directly on to the surface of rock-block mass, flowing by its own weight to fill all void spaces, as shown in Fig. 7.4 [31, 32]. Another method to achieve RFC is dump-type RFC, which consist of pouring the SCC before introducing the large-scale blocks of rocks (right side of Fig. 7.4). The mixture sets to

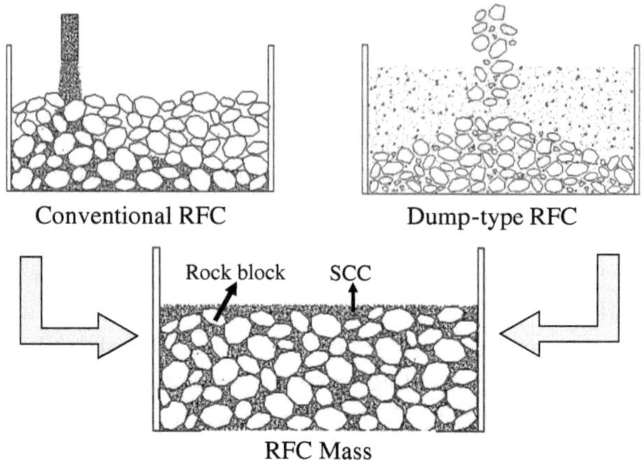

Fig. 7.4 Two types of RFC construction technologies

form the RFC mass. As a result, 55 to 60 % of the space can be filled by rock blocks in RFC mass, which leads to a great reduction of cement content in RFC. This is advantageous in massive concrete structures, in particular, concrete dams require a reduction of cement content to lower cost and heat of hydration.

7.3.1.1 Compressive Strength

RFC is expected to perform satisfactorily since the cement content for a strength grade of C15 is only about 80–90 kg/m^3 [33]. A RFC block measuring 2000 × 1000 × 1800 mm^3 was constructed in the laboratory [34]. The block was cast with three lifts with a thickness of 600 mm each. Cold and hot joints were formed between the various lifts. After curing, specimens with different sizes were cut from the RFC block to determine compression strength and permeability. Nine prismatic specimens measuring 150 × 150 × 300 mm^3 were cut from the RFC block for axial compression strength testing. The results of the compressive tests are shown in Fig. 7.5. They indicate a relationship between RFC and SCC compressive strengths. The average strength of SCC used in RFC construction was 13.2 MPa, while the average axial compression strength of nine RFC specimens was 16.7 MPa, which is 1.27 times the value of SCC. These results showed that the strength of most RFC samples was higher and only few specimens showed lower strength than the SCC alone.

7.3.1.2 Permeability Tests

Permeability is one of the most important properties concerned by engineers in hydraulic projects, since it is one of the most important potential weaknesses. Structural weaknesses, such as the cold and hot joints created by the limited thickness of concrete lifts, can significantly affect permeability of the dam body.

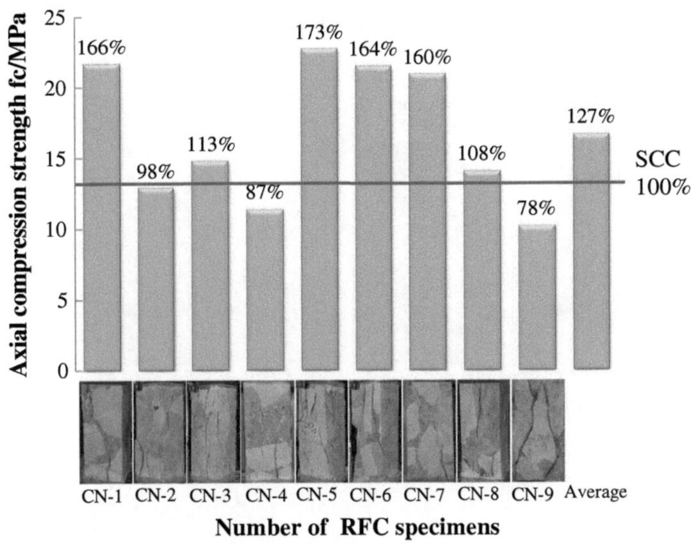

Fig. 7.5 Comparison of strength of RFC and SCC

Table 7.1 Permeation resistance index of RFC in different areas

Specimen position	RFC body	Hot joint	Cold joint
Permeation resistance index	W35	W31	W14

In order to study the permeability properties of RFC in different areas, 18 specimens (six specimens per case) were cut from a RFC body, including cold and hot joint areas. The permeation resistance indexes of RFC in different areas were tested and are reported in Table 7.1. Since the required permeation resistance index of the dam concrete in hydraulic engineering is from W2 to W10 in China [35] and the minimum index of RFC is W14, it is indicated the permeability of RFC is low enough to satisfy the requirement of the dam concrete in hydraulic engineering.

7.3.1.3 Field Applications

The construction and demonstration of RFC were carried out at the jobsites of Henan Baoquan pumped-storage project and Sichuan Xiangjiaba hydropower project, to simulate construction reality. As shown in Fig. 7.6, in the Baoquan project, the formwork is made of a masonry wall, and SCC was poured directly on the face of the rock-block mass by an excavator. After hardening, RFC with good compactness was found from the exploratory test.

Images of the Xiangjiaba in situ application are shown in Fig. 7.7 and RFC was constructed with dump-type RFC construction technology. SCC was poured into

Fig. 7.6 In-situ tests in baoquan pumped-storage projects

Fig. 7.7 In-situ application in xiangjiaba hydropower projects

the workspace directly from a mixture truck and then rock blocks were dumped with bulldozers and excavators. RFC presented satisfactory results for the apparent density test, the core compression test, the water pressure test, and the temperature rise of hydration heat test [36, 37]. Those results are shown in Table 7.2.

7.3.2 Preplaced Aggregate Filled with Self-Compacting Mortar

Preplaced aggregate filled with self-compacting mortar (PACSCM) involves the self-compacting mortar poured in a skeleton of the coarse aggregate that are preplaced in a formwork. The filling of the gaps between aggregate particles is achieved only by the self-weight of the mortar. The mortar needs to have high flowability and limited bleeding.

Table 7.2 Results of in situ tests at Baoquan and Xiangjiaba projects

Project	Apparent density test (kg/m^3)	Core compression test (MPa)	Water pressure test	Adiabatic thermal rising (°C)
Baoquan	2415	23.7	1.33 Lu	–
Xiangjiaba	2508	33.4	–	9.3

Fig. 7.8 Prototype before filling **(a)** and location of the mortar pouring **(b)**

The PACSCM concept was investigated in Portugal [38] in a cofferdam section of a hydroelectric power plant for mass concrete applications. Other application for PCA is under investigation in the Netherlands for the stabilization of railway bedding.

7.3.2.1 Mechanical Properties

The above study included, in a preliminary phase, the casting of a prototype with a cubic shape with inner edges of 1.97 m [38] as shown in Fig. 7.8a. The filling of the mould with mortar was carried out using 15 mixtures, resulting in a total volume of applied mortar of approximately 3.1 m^3, which results in mortar content of about 39 %. The pouring of the mortar was always carried out at the same location and ran for about 4 h (Fig. 7.8b).

The mortar composition had a paste content of 65 % and a sand/binder ratio of 1.2, by mass. The paste was composed with 120 kg/m^3 of cement, with 60 % fly ash of the total binder mass, and a W/CM of 0.42.

In-situ testing on hardened concrete and laboratory tests on samples taken from the prototype was carried out. The in situ tests were executed to asses durability related properties and included air permeability determination by the Torrent method, GWT water permeability, and ultrasonic pulse velocity. The laboratory tests were performed on cores taken from the prototype to determine the accessible porosity, water permeability, compressive and tensile strengths, elastic modulus,

Table 7.3 Mechanical properties of PACSCM

Properties	Result
Compressive strength	25.9 MPa
Tensile splitting strength	1.95 MPa
Direct tensile strength	1.41 MPa
Elastic modulus	25.3 GPa
Density	2410 kg/m^3

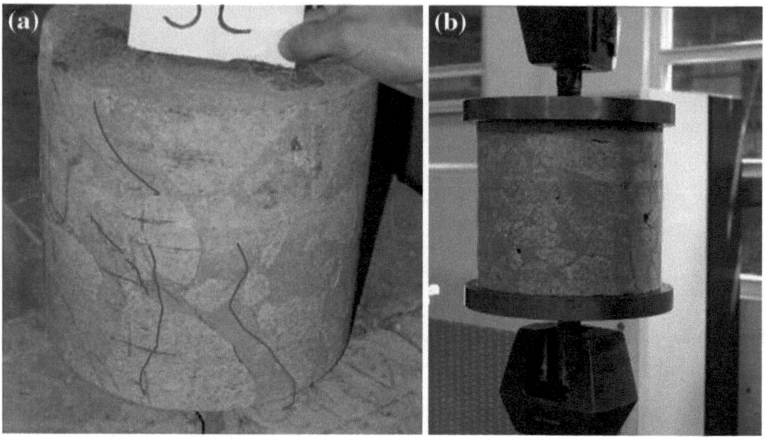

Fig. 7.9 Sample after compressive test (**a**) Direct tensile strength testing (**b**)

and density [39]. The average results of the mechanical properties at 90 days are presented in Table 7.3.

It is observed that the mortar had a significantly higher strength (36.8 MPa at 90 days) than the concrete. Part of this difference is due to the way of sample conservation, which in mortar is more favourable due to the availability of water. But it could also be due to the presence of voids in the mass of the concrete, which did not occur in the mortar, and possibly the influence of the interface mortar-aggregate. If bleeding occurred in this zone, the rupture lines tend to form through this interface (see drawn lines in Fig. 7.9), which, when the relationship between the aggregate size and the sample is large, as in this case, reduce the tensile strength. This phenomenon will have little impact on the performance of real size applications, given the large dimension of the elements on site.

Comparing the results of direct and splitting tensile tests, the first present lower values as expected. The ratio between the results of these two tests is about 75 % of the tensile splitting strength, which can be considered normal. Regarding the density the results are higher than the common SCC being similar to mass concrete applied in dam construction.

In general, the PACSCM presented a good performance compared to VC with regard to properties usually controlled for dam construction.

Table 7.4 Durability related properties of PACSCM

Testing	In-situ		Laboratory	
Property	Torrent ($\times 10^{-16}$ m^2)	GWT (mm^2 $s^{-1}bar^{-1}$)	Porosity[a] (%)	Permeability (m/s $\times$ 10^{-11})
Average	7.65	0.03	9.56	1.33

[a] Accessible porous volume under water, by volume

7.3.2.2 Durability

In-situ non-destructive testing was executed on the different faces of the prototype to determine the Torrent air permeability and GWT water permeability coefficients. At that time, the concrete was 28 days of age. Laboratory tests were also performed on cores taken from the block. Table 7.4 presents the average results obtained from in situ and laboratory testing. With the GWT a sealed pressure chamber is attached to the concrete surface, water is filled into the pressure chamber and a specified water pressure is applied to the surface. The surface permeability is then assessed by means of Darcy's law.

Although the average values of the GWT and Torrent tests lie within the expected range for the material used, they showed considerable variations of the results. These variations are due to the character of the tests as they only assess a limited and very small area, and so, they are sensitive to local defects. The values of the porosity are normal in view of the mixture used, so that one may conclude that the filling of gaps between the preplaced aggregates took place in a satisfactory way. Common coefficients of permeability for a 90–day old concrete with cement content of 200–400 kg/m^3 are about 10^{-13}–10^{-12} m/s. The cement content was less than 200 kg/m^3, but with the contribution to hydration of the fly ash one can use this value for comparison. Thus, the coefficient of water permeability can be considered to be slightly higher than for a VC used in dam construction. The results are still within the order of magnitude.

The water permeability of the PACSCM was higher than VC, which can be related to the quality of the coarse aggregate—mortar bond, which is affected by the bleeding of the mortar.

7.3.2.3 Field Applications

The implementation of the prototype discussed earlier was part of a preliminary trial preceding the full-scale implementation of a cofferdam in the Baixo Sabor hydroelectric power plant project. The cofferdam has a length of about 40 m and is 6 m high. Cast in three phases, the PACSCM approach was applied for the central section in three layers (Fig. 7.10).

The mortar was pumped to the same location during concrete placement (Fig. 7.11). The mortar had to be highly flowable with high segregation resistance to flow at least 11 m into place. The volume of mortar applied varied between

Fig. 7.10 Central section of the cofferdam under construction

40 and 42 % of the total volume of the concreted block, depending on the amount of lateral surface (wall-effect of the preplaced aggregates).

From this real-scale casting, the applicability of the PACSCM for mass concrete applications was confirmed. This construction method was applied to two other dams: the Baixo Sabor's Dam (Fig. 7.12a) and in a cofferdam of Salamonde's Dam (Fig. 7.12b) [40]. The PACSCM technology was shown to allow the reduction of concrete plant costs due to the absence of coarse aggregate stocking and handling as aggregate is transported directly from the quarry to the site, and also due to a reduction of concrete production where coarse aggregates can make up a large portion of concrete (up to 60 % of total volume in some cases). Additionally, there are lower risks of schedule non-fulfilment due to a more flexible production. This is done by dividing activities in the concrete placement procedure: pre-placement of coarse aggregate, production, and casting of the mortar.

7.4 Special Fiber Reinforced SCC

7.4.1 Self-Compacting SIFCON

Ordinary Slurry Infiltrated Fiber Concrete (SIFCON) is produced by a process in which fibers are placed into an empty mould, after which the fibers are infiltrated by cement slurry. Generally, the infiltration of the slurry into the layer of fibers is carried out under intensive vibration. Normally, fiber reinforced concrete contains

Fig. 7.11 Pumping of mortar into the cofferdam

Fig. 7.12 Baixo Sabor Dam (**a**) and the cofferdam of Salamonde's Dam power increase (**b**)

1 to 3 % fibers, by volume, while SIFCON can contain 4 to 25 % fibers [41]. The first introduction of cement composites with a high content of fibers infiltrated with Portland cement-based materials was reported in Haynes in 1968 [42]. The volume of fibers is influenced by the fiber type and procedure of fiber placing [43].

Naaman et al. proposed the use of SIFCON for joints in seismic resistant reinforced concrete frames [44]. SIFCON would be used only in small parts of the frames. These parts would be designed to be affected during seismic loading and form 'plastic hinges', which would absorb the fracture energy and protect the constructions against collapse.

Self-compacting SIFCON Concrete (SC-SIFCON) can be regarded as a special type of high-performance steel fiber reinforced cement composite. The development of SC-SIFCON, which does not require vibration. The mix design of slurries made with limestone powder or silica fume on the fresh properties was

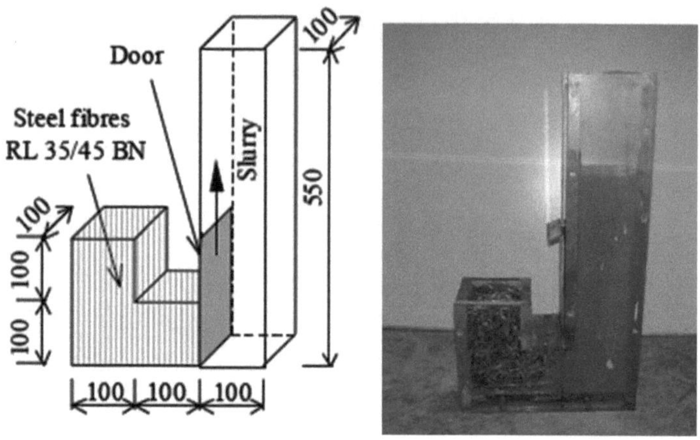

Fig. 7.13 J-penetration test for self-compacting SIFCON

investigated [45, 46] to optimize rheological parameters and compressive strength. Compressive strength of the SC-SIFCON slurry is shown to be affected by the W/C, dosage of silica fume, and the proportion of sand [47]. Another study showed that the increase in the proportions of limestone powder decreased the compressive strength of slurries for SC-SIFCON [46].

Two self-compacting slurries with different strengths were developed and applied for the production of SIFCON elements with fiber contents of up to 11 % with fine sand having 0.6 mm maximal size. Steel hooked-end fibers with a length of 35 mm and an aspect ratio (length/diameter) of 48 were used. Figure 7.13 shows an example of SCSIFCON tested for the penetrability of slurry through the fibers [46, 47]. A significant difference between the stress-strain response of samples loaded in a perpendicular direction to fiber orientation and samples loaded parallel to fiber orientation was observed. In three-point bending tests, a great difference of shapes of load–displacement curves for different orientation of fibers in relation to the plane of rupture was recorded. Fibers had a significantly lower effect on the results in the case of a parallel orientation to the initial plane of rupture when compared with a perpendicular orientation [48].

An "edge effect" and anisotropy of SIFCON influenced the preparation of test samples, and therefore compressive strength [48].

A significant anisotropy of SIFCON was underlined by samples with different fiber orientations. An 'edge effect' was investigated on 'cut' and 'cast' samples and a decrease in flexural strength of 'cut' samples was found. The samples which were cut from the blocks did not have the wall effect of fibers at the sides of the mould and the volume of fibers in these samples was higher than in the samples cast straight into the moulds. Figure. 7.14 shows a cross-section of a beam cut from a SC-SIFCON element. Whether the samples were cast in a mould or cut from a block of SC-SIFCON had no serious influence on the flexural strength (Fig. 7.15). The flexural strength of SC-SIFCON was between 3.7 times and

Fig. 7.14 Cross section of a beam cast with self-compacting SIFCON

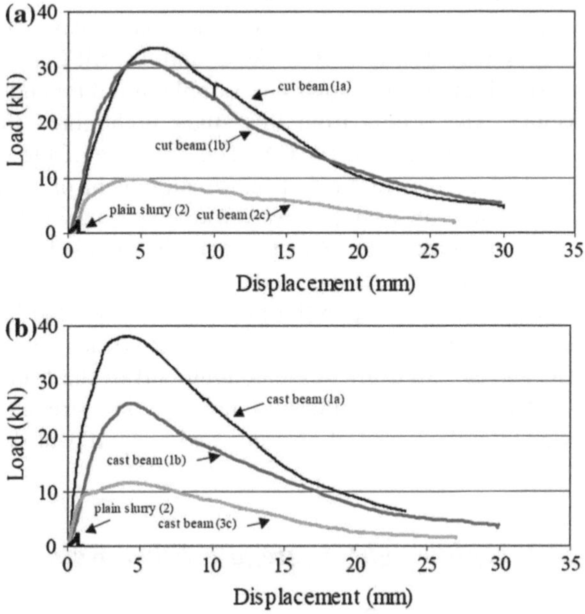

Fig. 7.15 Typical load–displacement curves: for beams cut from blocks of SIFCON (**a**); for straight cast beams (**b**)

18 times higher than that of plain cement slurry. This was depending on the orientation of fibers to the plane of rupture (Fig. 7.15). The 'post-peak' behaviour of all SC-SIFCON samples indicates low brittleness, high ductility, and high

Fig. 7.16 Casting Self-compacting SIFCON knee-joint in a frame made with SCC

energy absorption, which suggested the suitability of this material for use in seismic areas. The compressive strength of SC-SIFCON was affected by the placing of fibers and by the method of production (straight cast samples or samples cut from a block of SC-SIFCON).

Full-scale frames were produced and tested for loading using SC-SIFCON in the corner parts of the frames. The other parts of the frame were cast with SCC having a slump of flow of 650 mm and an average of compressive strength of 60 MPa at 28 days (Fig. 7.16). The compressive strength of plain SIFCON at 28 days was 30 MPa. Both SCC and SC-SIFCON were made with limestone powder. Figure 7.17 shows an example of failure and typical cracks of frame after loading. The results from this study highlighted the problems relating to the anisotropy of the SC-SIFCON. The presence of the reinforcing bars in the knee-joint had a significant effect on the alignment of fibers in the SC-SIFCON parts. Areas of low fiber content were formed directly below the bars [48].

The presence of SC-SIFCON parts in the test frames influenced the strength of the frames and the failure mode. In the case of SC-SIFCON, fibers being oriented perpendicularly to the initial plane of ruptures, the SC-SIFCON corner increased the maximum load. The most significant cracks were observed at the interfaces of SCC and SC-SIFCON in the column part of the frames. The presence of the SC-SIFCON did not significantly improve the strength of the frames. Moreover, unsuitable placement of fibers caused a significant decrease in strength [48].

7.4.2 Lightweight Aggregate Fiber Reinforced SCC

Hela and Hubertova [49] investigated fresh and hardened state properties of LSCC reinforced with different amounts of different types of fibers. Comparing two mixtures, one with only LWA ($\rho = 1430$ kg/m^3) and the other with blended normal and LWA ($\rho = 1740$ kg/m^3), the effect of fibers was found to be stronger

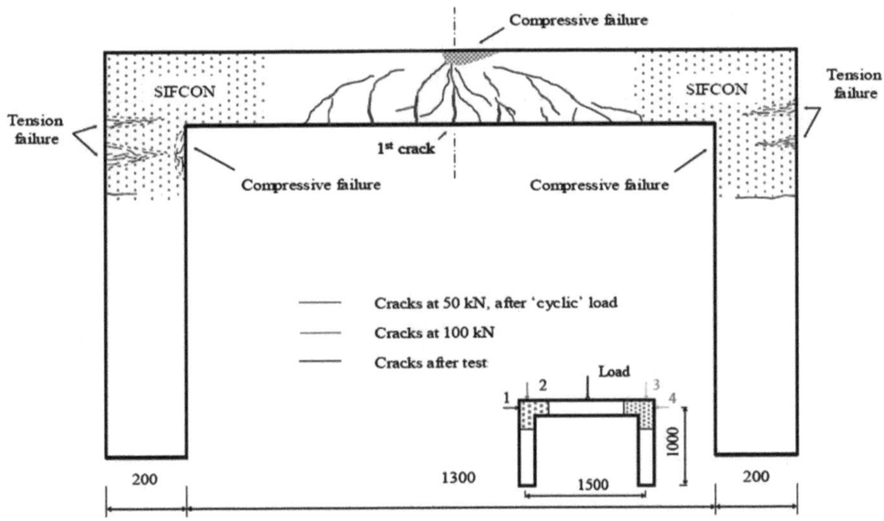

Fig. 7.17 Testing SC-SIFCON knee-joint frame

in the former, consistently with an expectable more brittle behaviour of the plain matrix due to the less effective crack arresting role of the larger amount of LWA.

An investigation on mechanical properties of LSCC reinforced with polypropylene fibers has been recently reported by Mazaheripour et al. [50]. Results confirmed well-known effects of fibers on tensile and flexural behaviour and their negligible influence on compressive strength and modulus of elasticity.

7.4.3 Self-Compacting High Performance Fiber Reinforced Cementitious Composites (SC-HPFRCCs)

High-performance fiber reinforced cementitious composites (HPFRCCs) are unique broad categories of fiber reinforced concretes which can exhibit significant level of stable multi-cracking system. This can be accompanied by an almost "plastic" or even strain hardening behaviour after the onset of the first crack and before unstable localization of the critical crack. In this unique composite material, after the formation of the first crack, the energy required to pull-out the fibers crossing the crack itself, is higher than that necessary to form a new crack at a prescribed distance from the former one. This multi-cracking process gives rise to significant stress redistribution, making it possible to achieve, after first cracking, the aforementioned "plastic" or even strain hardening behaviour. It stops once a percolation threshold is attained, i.e. when the spacing between the cracks has been completely saturated and unstable localization of the critical crack occurs.

Table 7.5 Mixture composition and properties of SC-UHPFRCC mixtures employed in [60, 61]

Mix design (dosages in kg/m^3), fresh and hardened state properties

Constituent	HPFRCC 100	HPFRCC 50	HPFRCC 0
Cement type I 52.5	600	600	600
Slag	500	500	500
Water	200	200	200
Superplasticiser	33 (l/m^3)	33 (l/m^3)	33 (l/m^3)
Sand 0–2 mm	983	1000	1017
Steel fibers (l_f = 13 mm—d_f = 0.16 mm)	100	50	0
Fresh state performance	–	–	–
Slump flow diameter (mm)	750	760	775
T_{50} (s)	6	6	5
V funnel flow time T_V (s)	30	29	25
Mechanical properties	–	–	–
Compressive strength (cube—N/mm^2)	116		

The mixture composition of this kind of cementitious composites is characterised by high dosage of cement and mineral additions, a low maximum aggregate size ($d_{a,max} \cong$ 2–4 mm), low W/B, high dosage of superplasticiser, and high volume of fibers (generally $V_f \geq 1$ %). Such material is can secure superior performance in the fresh state, such as self-compactability [51]. The superior workability characteristics can allow for ease of casting of structural elements with optimized geometries and dimensions (e.g. reduced thickness). Structural concepts can be developed that could be competitive with steel structures [52].

As mentioned in Chap. 6 of this report, the superior workability of SC-HPFRCC can be exploited to favour the orientation of steel fibers along the casting flow direction to maximize mechanical properties and structural performance [53, 54]. As such, strong orthotropic mechanical behaviour results, which cannot be disregarded since the same material can exhibit a strain—or deflection-hardening behaviour in the direction of preferential fiber alignment and a deflection and strain—softening in the orthogonal direction. This was illustrated in [55, 56] with SC-HPFRCC made with 50 or 100 kg/m^3 short steel fibers (Table 7.5).

The tensile performance was measured through indirect tests performed, according to the set-up shown in Fig. 7.18a, in the direction either parallel or orthogonal to the casting flow and hence to the preferential alignment of fibers (Fig. 7.18b). The latter was either evaluated through image analysis or non-destructively monitored by means of a magnetic method [57]. Fracture toughness parameters were effectively correlated to fiber dispersion and orientation (Fig. 7.18c).

Ferrara et al. [58] and Di Prisco et al. [59] have studied simply supported roof slabs, 1.2 m wide, 2.4 m long, and 25–30 mm thick made with SC-HPFRCC-100 detailed in Table 7.5, with no conventional reinforcement. The influence of the flow driven fiber orientation on the moment capacity and ductility of the cross section, as determined from the stress versus crack-opening relationships identified

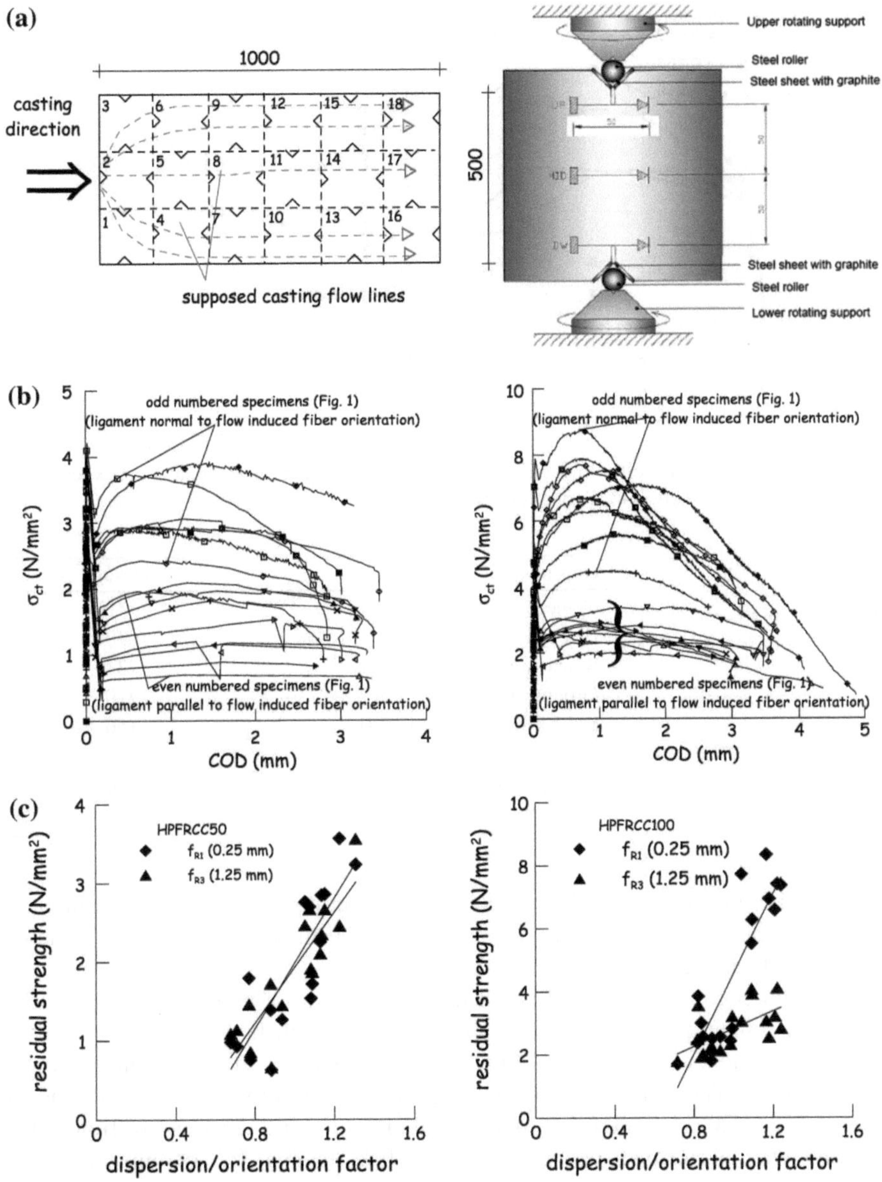

Fig. 7.18 Assessing the influence of fiber orientation on SC-HPFRCCs [55, 56]: scheme of casting and specimen cutting and test set-up for indirect tensile tests (Double Edge Wedge Splitting) (**a**); tensile stress verses. crack opening curves for HPFRCC-50 and HPFRCC-100 (**b**) and correlation of residual tensile stresses at different crack-opening levels (0.25 and 1.25 mm) to fiber dispersion/orientation factors inferred from magnetic non-destructive monitoring (**c**) [57]

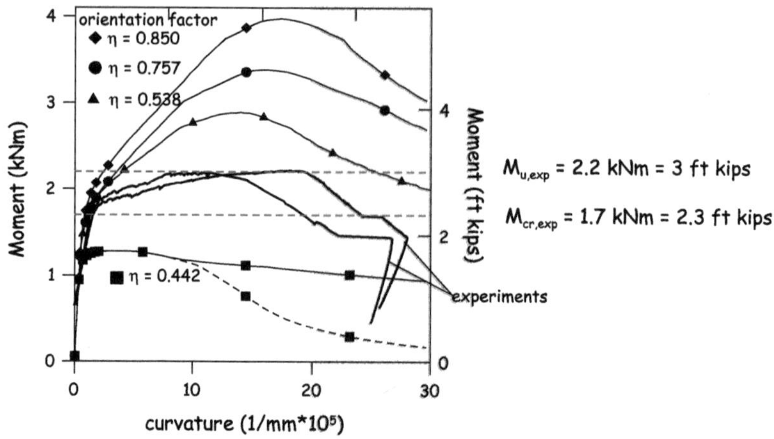

Fig. 7.19 Slabs cast with self-compacting HPFRCC. Moment–curvature behaviour: experiments and modelling as a function of fiber orientation [59]

Table 7.6 Mixture composition of SC-UHPFRCC mixture employed for cross-arms [60]

Constituent	Dosage (kg/m^3)
Cement I	712
Fine sand	1020
Silica fume	231
Ground quartz	210
Steel fibers $l_f/d_f = 12.7/0.2$	156
Water	109
Superplasticiser	30.7
Set accelerator	30

from results in Fig. 7.18c, highlights, through comparison with experimental results obtained from slabs cast with no preferential fiber orientation, the benefits of a tailored casting (Fig. 7.19).

El-Hacha et al. [60] reported about the use of an ultra-high performance fiber reinforced cementitious composite which "exhibits high flowability and is virtually self-placing" (Table 7.6), as quoted, for cross arms for high voltage transmission lines. The superior fresh state performance allowed for new-concept casting (Fig. 7.20). An extensive mechanical and durability characterization of the material was performed (Table 7.7), including testing for compression and bending strength, shrinkage and creep, chloride ion diffusion, carbonation depth, freeze-and-thaw resistance, salt-scaling resistance. The concrete was steam-cured at 90 °C and 95 % RH for 48 h.

Bending tests, about both the strong and weak axis on cross-arm mock-ups, were conducted and also investigations regarding the influence of the shear-span to depth ratio (between 1.6 and 3.6 for tests about the strong axis; between 2.2 and 5 for tests about the weak axis) were performed. Results were instrumental to assess how the different casting methods affected the fiber orientation and distribution,

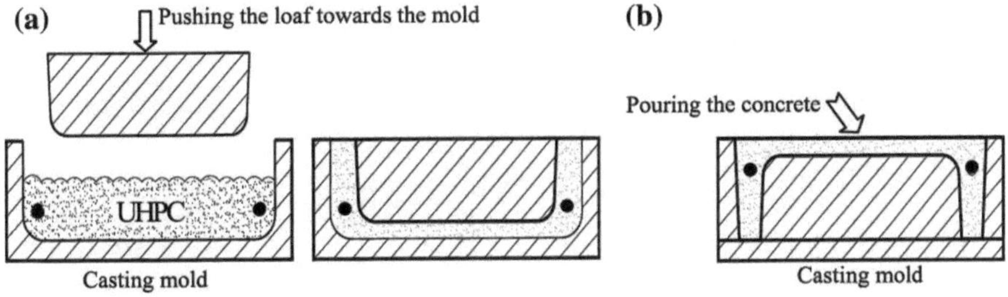

Fig. 7.20 Adopted casting procedures of cross-arm mock-ups [60]

Table 7.7 Mechanical properties of SC-UHPFRCC mixture employed for cross-arms [60]

Property	Results
f_c, (N/mm^2)	180–225
Bending strength (N/mm^2)	40–50
Young modulus (N/mm^2)	55000–58500
Creep coefficient	0.2–0.5
Shrinkage strain after curing	0

mainly around the passive reinforcing bars, and consequently the overall flexural behaviour (Fig. 7.21).

7.4.4 Glass Fiber Reinforced SCC

Different SCC mixtures reinforced with 0.6 kg/m^3 of glass fibers were tested by Suresh Babu et al. [61] and the effectiveness of the glass fiber reinforcement on the behaviour of beams in bending was assessed. The addition of fibers resulted in higher ultimate loads, deflection and curvatures and lower crack openings, either at service and ultimate state. Meanwhile, Sravana et al. [62] presented the following conclusions: the ultimate flexural strength of glass fiber reinforced self-compacting concrete beams at 0.03 % is on higher side when compared with other beams having glass fibers 0, 0.06, and 0.1 %; the presence of glass fibers in glass fiber reinforced self-compacting concrete slabs have not improved any flexural strength; development of multiple cracks and micro cracks is prevented with the use of glass fibers.

7.5 Application-Oriented SCC

7.5.1 Underwater Concrete

Technological progress in the field of underwater concrete placement and repair has proceeded through the development of improved methods of concrete

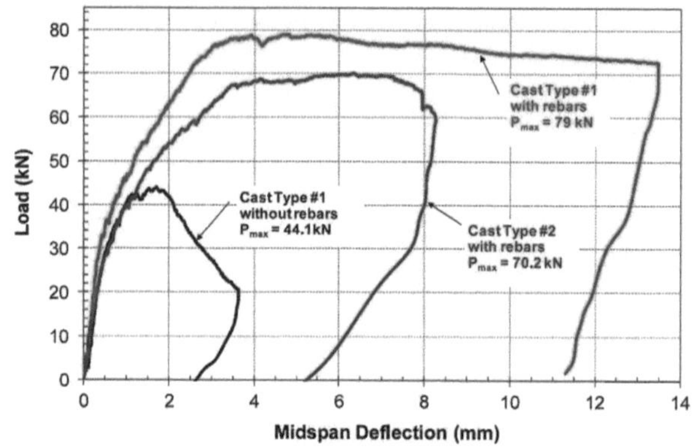

Fig. 7.21 Load-deflection curves (bending about strong axis) for differently cast cross-arms [60]

placement and growing use of chemical admixtures and supplementary cementitious materials to enhance engineering performance. Self-compacting, yet highly stable, concrete is required in underwater applications to ensure proper filling of the section without any external compaction and proper development of in situ properties, and in some cases the development of flat surfaces. The casting of relatively shallow lifts of concrete in repair applications can prevent the maintenance of bottom seal of the placement device, typically required for conventional tremie concrete to reduce segregation and water erosion. This necessitates greater care in proportioning the mixture to ensure high stability and flowability of the fresh concrete [63–67].

Underwater SCC (UW-SCC) has been used in many applications, including the construction of bases of massive piers of the Confederation Bridge in Canada. The bridge is a 13 km link between Prince Edward Island and New Brunswick. Such conditions necessitate that the concrete spreads readily into place and around obstacles and maintain minimum dilution with the surrounding water to develop adequate in situ strength, bond, and impermeability.

7.5.1.1 Effect of Mixture Composition of In Situ Performance of UW-SCC

Sonebi and Khayat [68] investigated the effects of the anti-washout admixture (AWA) concentration, W/CM, and binder composition on the variations of in situ compressive strengths on the washout resistance of UW-SCC. The investigated mixtures were prepared with W/CM of 0.41 and 0.47 and all mixtures incorporated a retarder to enhance fluidity retention. The concrete incorporated high contents of cementitious materials to reduce aggregate volume, thus enhancing flow characteristics. It is important to note that for the majority of the mixtures, a large part of the cement was substituted by fly ash (FA) or granulated blast-furnace slag to limit the heat of hydration and to enhance performance. The volume of coarse aggregate was limited to approximately 300 l/m^3, and the sand-to-total mass of aggregate

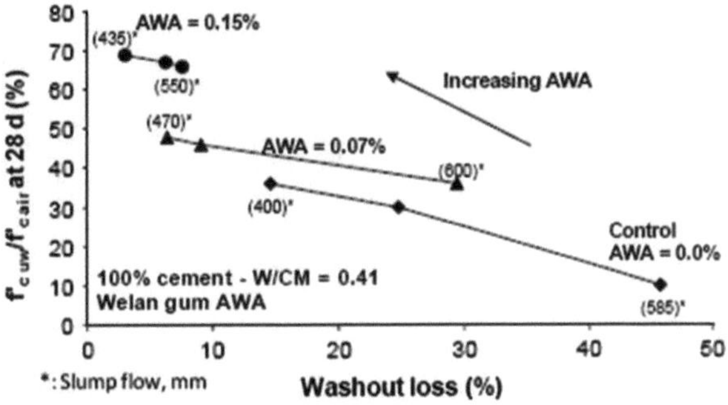

Fig. 7.22 Effect of AWA dosage on variations of residual compressive strength with washout loss (100 % C binder and a W/CM of 0.41) [68]

was fixed at 0.46. The concentrations of welan gum and cellulosic AWAs were set to 0.07–0.15 % (by mass of binder), respectively, which represent low and moderate dosages of these AWA for underwater concrete application. The dosages of the cellulosic AWA ranged between 1 l/100 and 1.65 l/100 kg of cementitious material, representing moderate to high contents of the admixture for underwater applications. Underwater strength was determined by casting concrete into 100 × 200 mm cylinders filled with water without any consolidation. These results were compared to strengths determined on cylinders cast and consolidated above water [68].

7.5.1.2 Effect of AWA Dosage on Variations in Residual Strength with Washout Loss

The effect of the dosage of welan gum, as an AWA, on the variations between washout mass loss and residual compressive strength of the 100 % cement (100 % C) concrete made with a W/CM of 0.41 is illustrated in Fig. 7.22 [68]. It was reported that in general the reduction of in situ strength compared to reference cylinders cast above water is due to the lack of full compaction coupled with water dilution of the cast concrete [68]. It was presented that a greater dosage of AWA resulted in net reduction in washout loss for any given fluidity and therefore the residual compressive strength was enhanced. For example, for the mixture with a W/CM of 0.41, at a constant washout loss of 6 %, the residual compressive strength has grown from 48 to 66 % with the increase in welan gum from 0 to 0.15 %, by mass of binder [68].

7.5.1.3 Effect of Binder Composition on Relationships Between Residual Strength and Washout Loss

The effect of the binder composition on the variations of residual compressive strength with washout for concrete mixtures (W/CM = 0.41) made with cellulosic AWAs are illustrated in Fig. 7.23 [68]. It was observed that for the majority of

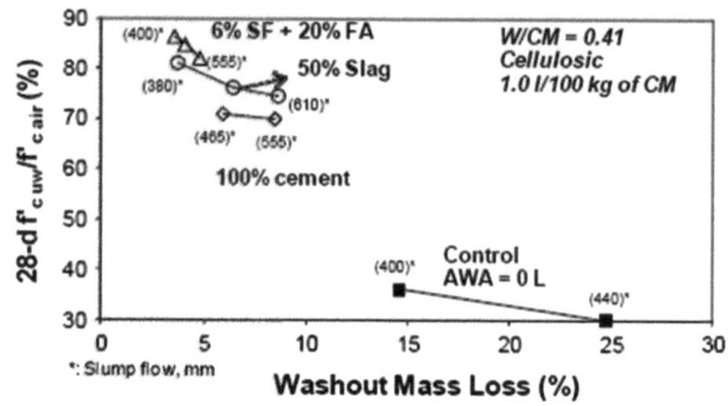

Fig. 7.23 Effect of binder composition on variations of residual compressive strength with washout loss (W/CM of 0.41) [68]

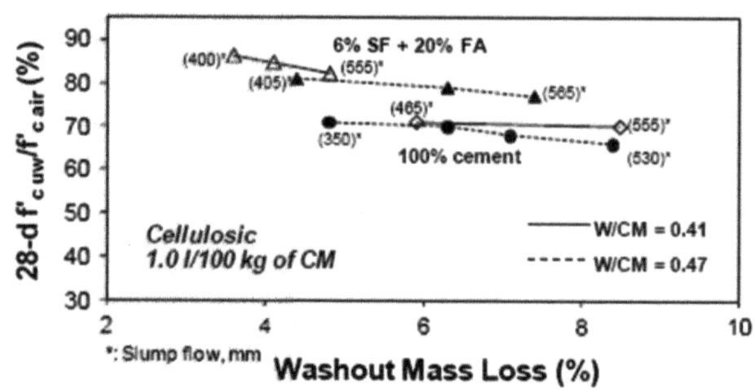

Fig. 7.24 Effect of W/CM on variations of residual compressive strength with washout loss [68]

mixtures, the use of the ternary binder (6 % silica fume (SF) and 20 % FA) and the binary mixtures with 10 % SF or 50 % slag resulted in net improvements in residual compressive strength.

7.5.1.4 Effect of W/CM on Variations in Residual Strength with Washout Loss

Figure 7.24 presents the effect of W/CM of mixtures made with 100 % C and 6 % SF combined with 20 % FA [68]. The increase in W/CM from 0.41 to 0.47 resulted in significant decrease in residual compressive strength. Figure 7.25 presents the residual compressive strength of a fixed slump flow of 500 mm [68]. It can be observed that for a fixed slump flow, the increase of W/CM led to a reduction of residual compressive strength and the increase of AWA and use of supplementary materials, such as silica fume or silica fume combined with fly ash, improved significantly the residual compressive strength [68].

7.5.1.5 Effect of Application Procedure

The casting of a UW-SCC in stagnant water can result in mean drops of approximately 20 and 10 % in compressive and splitting tensile strengths, respectively, over 5 m long blocks. Such drops were approximately 25 and 15 %,

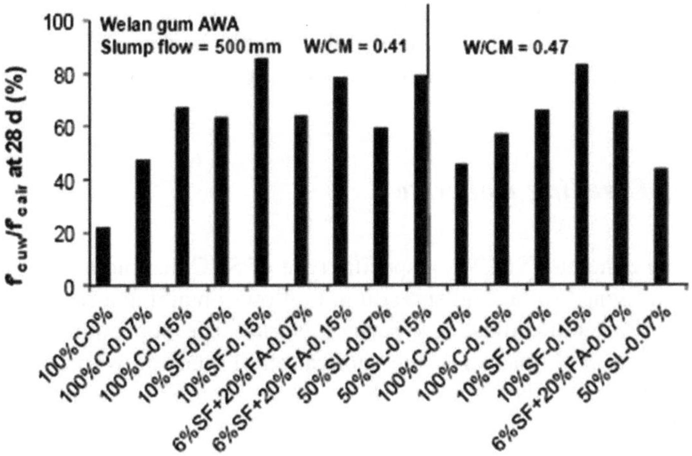

Fig. 7.25 Residual compressive strength of mixtures with slump flow of 500 mm containing welan gum [68]

respectively, when the concrete was cast in 0.1 m/s water velocity and approximately 40 and 25 %, respectively, when cast at 0.65 m/s. The investigated concrete contained 8 % silica fume and 0.15 % welan gum and had a slump flow of 500 mm and 5 % standard washout mass loss [69].

UW-SCC should be proportioned to flow readily into place with minimum materials separation. Unlike concrete cast for deep tremie seals, the use of concrete in repairs often necessitates some free-fall of the mixture through water. Such placement conditions lead to a greater risk of water erosion and segregation, and should be addressed in proportioning the concrete. UW-SCC can therefore be proportioned to secure adequate in situ properties. The incorporation of AWA and SCMs can enhance in situ characteristics of the concrete. A relationship between the extent of free-fall height (FFH) of UW-SCC in water and the influence of its washout resistance on in situ strength performance is already established [70].

7.5.1.6 Remarks on UW-SCC

The mixture composition is shown to play a significant role of the in situ mechanical performance of UW-SCC. The W/B as well as type and dosage of SCMs affect washout resistance and in situ properties of the hardened concrete. The use of silica fume or the combination of silica fume with fly ash can improve significantly the washout resistance and the residual compressive strength. The type and the dosage of AWA can affect workability, washout mass loss, and residual strength. Concrete containing 10 % of silica fume or 20 % of fly ash and 6 % of silica fume replacement can exhibit considerably greater washout resistance than similar concrete without any SCMs. Such binary and ternary mixtures can develop greater relative residual strengths and lower accumulations of laitance and washout deposits at upper surfaces of the cast blocks than concrete with 100 % cement, regardless of the initial free-fall height in water.

7.5.2 Self-Levelling Concrete

Self-levelling concrete (SLC) is a specific type of SCC, having a size distribution specifically designed for a fluid spreading. It allows obtaining a smooth surface of the pavement without almost any surface works. The application of all-in-one procedure, using SLC, instead of the application of concrete with a screed on top can lead to material savings up to 35 %, without counting the production time savings [71].

SLC and screeds of high resistance to frost were developed and discussed in [72]. It was concluded that SCC for screed applications can exhibit higher compressive and flexural strengths than standard screeds. A decrease in the nominal thicknesses of the indoor to the permissible values is thus technically feasible. The application of cements of the strength classes 32.5 is sufficient due to the high strength of the mixtures. In principle, the cement content of screeds for indoors can be significantly reduced to contents of about 100-150 kg/m^3 in comparison to customary contents. The required mechanical properties are satisfied by a particle size distribution optimization. A sufficient freeze–thaw resistance of SCC for floor applications is obtained by a dense package of the solid particles in combination with a typical cement content of outside steel-reinforced concrete building elements and low water content. Air entrainment increases freeze–thaw resistance, however, it is not compulsory.

7.5.3 Repair Mortars and Concretes

SCC may substitute normal concrete in most structural applications. Nevertheless, the interesting repair applications are those in which SCC can be effectively more competitive, both technologically and economically. Indeed, most repair works are done in locations where it is difficult to vibrate, for instance, in narrow places. These works require the use of small maximum size aggregate to allow flow in restricted spaces between parent concrete, reinforcement, and formwork [73]. This restriction leads to the conclusion that self-compacting mortars can be advantageously used in many repair situations [74].

Repair cases in which the application of SCC was considered as the most adequate repair method has been reported to go from strengthening existing structures to the replacement of degraded concrete.

One case of structural strengthening employs concrete overlay. Bonded concrete overlays constitute a versatile repair method that has been used successfully

Fig. 7.26 Cast in place of self-compacting concrete [76]

for concrete bridge decks, concrete pavements, and industrial concrete floors during several decades. There are experiences both in the field and in the laboratory indicating that SCC overlays have performed well [75]. One successful example was about slabs with a wrong design thickness, being necessary to increase that thickness by adding a new layer of concrete [76]. This new layer had a thickness of about 50–80 mm, and could hardly be vibrated (Fig. 7.26). One special requirement for the mix design of this concrete was the need to control shrinkage, since it was to be applied over an "old" concrete, made with shrinkage reducing admixtures. The major property of this repair application was the bond to old concrete which was assessed by pull-out tests 7 days after casting of the SCC. The old surface was prepared with water and sand blasted until total exposure of coarse aggregates. The rupture happened in the old concrete with an average result of tensile tension on pull-out test of about 2 MPa.

A problem often encountered during surface repair to existing concrete is the formation of cracks. Cracks are due to restrained shrinkage of the repair concrete. In such a case, given its special visco-plastic properties, the use of SCC represents a major advantage since it limits the formation of cracks [76].

The replacement of deteriorated concrete is the most common application of SCC technology for repair works. As an example, the replacing of the cover concrete on a wharf structure due to reinforcement corrosion in seashore structure was a case reported [77]. In this application, it was necessary to apply a high-performance concrete to cope with the hard environmental exposure conditions of the structure. Furthermore, a tight schedule was imposed and it was difficult to vibrate since the zone of application was underneath the wharf and above the water level of the river (Fig. 7.27). The mixture proportioning of the SCC was formulated to achieve both fresh characteristics to fill the forms and a high performance to durability requirements. As a result, the chloride diffusion coefficient at 28 days, determined by CTH method, was less than 4×10^{-12} m^2/s, fulfilling the high-performance requirements of repair applications.

Fig. 7.27 Wharf to be repaired (**a**) Application of SCC to repair the beams (**b**) [76]

Many repair/strengthening projects have a small formwork space that make concrete application challenging. These projects usually have tighter spacing between the reinforcement and the existing concrete surface, reinforcement, and formwork. SCC's flowability lends itself well to concrete beam, slab, and column "form and pump" repairs that commonly are found on structural restoration and strengthening projects. Frequently, these repairs involve enclosed formwork and require the material to be placed under extra pressure to ensure a good bond with the prepared, "open" substrate. This placement process and material alleviate problems associated with using traditional concrete where voids and honeycombs can occur and often are not revealed until the forms are stripped, if ever [78].

An experimental program was undertaken to evaluate the performance of self-compacting mortar designed for filling small annular spaces for the rehabilitation of underground water line or sewage pipelines [79]. The investigated mixtures had 7-day compressive strength higher than 20 MPa. In general, the investigated mortars exhibited higher 28 and 56 day compressive strengths than the neat cement grout. Another experimental project dealing with self-compacting repair mortars [80] involved the testing of six self-compacting mortars produced using combinations of limestone filler, slag from steel production, and crushed limestone sand. The mortars fulfilled the requirements of the standard and were characterized as R2 and R3 repair mortars according to European Standard EN 1504–3.

For the design of SCC for repair works the following has to be taken into account: the thickness of concrete; a good flowability in the restricted spaces; type of placement such that no air or water is trapped; a high rate of strength gain, particularly where a lot of small localised repairs have to be undertaken; to be dense and have a low permeability as often repairs result from a combination of low cover and chloride/carbonation attack of the original construction which must be overcome; to not suffer from any bleed that would decrease the bond strength of the repair concrete to the parent concrete, thus reducing the structure effectiveness and presenting a leakage path into the reinforcement [73]; a low shrinkage to avoid cracking and poor bonding with old concrete [81].

To avoid the occurrence of problems an efficient application of SCC requires a specific design, both from structural and material viewpoints, as well as qualified personnel involved in its application.

7.6 Final Remarks on Specialty SCC

SCC is already a special concrete. However the inclusion of special aggregates instead of standard ones, the use of fibers, and the development of different placement techniques requires tailor-design of SCC with specific performance characteristics. Specialty SCC materials can provide strong techniques to improve the efficiency of the construction process.

References

1. Kosmatka, S.H. et al.: Special types of concrete. In: PCA (ed.) Design and Control of Concrete Mixtures. Skokie, Illinois (2008)
2. Bartos, P.J.M.: Workability of special concrete mixes. Mater. Struct. **26**, 50–52 (1993)
3. European Committee for Standardization, CEN/TC 104: EN 206-1 Concrete—Part 1: Specification, performance, production and conformity. Brussels (2005)
4. Jensen, O.M., Lura, P. Kovler, K.: Materials and methods for internal curing. In: K. Kovler, K., Jensen, O.M. (eds.) Internal Curing of Concrete—State-of-the-Art Report of RILEM Technical Committee 196-ICC, pp. 45–55. RILEM, Bagneux (2007)
5. American Concrete Institute, ACI Committee 213: ACI 213R-87 Guide for Structural Lightweight Aggregate Concrete. In: ACI Manual of concrete practice, vol. **1**. ACI, Farmington Hills (1987)
6. Kowalsky, M.J., et al.: Shear and flexural behaviour of lightweight concrete bridge columns in seismic regions. ACI Struct. J. **96**(1), 136–148 (1999)
7. Müller, H.S. et al.: Development of self-compacting lightweight aggregate concrete. In: Ozawa, K., Ouchi, M. (eds) Proceedings of The Second International Symposium on SCC, Tokyo (2001)
8. Kwasny, J., et al.: Influence of the type of coarse lightweight aggregate on properties of semilightweight self-consolidating concrete. J. Mater. Civ. Eng. **24**, 1474–1483 (2012). doi:10.1061/(ASCE)MT.1943-5533.0000527
9. Papanicolaou, C., Kaffetzakis, M.: Pumice aggregate self-compacting concrete (PASCC). In: Khayat, K., Feys, D. (eds.) Proceedings of SCC2010, Montreal (2010)
10. Kaszynska, M.: Effect of aggregate mix composition on lightweight self-consolidating concrete. In: Khayat, K., Feys, D. (eds.) Proceedings of SCC2010, Montreal (2010)
11. Costa, H. et al.: Mix design and characterization of self-compacting lightweight aggregate concrete. In: Khayat, K., Feys, D. (eds.) Proceedings of SCC2010, Montreal (2010)
12. Topçu, I.B., Uygunoglu, T.: Effect of aggregate type on properties of hardened self-consolidating lightweight concrete (SCLC). Const. Build. Mater. **24**, 1286–1295 (2010)
13. Choi, Y.W., et al.: An experimental research on the fluidity and mechanical properties of high-strength lightweight self-compacting concrete. Cem. Concr. Res. **36**, 1595–1602 (2006)
14. Bentur, A., et al.: Prevention of autogenous shrinkage in high strength concrete by internal curing using wet lightweight aggregates. Cem. Concr. Res. **31**, 1587–1591 (2001)
15. Shi, C., Wu, Y.: Mix design and properties of self-consolidating lightweight concrete containing glass powder. ACI Mater. J. **102**, 355–363 (2005)

16. Shi, C., et al.: Use of self-consolidating lightweight concrete for insulated concrete form system. Concr. Int. **28**, 40–43 (2006)
17. Barros, J., et al.: Lightweight panels of steel fiber-reinforced self-compacting concrete. J. Mater. Civ. Eng. **19**, 295–304 (2007)
18. Collepardi, M. et al.: Recent developments of special SCC's. In: Malhotra, M.H. (ed.) Proceedings of Seventh CANMET/ACI International Conference on Recent Advances in Concrete Technology. ACI, Las Vegas (2004)
19. Molina, J., Gimeno, J.J.: Hormigon autocompactante (ligero y de densidad normal) para la base de la Copa América del equipo suizo. In: Barragan, B.E. et al. (eds) Proceedings of the 1st Spanish Congress on SCC, Valência (2008)
20. Foster, B.E.: Attenuation of X-rays and gamma rays in concrete. Mat. Res. Stand. **8**, 19–24 (1968)
21. Revuelta, D., et al.: Measurement of properties and of the resistance to segregation in heavyweight, self-compacting barite concrete. Mat. Const. **59**, 31–44 (2009)
22. Revuelta, D. et al.: Experimental study on a barite heavyweight self-consolidating concrete. In: Gupta, P. et al. (eds) Proceedings of the Tenth ACI International Conference on Recent Advances in Concrete Technology and Sustainability Issues. ACI, Seville (2009)
23. Nishikawa, K., et al.: Mechanical properties of SCC using recycled materials from demolished concrete structure as aggregate and powder. In: De Schutter, G., Boel, V. (eds.) Proceedings PRO054: Self-Compacting Concrete—SCC 2007. RILEM, Ghent (2007)
24. Kou, S.C., Poon, C.S.: Properties of self-compacting concrete prepared with coarse and fine recycled concrete aggregates. Cem. Concr. Comp. **31**, 622–627 (2009)
25. Corinaldesi, V., et al.: SCC: A way to sustainable construction development. In: Yu, Z., et al. (eds.) Proceedings PRO42 SCC 2005—China 1st International Symposium on Design. Performance and Use of SCC. RILEM, Changsha (2005)
26. Grdic, Z.J., et al.: Properties of self-compacting concrete prepared with coarse recycled concrete aggregate. Const. Build. Mater. **24**, 1129–1133 (2010)
27. Kubens, S., Wallevik, O.: Self-compacting mortar for concrete furniture and design objects. In: Khayat, K., Feys, D. (eds.) Proceedings of SCC2010, Montreal (2010)
28. Lappa, E.S., et al.: Self-compacting, high strength steel fibre reinforced mortar for pre-cast sheet piles. In: Wallevik, O., Nielsson, I. (eds.) Proceedings PRO033: Self-Compacting Concrete—SCC 2003. RILEM, Reykjavik (2003)
29. Sahmaran., M. et al: The effect of chemical admixtures and mineral additives on the properties of self-compacting mortars. Cem. Concr. Comp. **28**, 432–440 (2006)
30. ACI Committee 116: ACI 116R-00 cement and concrete terminology. In: ACI Manual of concrete practice, Vol.1. Farmington Hills, American Concrete Institute (2005)
31. Jin. F, et al.: Construction method of conventional rock-filled concrete 200710100315.3 (2007) (in Chinese)
32. An, X. et al.: Construction method of dump-type rock-filled concrete. Chinese Patent No 200710121791.3. (2007) (in Chinese)
33. Jin, F., et al.: Study on rock-filled concrete dam. J. Hydraul. Eng. **36**, 1348–1351 (2005). (in Chinese)
34. Huang, M., et al.: A pilot study on integrated properties of rock-filled concrete. J. Build. Mater. **11**, 206–211 (2008). (in Chinese)
35. Design specification for concrete gravity dams. SL319-2005, China. (in Chinese) (2005)
36. Song, D., Liu, J.: Application of self-compacted rockfill concrete in baoquan pump-storage power station. Water Power **33**, 26–27 (2007). (in Chinese)
37. China Gezhouba (Group) Corporation. Experimental Report of RFC.. 2 (2007).(in Chinese)
38. Vieira, M. et al.: Self-compacting mortar for mass concrete application with PAC technology. In: Khayat, K., Feys, D. (eds.) Proceedings of SCC2010, Montreal (2010)
39. Vieira, M., Bettencourt, A.: Study for the implementation of preplaced aggregate concrete in dams-Report of 1st phase. LNEC, Lisbon (2009) (Private report in Portuguese)
40. Vieira, M., et al.: Preplaced-aggregate concrete with self-compacting mortar. on site world premier applications. Proceedings of SCC2013, Paris (2013)

41. Naaman, A.E., Baccouche, M.R.: Shear response of dowel reinforced SIFCON. ACI Struct. J. **92**, 587–596 (1995)
42. Haynes H Investigation of fibre reinforcement methods for thin shell concrete. Naval Civil Engineering Laboratory, Port Hueneme, CA, N-979, 1-26 (1968)
43. Reinhardt, H.W., Fritz, C.: Optimization of SIFCON mix. In: Swamy, R.N., Barr, B. (eds.) Proc. Int. Symp. Fibre reinforced concrete—recent developments, Cardiff (1989)
44. Naaman, A.E., et al.: SIFCON connections for seismic resistant frames. Concr. Int. **9**, 34–49 (1987)
45. Bartos, P.J.M., Marrs, D.L.: Development and testing of self-compacting grout for the production of SIFCON. In: Reinhardt, H.W., Naaman, A.E. (eds) Proc. Int. Work. On High performance fibre reinforced cement composites. RILEM, Mainz (1999)
46. Sonebi, M., et al.: Factorial design for cement slurries containing limestone powder for self-consolidating SIFCON. ACI Mater. J. **101**, 136–145 (2004)
47. Sonebi, M., et al.: Statistical modelling of cement slurries for self-compacting SIFCON containing silica fume. Mater. Struct. **38**, 79–86 (2005)
48. Svermova, L. et al. Design and testing of selfcompacting sifcon produced with low strength slurry. In: Proceedings of the 7th International Symposium on Brittle Matrix Composites, Warsaw (2007)
49. Hela, R., Hubertova, M.: Research of effect of fibre reinforcement at characteristics of lightweight self-compacting concrete. In: De Schutter, G., Boel, V. (eds.) Proceedings of the SCC2007, 5th International RILEM Symposium on Self-Compacting Concrete, Gent (2007)
50. Mazaheripour, H., et al.: The effect of polypropylene fibers on the properties of fresh and hardened lightweight self-compacting concrete. Const. Build. Mater. **25**, 351–358 (2011)
51. Olsen, E.C., Billington, S.L.: Cyclic response of precast high-performance fiber rein-forced concrete infill panels. ACI Struct. J. **108**(1), 51–60 (2011)
52. Camacho, E., Serna, P.: Design and experimental verification of self-compacting ultra-high performance hybrid fiber-reinforced concrete ties. In: Khayat, K., Feys, D. (eds.) Proceedings of SCC2010, Montreal (2010)
53. Ferrara, L., e al.: High mechanical performance of fiber reinforced cementitious composites: the role of "casting-flow" induced fiber orientation. Mater. Struct. **44**, 109–128 (2011)
54. Pansuk, W., et al.: Tensile behaviors and fiber orientation of UHPC. In: Fehling et al. (eds.) Ultra High Performance Concrete, Kassel, Germany, March 2008. Proceedings 2nd International Symposium on UHPC, pp. 161–168. Kassel University Press (1996)
55. Ferrara, L., et al.: A magnetic method for non-destructive monitoring of fiber dispersion and orientation in Steel Fiber Reinforced Cementitious Composites—part 1: method calibration. Mater. Struct. **45**, 575–589 (2012)
56. Ferrara, L., et al.: A magnetic method for non-destructive monitoring of fiber dispersion and orientation in steel fiber reinforced cementitious composites—part 2: correlation to tensile fracture toughness. Mater. Struct. **45**, 591–598 (2012)
57. Faifer, M., et al.: Non-destructive testing of steel fiber reinforced concrete using a magnetic approach. IEEE Trans. Instrum. Meas. **60**(5), 1709–1717 (2011)
58. Ferrara, L. et al.: Self consolidating high performance SFRC: an example of structural application in Italy. In: Aldea, C.M., Ferrara, L. (eds.) Fiber reinforced Self Consolidating concrete: research and application", pp. 109–128. ACI-SP 274, Farmington Hills (2010)
59. di Prisco, M., et al.: HPFRCC thin plates for precast roofing. In: Fehling et al. (eds.) Ultra High Performance Concrete, Kassel, Germany, March 2008. Proceedings 2nd International Symposium on UHPC. Kassel University Press, (1996)
60. El-Hacha, R., et al.: Effect of casting method and shear span-to-depth ratio on the behaviour of Ultra-High Performance Concrete cross arms for high voltage transmission lines. Eng. Struct. **32**, 2210–2220 (2010)
61. Suresh Babu, T., et al.: A study on the flexural behaviour of glass fiber reinforced self-compacting concrete. In: Gettu, R. (ed.) Fiber Reinforced Concrete: Design and Applications, Proceedings of the 7th International RILEM Symposium BEFIB 2008, Chennai, India, September 2008. RILEM PRO 60, pp. 793–802. RILEM Pubs (2008)

62. Sravana, P., et al.: Flexural behaviour of glass fibre reinforced self-compacting concrete slabs. 35th Conference on "our world in concrete & structures", Singapore (2010)
63. Khayat, K.H., et al. : High quality tremie concretes for underwater repairs. In: Whitting, D. (ed.) Symposium on Performance of Concrete, pp. 125–138. ACI SP-122, Farmington Hills (1989)
64. Khayat, K.H., Hester, H.T.: Evaluation of concrete mixtures for underwater pile repairs. ASTM Cem. Concr. Aggreg. J. **13**(1), 32–41 (1991)
65. Khayat, K.H.: In-situ properties of concrete piles repaired under water. Concr. Int. **14**(3), 42–49 (1992)
66. Khayat, K.H., et al.: Self-leveling and stiff consolidated concretes for casting high-performance flat slabs in water. Concr. Int. **15**(8), 36–43 (1993)
67. Sonebi, M., Khayat, K.H.: Performance of underwater concrete cast in still and flowing water. Concr. **37**(3), 29–31 (2003)
68. Sonebi, M., Khayat, K.H.: Effect of mixture composition on relative strength of highly flowable underwater concrete. ACI Mater. J. **98**(3), 233–239 (2001)
69. Sonebi, M., Khayat, K.H.: Effect of water velocity on performance of selfconsolidating underwater-cast concrete. ACI Mater. J. **96**(5), 519–528 (1999)
70. Sonebi, M., Khayat, K.H.: Effect of free-fall height in water on performance of flowable concrete. ACI Mater. J. **98**(1), 72–78 (2001)
71. Smeplass, S., Pedersen, B.: SCC-Economic benefits? In: Nordic SCCNet, Oslo (2006)
72. Uebachs, S.: Self-consolidating concrete for floor slab and screed applications. In: The Third North American Conference on the Design and Use of Self- Consolidating Concrete, Chicago (2008)
73. McLeish, A.: Flowable concrete for structural repairs. In: Bartos, P.J.M., Marrs, D.L. and Cleland, D.J. (eds.) Proc. Int. RILEM Conf. On Production methods and workability of concrete (1996)
74. Courard, L., Bissonnette, B.: Compatibility performance as a fundamental requirement for the repair of concrete structures with self-compacting repair mortars. In: De Schutter, G., Boel, V. (eds.) Proceedings PRO054: Self-Compacting Concrete—SCC 2007. RILEM, Ghent (2007)
75. Silfwerbrand, J.L.: Use of self-compacting concrete for bonded overlays. In: De Schutter, G., Boel, V. (eds.) Proceedings PRO054: Self-Compacting Concrete—SCC 2007. RILEM, Ghent (2007)
76. Vieira M, Bettencourt A (2005) Repair applications of self-compacting concrete. In: Proceedings of SCC2005. RILEM, Chicago
77. Khayat, K.H. et al.: Rehabilitation strategies and material performance of SCC used for the repair of Jarry/Querbes underpass in Montreal. In: Alexander et al. (eds) Concrete Repair, Rehabilitation and Retrofitting. London (2006)
78. Miller, M. et al.: SCC Proves Successful in Repair and Strengthening Projects. http://www.concreteconstruction.net/. (2009). Accessed 26 June 2012
79. Hwang, S.D. et al.: Specifications and quality control testing of self-consolidating mortar designated for annular space grouting. In: Khayat, K., Feys, D. (eds.) Proceedings of SCC2010, Montreal (2010)
80. Vitsios, I., et al.: Self Compacting Mortars for Repair Applications According to EN 1504–3. Democritus University of Thrace, Diploma Thesis (2012). (in Greek)
81. Pistolesi C et al.: Low shrinking self-compacting concretes for concrete repair. In: Alexander et al. (eds) Concrete Repair, Rehabilitation and Retrofitting II. London (2009)

Chapter 8
Summary and Conclusions

Geert De Schutter and Kamal H. Khayat

8.1 Scope of the Report

The State-of-the-Art Report of RILEM Technical Committee 228-MPS on Mechanical Properties of Self-Compacting Concrete gathers available information related to mechanical properties and mechanical behaviour of SCC. Due attention is given to the fact that the composition of SCC might be significantly different in different regions. Furthermore, it is not the intention to review the mechanical behaviour of conventional or vibrated concrete. The committee focused on specific mechanical aspects related to self-compacting concrete. All relevant mechanical issues are considered, including compressive strength, stress–strain relationship, tensile and flexural strength, modulus of elasticity, shear strength, effect of elevated temperature, in situ properties, creep, shrinkage, bond properties, and structural behaviour. A chapter on fiber reinforced SCC is included, as well as a chapter on special SCC (light-weight SCC, heavy-weight SCC, preplaced aggregate SCC, special fiber reinforced SCC, and underwater concrete).

8.2 Mechanical Properties

The intrinsic mechanical properties of SCC are reported in Chap. 2. Based on an extended data base (more than 1,500 mixtures) containing mechanical information on a wide range of SCC types (Powder-type SCC, VMA-type SCC, and combined

G. De Schutter (✉)
Magnel Laboratory for Concrete Research, Ghent University, Ghent, Belgium
e-mail: geert.deschutter@ugent.be

K. H. Khayat
Civil, Architectural and Environmental Engineering, Missouri University of Science and Technology, Rolla, MO, USA
e-mail: khayatk@mst.edu

K. H. Khayat and G. De Schutter (eds.), *Mechanical Properties of Self-Compacting Concrete*, RILEM State-of-the-Art Reports 14,
DOI: 10.1007/978-3-319-03245-0_8, © RILEM 2014

types), the mechanical properties are discussed and analysed in relation to several parameters, of which some parameters specifically related to the SCC mix design (e.g. W/P, C/P, etc.).

8.2.1 Compressive Strength

The compressive strength is studied in relation to existing standard formulations and in comparison with vibrated concrete (VC).

While in case of VC, the ***influence of specimen size and shape*** on the measured compressive strength is well documented, it is not the case for SCC. Based on literature data, a tendency for the development of higher conversion factors (f_{ccyl}/f_{ccub}) in the case of SCC is noticed. Several factors might be responsible for the different strength conversion factors for SCC in comparison with VC. The denser microstructure of SCC and enhanced bonding to the aggregates may lead to a more uniform stress distribution during compression. Lower stress concentrations could reduce the probability of premature failure. This might influence the effect of specimen shape and size on the compressive strength. The considerable difference in mixture composition between SCC and VC might also lead to a different effect of the multi-axial stresses in cylinders and cubes while loaded in the testing machine. SCC contains less coarse aggregates compared to VC. For VC it is known that changing the aggregate grading affects cube strength more than cylinder strength. This effect might be more pronounced in case of SCC. The lower coarse aggregate content in SCC might also have an effect through a reduced aggregate interlock in sheared and cracked sections. During compression testing, shear movement has a more significant effect in cubes in comparison with cylinders. This can also contribute to higher f_{ccyl}/f_{ccub} in case of SCC.

The ***influence of W/C and cement strength class*** on the resulting compressive strength of SCC does not seem to be different in comparison with VC.

The most common used ***addition types*** to produce SCC are limestone filler, fly ash, blast furnace slag, and silica fume. Literature data shows that larger strength (at 28 days) will be obtained with fly ash and natural pozzolan compared to limestone filler, while maintaining the cement content and the C/P. No significant differences have been reported between limestone filler, marble powder and basalt powder, although some differences might be noticed depending on the fineness of the powder and the resulting packing density of the binder. Limestone filler and marble powder both have a significant physical effect on the early hydration process of cement, due to the improved nucleation provided by the high specific surface. This positively influences the early age strength development of SCC.

The ***influence of C/P*** was analysed in case of the addition of limestone filler, fly ash, slag and silica fume. At a constant W/C a higher C/P generally leads to lower strengths. This is probably due to the fact that increasing the cement content also needs increasing the water content in order to maintain the W/C, meaning a higher

W/P. More water in the mixture leads to a higher capillary porosity of the concrete and a lower compressive strength.

An increase in ***air content*** decreases the compressive strength of SCC with about 4 MPa per 1 % increase in air content. No significant difference in comparison with VC has been noticed.

Coarse aggregates could have an influence on the compressive strength due to their shape, size, surface, texture and origin. However, no significant effects could be noticed, except some general trends. The higher the compressive strength, the more important becomes the aggregate compressive strength and as such also the origin of the coarse aggregate. At normal strengths, the compressive strength of the concrete is rather governed by the strength of the cement matrix. Consequently, as most reported concrete strength results are lower than 100 MPa, no effect of the aggregate origin is noticed on the compressive strength. Moreover, the effect of the ***maximum grain size*** of the aggregates has been analysed, but no clear influence was observed.

Existing ***strength prediction models*** (Eurocode, Model Code, ACI, etc.), valid for VC, can be extended to better cover SCC. This can be achieved by e.g. implementing new parameters like W/P or C/P or specific modification factors to modulate for the specific powder materials used in SCC mix designs.

8.2.2 Stress–Strain Relationship

Accurate measurements of the ***stress–strain relationship*** require a high level of care and an appropriate steering signal during testing. This can lead to different conclusions in the literature concerning the stress-strain relationship of SCC. Nevertheless, it can be seen that some differences exist between SCC and VC, also depending on the mineral addition type. For limestone filler-based SCC, the ***peak strain*** at different ages (3 days up to 3 months) and concrete strengths (range of 20–70 MPa) is higher than that of VC. The use of blast furnace slag, fly ash and silica fume has been investigated as well. For almost all tested combinations, the peak strain of SCC was higher than for VC. The largest normalised peak strains were measured for SCC containing limestone filler. The values for the SCC mixtures with fly ash, silica fume, or combinations of these supplementary cementitious materials (SCMs) were slightly lower. The lowest values have been found when blast furnace slag was used as addition.

SCC with lower compressive strength (40–50 MPa) showed a more ductile behaviour than higher strengths (50–60 MPa), as is the case for VC. When high amounts of fly ash were combined with blast furnace slag, the behaviour was more brittle. The peak strain is increasing with increasing compressive strength.

The ***strain softening behaviour*** of SCC and VC are comparable for the same strength level. The ***toughness*** (area under the entire stress–strain diagram) of the limestone filler-based SCC was slightly higher than for VC. The largest difference was found for the ascending branch, whereas the surface under the descending branch was comparable. However, the difference in toughness does not seem to be significant.

8.2.3 Tensile Strength

An increasing trend of ***direct tensile strength*** is noticed for increasing cube compressive strength, as can be expected. The available literature results follow the mean relationship proposed by EC2 and MC2010, especially when crushed coarse aggregates are used. In case of uncrushed coarse aggregates (gravel), the direct tensile strength results lie between the 5 % percentile and the mean value proposed by EC2 and MC2010.

Based on limited results, no significant effect of paste volume or filler type on the correlation between direct tensile strength and compressive strength is found. However, the C/P seems to have some visible effect, yielding lower direct tensile strength values in case of C/P < 75 %, for a given strength.

An increasing trend of ***splitting tensile strength*** is also noticed for increasing compressive strength, following the mean relation as proposed by EC2, and following the upper range as proposed by MC2010. The improved bond between paste and aggregates can contribute to somewhat higher values, closer to the upper range in comparison with VC. Splitting tensile strength results of SCC using granite coarse aggregates lay in the upper half or above the ranges proposed by EC2 and MC2010. When limestone or gravel aggregates are used, this is not the case, as the data tend to follow the mean relation proposed by EC2. The ratio between the splitting tensile strength of SCC and VC for equal W/C seems to be influenced by the aggregate type.

Further analysis of available literature data shows that there is no significant effect of coarse aggregate size or paste volume on the correlation between splitting tensile strength and compressive strength. The influence of the C/P or filler type is also not significant according to the available data.

It can be concluded that the existing models EC2 and MC2010 for estimating the tensile strength of SCC are safe and realistic.

For the evaluation of the ***flexural tensile strength*** of SCC, only a limited amount of test results could be found in literature. A logically increasing trend of flexural strength is noticed for increasing cube compressive strength. As could be expected, higher values are found in case of 3-point bending tests, with most of the results lying above the mean value and even above the upper range proposed by EC2 and MC2010. In case of 4-point bending, the test results follow the mean value proposed by EC2 and MC2010, and all data are between the lower and upper range as proposed by the codes.

8.2.4 Modulus of Elasticity

Due to the considerable contribution of aggregates to the overall stiffness of concrete, it is often assumed that SCC, with its higher paste content, is characterised by a lower ***modulus of elasticity***. Nevertheless, analysis of the data available in the RILEM database shows that the available results fit well into the range acceptable for design using the CEB-fib Model Code. The modulus of elasticity of SCC thus seems to be very similar to that of VC, with an important but similar scattering of the results for both types of concrete. ACI 318-08 captures the relationship between compressive strength and E-modulus in a better way, but seems to provide an underestimation of the modulus of elasticity. No significant influence of the paste volume on the elastic modulus was found.

8.2.5 Mechanical Properties of SCC at Elevated Temperatures

The denser microstructure of SCC in comparison with VC seems to be a disadvantage when exposed to fire, as it is the case for high-performance concrete (HPC). The few studies on SCC subjected to high temperatures show that some differences exist between SCC and VC after exposure to high temperatures.

On average, the strength loss after heating of SCC is comparable to VC. The strength of SCC at fire temperature (hot strength) and the residual or post-cooling strength of SCC more or less follow the strength decrease as predicted by EC2. The strength decrease of SCC with temperature increase is less than what is observed for HPC.

By gradually replacing Portland cement by blast furnace slag or fly ash, the residual or hot compressive strength loss increases, especially for temperatures exceeding 400 °C. Higher weight losses have been observed for fly ash in comparison with blast furnace slag. On the other hand, in case of blast furnace slag a higher reduction in pulse velocity was observed in comparison with fly ash. SCC with blast furnace slag showed better performance than with fly ash, for all heating cycles. Nevertheless, compared to EC2 predictions, the residual compressive strengths tend to follow the curves proposed for VC. However, a prediction with EC2 of the residual tensile strength of SCC exposed to temperatures up to 300 °C could lead to an overestimation of the actual losses. The effect of temperature on the tensile strength of SCC becomes more pronounced at higher temperatures due to development of pore pressures.

In concrete mixtures with finer pore structure, such as HPC and SCC, the internal pore pressure upon heating is not released by sufficiently fast enough and can lead to severe spalling of the surface layers. The probability of spalling of SCC is higher compared to VC, although sometimes conflicting results are given in the literature. The explosive fire spalling is mainly dependent on the stress in the concrete, the C/P, and the W/C. One way to limit the amount of explosive fire spalling is to introduce polypropylene fibers. Another way is to keep the combination of C/P and W/C sufficiently high (which of course will have an impact on other material properties as well, e.g. reducing strength…). Due to the complex physics behind the spalling behaviour of concrete in general, no reliable prediction models are available for now.

8.2.6 In-situ Properties

In the literature, several studies concerning the in situ properties of SCC can be found. Overall, it can be confirmed that SCC cast in situ can provide similar or even slightly better properties compared to properly compacted VC. The robustness of SCC, i.e. the uniformity of its properties due to small changes in the original mixture proportioning, mixing conditions, and temperature by errors in weighing the constituents, is an important parameter that should not be overlooked.

8.3 Creep and shrinkage of SCC

The most prominent changes in mix design when shifting from VC to SCC are the higher paste volume, the substantial use of mineral additions and high dosages of superplasticiser, often in combination with a viscosity-modifying agent (VMA). These changes in paste volume and binder composition could influence the viscoelastic properties of the concrete.

Based on the published literature, no general conclusion can be drawn concerning the influence of paste volume and binder composition on *creep*. In general, SCC shows a slightly higher creep coefficient than VC, and the influence of limestone powder on creep coefficient seems to be small. However, the use of slag and fly ash can result in a decrease of the creep coefficient. The somewhat inconsistent results might be caused by differences in kinetics of both cement hydration and compressive strength development; the higher the strength increase after the load is applied, the lower the resulting creep coefficient. The current situation demands further research. In particular, data about creep in tension that can directly be compared with creep in compression are needed.

Autogenous shrinkage of SCC follows the same pattern as autogenous shrinkage of VC and HPC. The autogenous shrinkage increases with increasing cement content and decreasing W/P. Since the higher paste volume of SCC is usually achieved by using mineral additions and not by increasing cement content, self-desiccation might be reduced depending on the type of mineral additions. Therefore, autogenous shrinkage of SCC with mineral additions is usually not higher than the one of VC of similar compressive strength produced with OPC.

However, the higher paste volume results in a higher ***drying shrinkage*** of SCC compared to VC. This increase is slightly reduced by the use of mineral additions.

Due to reduced W/P, finer cements, and higher volume of filler, SCC should be more susceptible to ***plastic shrinkage*** cracking than VC. However, the scarce results in the literature appear contradictory and more research is needed to clarify this issue. The differences between VC and SCC in ***cracking risk due to restrained shrinkage*** of hardened concrete depend on the degree of restraint and the drying velocity. When the degree of restraint and drying velocity are high, SCC exhibits a higher cracking risk than VC as a result of higher shrinkage. However, the importance of creep and E-modulus is increased in case of slow drying and/or low degree of restraint resulting in a relaxation of stress. Consequently, the cracking risk of SCC is decreased and can be equal to or even lower than the one of VC. The influence of mineral additions on the cracking risk is not clear. It seems to depend on the duration of curing.

Several ***models*** enable a calculation of creep and shrinkage. Their accuracy with regard to creep and shrinkage differs. The main problem is that the models are designed for VC and do not take into account the influence of paste volume. Therefore, a modification of the existing models is necessary. However, the wide variety of mix designs and characteristics of mineral additions used could restrict the accuracy of models or demand models adapted for particular types of SCC and mineral additions.

8.4 Bond Strength in SCC

Bond between SCC and reinforcement is not less than that of VC, and in some cases higher values are reported. This may be attributed to the superior stability and filling ability of SCC that can lead to better encapsulation of the reinforcement and contact with existing surfaces. The enhanced quality of the ITZ with embedded reinforcement in SCC can also lead to higher bond strength.

Despite the high fluidity of SCC, high static stability after placement and until the onset of setting is necessary to secure more homogenous in situ properties and denser matrix at the interface between the cement paste and reinforcement. Such bond can be significantly affected by excessive bleeding or segregation found in poorly designed SCC.

Static stability of the SCC is critical in reducing the ***top-bar effect*** to embedded reinforcing steel and prestressing strands. Static stability can be expressed in terms of the maximum surface settlement percentage and static segregation resistance determined from the column segregation test. Such values should be limited to 0.5 and 15 %, respectively, particularity in deep elements, in order to ensure relatively low top-bar effect of the SCC. Highly flowable SCC can develop at least 90 % in situ relative compressive and modification factor (or top-bar effect to prestressing strands) of 1.4 for horizontally embedded prestressing strands. The increase in plastic viscosity of SCC at a given yield stress (or slump flow value) can lead to greater resistance to surface settlement and segregation. However, high plastic viscosity can hinder the self-consolidation of the concrete and can lead to entrapment of air-voids during casting, with negative implication on bond strength. Upper limits to plastic viscosity and t_{50} to ensure adequate self-consolidation have been reported in the literature.

The incorporation of SCMs and fillers can also enhance bond strength of SCC. Incorporating additions, such as silica fume, fly ash, and limestone powder are shown to enhance bond properties of SCC.

Thixotropy and surface roughness seem to be the most important concrete properties affecting the multi-layer casting of SCC. An optimized combination of low thixotropic behaviour of the SCC, in the case of horizontal casting, and the limitation of the delay time between casting different layers are essential in enhancing the ***bond strength in multi-layer casting***. The residual strength in shear is especially affected by the presence of such defect, compared to flexural and compressive modes of stress application. Water permeability across lift lines is also highly affected by the thixotropy of the lower concrete lift at the time of casting and time lag between successive layers.

Because of its enhanced filling ability and self-consolidating properties, properly proportioned SCC used as a ***repair material*** is shown to develop greater ***bond strength to existing surfaces*** than repair overlay made with VC. Such bond is also affected by the roughness of the substrate and the moisture condition of the existing concrete at the time of casting, as is the case for conventional repair materials.

8.5 Structural Behaviour of SCC

According to the available literature, SCC columns under uniaxial compression showed slightly different behaviour compared to VC columns in terms of maximum axial strength, strain at maximum load, stiffness, and ductility.

Maximum axial strength (P_{max}) for both SCC and VC columns was dependent on the concrete compressive strength, and was always higher than the corresponding theoretical column capacity (P_0). With the increase in concrete strength, some researchers reported a slight reduction of the P_{max}/P_0 in SCC columns compared to VC columns.

SCC usually contains SCMs, and often has a lower W/C. As this is warranting a stronger and denser matrix, ***the strain associated with the maximum load*** for SCC columns tested under uniaxial compression could be expected to be lower than that of VC. At the same time, however, SCC mixtures contain a relatively lower amount of coarse aggregates compared to VC. The overall effect seems to be that SCC columns in compression are characterized by a higher strain at peak load.

The ***stiffness of concrete columns*** under uniaxial compression depends on several factors including W/C, the use of SCM, and the amount of coarse aggregate. When a comparable amount of coarse aggregate is maintained in both

VC and SCC mixtures, SCC columns are expected to exhibit higher stiffness than VC columns before peak load when tested under uniaxial compression. Meanwhile, using a lower amount of coarse aggregate in the SCC mixture may significantly reduce column stiffness.

Ductility is the ability of concrete to sustain significant inelastic deformation prior to failure. It is a desirable structural property as it allows absorption of energy for dynamic load due to earthquake, explosion, impact load, harmonic load, etc., and provides warning before structural failure. Members constructed with SCC are expected to have similar or even better seismic behaviour than those constructed with VC, due to better particle gradation, fewer voids, and a denser matrix structure. This has been confirmed in several experimental studies. Results typically show that the ductility of columns is higher in SCC compared to VC mixtures for columns made with normal strength concrete. However, for high-strength concrete, no notable differences have been observed between SCC and VC.

Since no significant differences in the stress–strain curve distribution were reported between SCC and VC, the computation of the ***ultimate flexural resistance*** of SCC beams should not be different from that of VC beams, and can be calculated using the same techniques provided in the related design codes. As to the ***cracking behaviour***, at service load, more and slightly wider cracks with greater penetration have been reported for VC beams in some experimental studies in comparison with SCC beams. The mode of failure and load deflection response of beams cast with SCC and VC were similar. It was also observed that the ultimate moment capacity of the SCC beam was comparable with that of the VC beam and the maximum deflection of the SCC beam was slightly higher.

Since SCC is normally produced by reducing the coarse aggregate content in the mixture, its structural ***shear strength*** is expected to be less than that of VC. Several experimental studies have been performed, with sometimes conflicting results. However, it could be summarized that SCC, with the same maximum size of coarse aggregate but a lower coarse aggregate content, has similar shear resistance in the pre-cracking stage compared with VC. However, a reduced post-cracking shear resistance in SCC can also be noticed, due to the reduced aggregate interlock.

8.6 Fiber Reinforced SCC

Fiber reinforced self-compacting concrete (FR-SCC) has been used in several cases in different types of structural elements (slabs, beams, piles, etc.). In most cases, steel fibers have been incorporated, although the use of non-metallic fibers has also been reported (polypropylene, glass, sisal, cellulose, etc.). One of the main advantages for the implementation of fibers in SCC, as it is for VC, is to partially or even completely substitute traditional reinforcement. Given the special rheological properties of SCC and the absence of vibration during casting, a random and uniform dispersion of the fibers can be obtained, not affected by settlement or segregation. It is also recognized that, depending on a well-balanced mix design and performance of the FR-SCC, fibers can be aligned along the streamlines of the concrete flow while casting the concrete element. A tailor-made casting procedure based on a well-designed FR-SCC, duly considering the target alignment of the fibers, can lead to superior mechanical and structural performance. The influence of the flow-driven fiber dispersion and orientation on the mechanical performance represents a key distinctive feature of self-compacting fiber reinforced concrete as compared to traditional vibrated fiber reinforced concrete, which cannot be disregarded when analysing engineering, mechanical, and structural properties.

Within this state-of-the-art report, a literature review has been given of the main findings with reference to the correlation among the fresh state performance, the fiber dispersion and the mechanical properties of FR-SCC and the possibility of governing the aforementioned correlations to conceive, design and make high end engineering and structural applications employing this category of advanced cementitious composites. Some interesting conclusions can be highlighted:

- Tailored alignment of fibers in structural elements made with FR-SCC can be obtained due to the superior fresh state performance of the material and as the outcome of a suitably designed casting process;
- Non-destructive test methods for fiber dispersion in full scale castings have been developed, and seem ready for application in construction practice;
- Tools to design the casting process and predict fiber alignment are still at an embryonic stage, due to the required computational time for a real scale casting flow simulation. Interesting results however have already been obtained for small volume castings. Furthermore, simple empirical and heuristic approached have been developed and calibrated;
- Fiber orientation dependent design procedures for FR-SCC element represent the next step of the knowledge transfer into engineering construction practice;
- Experimental characterization and modelling of the "flow induced" anisotropy of the material behaviour, under both static and cyclic loadings, have to be urgently tackled also in the sight of emerging applications of FR-SCCs.

A good understanding, modelling and control of the fiber orientation on the final structural behaviour would enable a "holistic" design approach which tailors both the material composition and the casting process to the anticipated structural performance.

8.7 Specialty SCC

Although SCC in itself is considered as a specialty concrete, even more special applications can be obtained by considering unusual techniques or materials leading to the definition of specialty SCC. From the large range of specialty SCC that can be used, this state-of-the-art report presents a review on the mechanical properties of SCC with uncommon aggregates, self-compacting mortars, preplaced aggregate SCC, special fiber reinforced SCC, and application-oriented SCC.

Lightweight SCC (LSCC) combines the advantages of lightweight concrete with those of SCC fresh properties. The strength of LSCC depends on the aggregate characteristics, W/C, cement content, and type and content of mineral powder, including fillers and SCMs. Generally speaking, the difference in strength development between LSCC and vibrated lightweight concrete is similar to that between SCC and VC. The elastic modulus of LSCC is affected by the type and properties of the aggregate, as well as the compressive strength of the concrete, ranging from 40 to 70 % of the equivalent normal weight concrete.

Heavyweight SCC (HSCC) presents an oven-dry density larger than 2,600 kg/m^3, and can be applied for counterbalance elements, or for radiation shielding. Application of HSCC requires formwork that can withstand the higher lateral pressures. Moreover, segregation due to the high density of the aggregate should be controlled.

As recycling is one of the measures to reduce the deposit of construction and demolition materials, recycled aggregates also find application in SCC. Although several studies exist concerning ***recycled aggregate SCC***, there is no general agreement on the influence of the coarse recycled aggregates on the concrete strength.

Given knowledge related to SCC, the special technique of ***preplaced aggregate concrete*** (PAC) has developed into two variants, rock-filled concrete and PAC filled with self-compacting mortar. ***Rock-filled concrete*** (RFC) is produced by filling the volume with large-scale blocks of rock, after which the empty space is filled with SCC, either by pumping or by pouring. Another method to achieve RFC is dump-type RFC, which consists of pouring the SCC before launching large-scale blocks of rock into it. As a result, 55–60 % of the volume of the RFC is filled by rocks, and thus only about 40–45 % of the volume needs to be filled with SCC, leading to a significant reduction in cement content. Experimental results show that the strength of RFC is typically higher than the strength of the SCC solely.

Preplaced aggregate filled with self-compacting mortar (PACSCM) is similar to more classical PAC, however with the difference that the mortar is simply poured through the aggregate instead of being injected. The filling of the gaps between the aggregates is achieved only by the self-weight of the mortar. Some field cases have shown good potential for the special method.

Self-compacting slurry infiltrated fiber concrete (SCSIFCON) can be regarded as a special type of high performance steel fiber reinforced cement composite, as a further development of SIFCON for which intensive vibration is needed. This special material could be used in small parts of frames, which are expected to be affected during seismic loading and form "plastic hinges", which would absorb the fracture energy and protect the construction against collapse.

High-performance, fiber reinforced cementitious composites (HPFRCC) are a unique broad category of fiber reinforced concretes which can exhibit significant stable multi-cracking accompanied by an almost "plastic" or even strain hardening behaviour after the onset of the first crack and before unstable localisation of the critical crack. This is obtained thanks to a micromechanics-based mix design. Self-compactability is often obtained, given the high dosage of cement, powder, and superplasticiser in these specialty mixtures.

Underwater self-compacting concrete (UW-SCC) can be stabilized by the application of anti-washout agents (AWA). The mixture composition is crucial for the in situ mechanical performance of UW-SCC. The water-to-binder ratio, the type and dosage of SCMs affect the washout resistance, and thus influence the in situ mechanical performance. The use of silica fume or the combination of silica fume with fly ash can significantly improve the washout resistance and the residual compressive strength. The type and the dosage of AWA affect the workability and the washout mass loss, and the residual strength.

Self-levelling concrete is a specific type of SCC specifically designed for a fluid spreading. It allows obtaining a smooth surface of the pavement without almost any surface works. For concrete slabs, the application of an all-in-one procedure, using self-levelling concrete, instead of the application of concrete with a screed on top can lead to material savings up to 35 %, without counting the production time savings. Higher compressive and flexural strengths can be obtained in comparison with traditional screeds.

For many repair situations, ***self-compacting repair mortars and concretes*** can be advantageously, and sometimes even offer the only proper solution, as repair works are typically done in locations where it is difficult or even impossible to vibrate. Applications can range from strengthening existing structures to replacement of degraded concrete. To avoid problems, application of self-compacting repair mortars or concretes requires a specific design, both from structural and materials viewpoints, as well as qualified personnel.

8.8 Final Remark

SCC is a new type of high-performance cementitious material, and at the same time a new type of production method for casting concrete structures. However, SCC mainly remains a cement-based material, which means that most of our knowledge and understanding based on VC is not obsolete. SCC pushes the limits of classical concrete technology. However, the main driving forces and the fundamental chemical, physical, and mechanical laws remain unchanged. Nevertheless, due to its specific mix design, SCC can sometimes behave differently in comparison with VC. In most cases, classical material laws should be extended to cover the specific situation of SCC. This also holds for the mechanical properties. Generally speaking, design codes and material mechanical properties for traditional concrete structures in most cases apply when using SCC, although creep and shrinkage models should be evaluated with care as most of them do not include the effect of paste volume.

However, in many cases a significant structural benefit can be obtained by considering the specifics advantages of SCC, such as the improvement of the ITZ with coarse aggregate and embedded reinforcement that can enhance bond and structural behaviour. Unfortunately, these specific advantages are not yet covered in traditional design codes. Hopefully, this RILEM state-of-the-art report can contribute to further development of research needs, standards, and code regulations for SCC.

Index

K. H. Khayat and G. De Schutter (eds.), *Mechanical Properties of Self-Compacting Concrete*, RILEM State-of-the-Art Reports 14, DOI: 10.1007/978-3-319-03245-0, © RILEM 2014

MIX
Papier aus verantwortungsvollen Quellen
Paper from responsible sources
FSC® C105338

If you have any concerns about our products,
you can contact us on
ProductSafety@springernature.com

In case Publisher is established outside the EU,
the EU authorized representative is:
Springer Nature Customer Service Center GmbH
Europaplatz 3, 69115 Heidelberg, Germany

Printed by Libri Plureos GmbH
in Hamburg, Germany